Essential Algebraic Number Theory

Series on Number Theory and Its Applications* ISSN 1793-3161

Published

Vol. 18 *Essential Algebraic Number Theory*
by Ivan Fesenko

Vol. 17 *Smooth-Automorphic Forms and Smooth-Automorphic Representations*
by Harald Grobner

Vol. 16 *Elementary Modular Iwasawa Theory*
by Haruzo Hida

Vol. 15 *Derived Langlands: Monomial Resolutions of Admissible Representations*
by Victor Snaith

Vol. 14 *Problems and Solutions in Real Analysis (Second Edition)*
by Masayoshi Hata

*For the complete list of the published titles in this series, please visit:
www.worldscientific.com/series/sntia

Series on Number Theory and Its Applications Vol. 18

Essential Algebraic Number Theory

Ivan Fesenko

World Scientific

NEW JERSEY • LONDON • SINGAPORE • BEIJING • SHANGHAI • HONG KONG • TAIPEI • CHENNAI • TOKYO

Published by

World Scientific Publishing Co. Pte. Ltd.
5 Toh Tuck Link, Singapore 596224
USA office: 27 Warren Street, Suite 401-402, Hackensack, NJ 07601
UK office: 57 Shelton Street, Covent Garden, London WC2H 9HE

Library of Congress Control Number: 2025054220

British Library Cataloguing-in-Publication Data
A catalogue record for this book is available from the British Library.

Cover image: *Delicate Tension, No. 85* (1923) by Wassily Kandinsky. Public domain.

Series on Number Theory and Its Applications — Vol. 18
ESSENTIAL ALGEBRAIC NUMBER THEORY

ISBN 978-981-98-2571-4 (hardcover)
ISBN 978-981-98-2681-0 (paperback)
ISBN 978-981-98-2572-1 (ebook for institutions)
ISBN 978-981-98-2573-8 (ebook for individuals)

For any available supplementary material, please visit
https://www.worldscientific.com/worldscibooks/10.1142/14658#t=suppl

Desk Editors: Nambirajan Karuppiah/Lai Fun Kwong

Typeset by Stallion Press
Email: enquiries@stallionpress.com

Preface

This book is based on courses given in Russia, the UK and China, as well as numerous talks delivered in Germany, Japan, France, and the USA. The most recent lecture courses were given at Tsinghua University in 2023–2024 and at Westlake University in 2025. The material of the first chapter was lectured for over 10 years in the UK.

The book offers a fast and relatively easy introduction into the main aspects of algebraic number theory, including its basic aspects, main results about local and global fields and of explicit class field theory. The emphasise is on a clear presentation, using as little auxiliary tools from algebra or analysis as possible. The presentation of this book was motivated by the author's study and research in generalisations of class field theory such as higher class field theory, anabelian geometry and some topics in the Langlands program. The main aim of the book is the enable the reader to learn central key aspects of the theory and to prepare for research work in modern number theory.

The first chapter is a compact simplified presentation of core features of algebraic number fields, without using analytic tools. Various properties of algebraic integers are deduced from properties of Dedekind rings. The reader is assumed to have a modest knowledge of elementary number theory and some first knowledge of commutative algebra. The chapter also includes a first introduction to p-adic numbers, which are studied in much more detail in Chapter 2, and to class field theory of rational numbers. This chapter can be used by readers who do not have time to learn more advanced areas of algebraic number theory.

In view of the important role of local theories in modern research, the second chapter includes various results about complete discrete valuations fields, which are used in modern research but typically not presented in older books on algebraic number theory. Complete discrete valuations fields with perfect residue fields are called local fields in this book. Algebraic number fields and the function fields of irreducible curves over finite fields are called global fields. The study of local fields can be viewed as a local study of global fields. Included are topics on Artin–Schreier extensions of local fields with residue field of positive characteristic, and an approach to the Hasse–Herbrand function of finite separable extensions of local fields by using the behaviour of the norm map. The chapter also includes a more advanced Fontaine–Wintenberger's theory of fields of norms, demonstrating that certain infinite extensions of local fields of characteristic zero with residue field of positive characteristic behave like local fields of positive characteristic. Beginners can omit the latter theory. Chapter 2 ends with a property of local fields that leads to abstract class field theory mechanism in the next chapter. This mechanism is applicable, as explained in Chapter 3, not only to class field theory of local fields but of global fields as well.

How to describe finite subextensions of the maximal separable extension, maximal abelian extension, maximal nilpotent extension of fixed nilpotent class of global and local fields, in terms of objects associated to the ground field and in ways which allow a good range of applications? This problem for abelian extensions is solved by various types of class field theory. For nilpotent extensions of local fields it is partially solved by arithmetic non-abelian local class field theory. The Langlands correspondence provides a different representation theoretic point of view, with mostly still conjectural expectations for arbitrary number fields. Anabelian geometry provides yet another very different insight into how the full structure of the absolute Galois group of a global field can be used to algorithmically recover the field. These generalisations of class field theory, as well as higher class field theory, also shed a new light on class field theory, helping to distinguish essential features of class field theory, as presented in this book.

Class field theory was initiated in the 19th century by German mathematicians as special class field theory of few small algebraic number fields such as the field of rational numbers and its imaginary quadratic fields. Takagi's work on general existence theorem and then Artin's introduction of global reciprocity map, published 100 years ago, started general global class field theory. Special class field theory was upgraded and fundamentally changed in general class field theory that works for all global fields. The latter theory became a key part of algebraic number theory of the 20th century, with a huge number of applications. As with many fundamental developments, first presentations of class field theory were very far from easy to study. It had taken a few decades for clearer presentations of the theory to emerge. One of popular presentations of class field theory used Galois cohomology; it was not quite explicit. We should also keep in mind the opinion of Hasse about the edifice of class field theory, "the sharply profiled lines and individual features of this magnificent edifice seem to me to have lost somewhat of their original splendour and plasticity by the penetration of class field theory with cohomological concepts and methods", [**15**]. Comparing different approaches to general class field theory that work for all local and global fields, the clearest and easiest approach to class field theory is presented in Neukirch's books [**29**] and [**30**]. This approach was motivated by anabelian geometry results for number fields. Later, it was found that this approach it is more directly generalisable to higher local and global theories, unlike the cohomological approach.

The third chapter presents abstract class field theory and its applications to class field theory for local fields with finite residue field and for global fields. This succinct presentation does not involve tools typical for other modern presentations of class field theory such as central division algebras nor group cohomology groups, nor formal Lubin–Tate groups. The approach to abstract class field theory uses slightly simplified Neukirch's class field theory mechanism, a topological group theoretical algorithm to deduce almost all main theorems of class field theory from its two axioms. Chapter 3 also includes an important adelic theory of zeta integrals, discovered by Iwasawa and Tate, and later generalised in the Langlands correspondence and higher adelic zeta integrals theory. Remarks on more recent developments and the three generalisations of class field theory are included in Chapters 2 and 3; more details and references can be found in [**9**].

The three chapters are written in different styles. The first chapter has a slow pace accessible to 3rd–4th year level undergraduate students. The second chapter is suitable for the level of senior undergraduate students or master students. The third chapter proceeds with the speed of textbooks for graduate students. The reader interested in class field theory and who already feels confident about local fields, can start with Section 18 of Chapter 2.

During preparation of this book, a number of mistakes in some of the main textbooks on class field theory were detected, they are corrected in this text if a relevant piece is included.

Sets of exercises can be found at the end of the book. The reader able to solve all of them may consider oneself suitable for research work in modern number theory.

A reference in a chapter to an assertion in the same chapter does not state the chapter number.

Hangzhou
September 10, 2025

About the Author

Ivan Fesenko is a mathematician whose career spans several countries and decades, marked by pioneering contributions in number theory and proactive leadership in the mathematical sciences. His major areas of specialization include the core of algebraic number theory: class field theory and its three generalizations. Fesenko introduced higher Haar measure and integration on higher local fields and higher adeles, and associated harmonic analysis. He launched a research program in higher adelic analysis and geometry, which extended the classical Iwasawa–Tate theory. This work involved the development of higher adelic integrals to study the zeta functions of elliptic surfaces using topological and measure-theoretic structures. He has studied Shinichi Mochizuki's IUT theory, co-authoring a 2022 paper that established effective abc-inequalities and offered a new proof of Fermat's Last Theorem as an application. In recent years, his interdisciplinary interests have expanded to encompass topics at the intersection of arithmetic geometry and quantum theory and computing, as well as new epidemic modelling. His advisory work in science policy has informed national strategies, contributing to a significant rise in funding for modern mathematical research.

Contents

Chapter 1

Algebraic Number Fields

This chapter presents in its Sections 2 and 3 basic features of algebraic number fields, typical for undergraduate courses on algebraic number theory. Section 4 includes a brief introduction to p-adic numbers whose properties are studied in much greater extent in Chapter 2. Chapter 1 ends with a brief presentation of cyclotomic class field theory; a full presentation of class field theory and many further results on algebraic number fields obtained via the adelic method are included in Chapter 3.

1. Algebraic Prerequisites

1.1. *Roots of polynomials and field extensions*

1.1.1. DEFINITION. For a field F define the ring homomorphism $\mathbb{Z} \longrightarrow F$ by $n \mapsto n \cdot 1_F$. Its kernel I is an ideal of $\mathbb{Z}$ such that $\mathbb{Z}/I$ is isomorphic to the image of $\mathbb{Z}$ in F. The latter is an integral domain, so I is a prime ideal of $\mathbb{Z}$, i.e. $I = 0$ or $I = p\mathbb{Z}$ for a prime number p. In the first case F is said to have *characteristic 0*, in the second – *characteristic p*.

DEFINITION. Let F be a subfield of a field L. An element $a \in L$ is called *algebraic over F* if one of the following equivalent conditions is satisfied:
(i) $f(a) = 0$ for a non-zero polynomial $f(X) \in F[X]$;
(ii) elements $1, a, a^2, \dots$ are linearly dependent over F;
(iii) F-vector space $F[a] = \{\sum a_i a^i : a_i \in F\}$ is of finite dimension over F;
(iv) $F[a] = F(a)$.

Proof. (i) implies (ii): if $f(X) = \sum_{i=0}^{n} c_i X^i$, $c_0, c_n \neq 0$, then $\sum c_i a^i = 0$.
(ii) implies (iii): if $\sum_{i=0}^{n} c_i a^i = 0$, $c_n \neq 0$, then $a^n = -\sum_{i=0}^{n-1} c_n^{-1} c_i a^i$,

$$a^{n+1} = a \cdot a^n = -\sum_{i=0}^{n-1} c_n^{-1} c_i a^{i+1} = -\sum_{i=0}^{n-2} c_n^{-1} c_i a^{i+1} + c_n^{-1} c_{n-1} \sum_{i=0}^{n-1} c_n^{-1} c_i a^i,$$

etc.

(iii) implies (iv): for every $b \in F[a]$ we have $F[b] \subset F[a]$, hence $F[b]$ is of finite dimension over F. So if $b \notin F$, there are d_i such that $\sum d_i b^i = 0$, and $d_0 \neq 0$. Then $1/b = -d_0^{-1} \sum_{i=1}^n d_i b^{i-1}$ and hence $1/b \in F[b] \subset F[a]$.

(iv) implies (i): if $1/a$ is equal to $\sum e_i a^i$, then a is a root of $\sum e_i X^{i+1} - 1$. □

For an element a algebraic over F denote by

$$f_a(X) \in F[X]$$

the monic polynomial of minimal degree such that $f_a(a) = 0$.

This polynomial is irreducible: if $f_a = gh$, then $g(a)h(a) = 0$, so $g(a) = 0$ or $h(a) = 0$, contradiction. It is called *the monic irreducible polynomial of a over F*.

For example, $f_a(X)$ is a linear polynomial if and only if $a \in F$.

LEMMA. *Define a ring homomorphism $F[X] \longrightarrow L$, $g(X) \mapsto g(a)$. Its kernel is the principal ideal generated by $f_a(X)$ and its image is $F(a)$, so*

$$F[X]/(f_a(X)) \simeq F(a).$$

Proof. The kernel consists of those polynomials g over F which vanish at a. Using the division algorithm write $g = f_a h + k$ where $k = 0$ or the degree of k is smaller than that of f_a. Now $k(a) = g(a) - f_a(a)h(a) = 0$, so the definition of f_a implies $k = 0$ which means that f_a divides g. □

DEFINITION. A field L is called *algebraic over* its subfield F if every element of L is algebraic over F. The extension L/F is called *algebraic*.

DEFINITION. Let F be a subfield of a field L. The dimension of L as a vector space over F is called the *degree* $|L:F|$ of the extension L/F.

If a is algebraic over F then $|F(a):F|$ is finite and it equals the degree of the monic irreducible polynomial f_a of a over F.

Transitivity of the degree $|L:F| = |L:M||M:F|$ follows from the observation: if α_i form a basis of M over F and β_j form a basis of L over M then $\alpha_i\beta_j$ form a basis of L over F.

Every extension L/F of finite degree is algebraic: if $\beta \in L$, then $|F(\beta):F| \leqslant |L:F|$ is finite, so by (iii) above β is algebraic over F. In particular, if α is algebraic over F then $F(\alpha)$ is algebraic over F.

If α, β are algebraic over F then the degree of $F(\alpha, \beta)$ over F does not exceed the product of finite degrees of $F(\alpha)/F$ and $F(\beta)/F$ and hence is finite. Thus all elements of $F(\alpha, \beta)$ are algebraic over F.

In particular, for two algebraic over F non-zero elements α, β the elements $\alpha + \beta$, $\alpha - \beta$, $\alpha\beta$, $\alpha\beta^{-1}$ are algebraic over F.

An algebraic extension $F(\{a_i\})$ of F is the composite of extensions $F(a_i)$, and since a_i is algebraic $|F(a_i):F|$ is finite, thus *every algebraic extension is the composite of finite extensions.*

1.1.2. DEFINITION. An extension F of $\mathbb{Q}$ of finite degree is called an *algebraic number field*, the degree $|F:\mathbb{Q}|$ is called the *degree of F*.

EXAMPLES.

1. Every quadratic extension L of $\mathbb{Q}$ can be written as $\mathbb{Q}(\sqrt{e})$ for a square-free integer e. Indeed, if $1,\alpha$ is a basis of L over $\mathbb{Q}$, then $\alpha^2 = a_1 + a_2\alpha$ with rational a_i, so α is a root of the polynomial $X^2 - a_2X - a_1$ whose roots are of the form $a_2/2 \pm \sqrt{d}/2$ where $d \in \mathbb{Q}$ is the discriminant. Write $d = f/g$ with integer f,g and notice that $\mathbb{Q}(\sqrt{d}) = \mathbb{Q}(\sqrt{dg^2}) = \mathbb{Q}(\sqrt{fg})$. Obviously we can get rid of all square divisors of fg without changing the extension $\mathbb{Q}(\sqrt{fg})$.

2. Cyclotomic extensions $\mathbb{Q}^m = \mathbb{Q}(\zeta_m)$ of $\mathbb{Q}$ where ζ_m is a primitive mth root of unity. If p is prime then the monic irreducible polynomial of ζ_p over $\mathbb{Q}$ is $X^{p-1} + \cdots + 1 = (X^p - 1)/(X-1)$ of degree $p-1$.

One way to show the irreducibility over $\mathbb{Q}$ of this polynomial is to make change of variable $Y = X+1$ and show that the polynomial in Y is irreducible over $\mathbb{Q}$ (applying the Eisenstein's criteria of irreducibility).

1.1.3. DEFINITION. Let two fields L, L' contain a field F. A homo(iso)morphism $\sigma\colon L \longrightarrow L'$ such that $\sigma|_F$ is the identity map is called an F-homo(iso)morphism of L into L'.

The set of all F-homomorphisms from L to L' is denoted by $\mathrm{Hom}_F(L,L')$. Notice that every F-homomorphism is injective: its kernel is an ideal of F and 1_F does not belong to it, so the ideal is the zero ideal. In particular, $\sigma(L)$ is isomorphic to L.

The set of all F-isomorphisms from L to L' is denoted by $\mathrm{Iso}_F(L,L')$.

Two elements $a \in L, a' \in L'$ are called *conjugate over F* if there is a F-homomorphism σ such that $\sigma(a) = a'$. If L, L' are algebraic over F and isomorphic over F, they are called *conjugate over F*.

LEMMA.

(1) *Any two roots of an irreducible polynomial over F are conjugate over F.*

(2) *An element a' is conjugate to a over F if and only if $f_{a'} = f_a$.*

(3) *The polynomial $f_a(X)$ is divisible by $\prod(X - a_i)$ in $L[X]$, where a_i are all distinct conjugate to a elements over F, L is the field $F(\{a_i\})$ generated by a_i over F.*

Proof. (1) Let $f(X)$ be an irreducible polynomial over F and a, b be its roots in a field extension of F. Then $f_a = f_b = f$ and we have an F-isomorphism

$$F(a) \simeq F[X]/(f_a(X)) = F[X]/(f_b(X)) \simeq F(b), \quad a \mapsto b$$

and therefore a is conjugate to b over F.

(2) $0 = \sigma f_a(a) = f_a(\sigma a) = f_a(a')$, hence $f_a = f_{a'}$. If $f_a = f_{a'}$, use (1).

(3) If a_i is a root of f_a then by the division algorithm $f_a(X)$ is divisible by $X - a_i$ in $L[X]$. □

1.1.4. DEFINITION. A field is called *algebraically closed* if it does not have algebraic extensions.

THEOREM. (without proof) *Every field F has an algebraic extension C which is algebraically closed. The field C is called an algebraic closure of F. Every two algebraic closures of F are isomorphic over F.*

EXAMPLE. The field of rational numbers $\mathbb{Q}$ is contained in algebraically closed field $\mathbb{C}$. The subfield of complex numbers generated by all algebraic elements over $\mathbb{Q}$ is an algebraic closure $\mathbb{Q}^{\mathrm{alg}}$ of $\mathbb{Q}$. This field $\mathbb{Q}^{\mathrm{alg}}$ is algebraically closed: if $\alpha \in \mathbb{C}$ is algebraic over $\mathbb{Q}^{\mathrm{alg}}$ then it is a root of a non-zero polynomial with finitely many coefficients, each of which is algebraic over $\mathbb{Q}$. Therefore α is algebraic over the field M generated by the coefficients. Then $M(\alpha)/M$ and $M/\mathbb{Q}$ are of finite degree, and hence α is algebraic over $\mathbb{Q}$, i.e. belongs to $\mathbb{Q}^{\mathrm{alg}}$. The degree $|\mathbb{Q}^{\mathrm{alg}} : \mathbb{Q}|$ is infinite, since $|\mathbb{Q}^{\mathrm{alg}} : \mathbb{Q}| \geqslant |\mathbb{Q}(\zeta_p) : \mathbb{Q}| = p - 1$ for every prime p.

The field $\mathbb{Q}^{\mathrm{alg}}$ is is much smaller than $\mathbb{C}$, since its cardinality is countable whereas the cardinality of complex numbers is uncountable).

Everywhere below we denote by C an algebraically closed field containing F.

Elements of $\mathrm{Hom}_F(F(a), C)$ are in one-to-one correspondence with distinct roots of $f_a(X) \in F[X]$: for each such root a_i, as in the proof of (i) above we have $\sigma\colon F(a) \longrightarrow C$, $a \mapsto a_i$; and conversely each such $\sigma \in \mathrm{Hom}_F(F(a), C)$ maps a to one of the roots a_i.

1.2. *On Galois extensions*

1.2.1. DEFINITION. A polynomial $f(X) \in F[X]$ is called *separable* if all its roots in C are distinct.

Recall that if a is a multiple root of $f(X)$, then $f'(a) = 0$. So a polynomial f is separable if and only if the polynomials f and f' don't have common roots.

LEMMA. *Irreducible polynomials over fields of characteristic zero and irreducible polynomials over finite fields are separable polynomials*

Proof. If f is an irreducible polynomial over a field of characteristic zero, then its derivative f' is non-zero and has degree strictly smaller than f; and so if f has a multiple root, than a g.c.d. of f and f' would be of positive degree strictly smaller than f which contradicts the irreducibility of f. For the case of irreducible polynomials over finite fields see 1.3. □

DEFINITION. Let L be a field extension of F. An element $a \in L$ is called *separable* over F if $f_a(X)$ is separable. The extension L/F is called *separable* if every element of L is separable over F.

EXAMPLE. Every algebraic extension of a field of characteristic zero or a finite field is separable.

1.2.2. LEMMA. *Let M be a field extension of F and L be a finite extension of M. Then every F-homomorphism $\sigma\colon M \longrightarrow C$ can be extended to an F-homomorphism $\sigma'\colon L \longrightarrow C$.*

Proof. Let $a \in L \setminus M$ and $f_a(X) = \sum c_i X^i$ be the minimal polynomial of a over M. Then $(\sigma f_a)(X) = \sum \sigma(c_i) X^i$ is irreducible over σM. Let b be its root. Then $\sigma f_a = f_b$. Consider an F-homomorphism $\phi\colon M[X] \longrightarrow C$, $\phi(\sum a_i X^i) = \sum \sigma(a_i) b^i$. Its image is $(\sigma M)(b)$ and its kernel is generated by f_a. Since $M[X]/(f_a(X)) \simeq M(a)$, ϕ determines an extension $\sigma''\colon M(a) \longrightarrow C$ of σ. Since $|L : M(a)| < |L : M|$, by induction σ'' can be extended to an F-homomorphism $\sigma'\colon L \longrightarrow C$ such that $\sigma'|_M = \sigma$. □

1.2.3. THEOREM. *Let L be a finite separable extension of F of degree n. Then there exist exactly n distinct F-homomorphisms of L into C, i.e.*

$$|\operatorname{Hom}_F(L,C)| = |L : F|.$$

Proof. The number of distinct F-homomorphisms of L into C is $\leqslant n$ is valid for any extension of degree n. To prove this, argue by induction on $|L : F|$ and use the fact that every F-homomorphism $\sigma\colon F(a) \longrightarrow C$ sends a to one of roots of $f_a(X)$ and that root determines σ completely.

To show that there are n distinct F-homomorphisms for separable L/F consider first the case of $L = F(a)$. From separability we deduce that the polynomial $f_a(X)$ has n distinct roots a_i which give n distinct F-homomorphisms of L into C: $a \mapsto a_i$.

Now argue by induction on degree. For $a \in L \setminus F$ consider $M = F(a)$. There are $m = |M : F|$ distinct F-homomorphisms σ_i of M into C. Let $\sigma_i' \colon L \longrightarrow C$ be an extension of σ_i which exists according to 1.2.2. By induction there are n/m distinct $F(\sigma_i(a))$-homomorphisms τ_{ij} of $\sigma_i'(L)$ into C. Now $\tau_{ij} \circ \sigma_i'$ are distinct F-homomorphisms of L into C. □

1.2.4. PROPOSITION. *Every finite subgroup of the multiplicative group $F^\times$ of a field F is cyclic.*

Proof. Denote this subgroup by G, it is an abelian group of finite order. From the well known result about the structure of finitely generated abelian groups we deduce that

$$G \simeq \mathbb{Z}/m_1\mathbb{Z} \oplus \cdots \oplus \mathbb{Z}/m_r\mathbb{Z}$$

where m_1 divides m_2, etc. We need to show that $r = 1$ (then G is cyclic). If $r > 1$, then let a prime p be a divisor of m_1. The cyclic group $\mathbb{Z}/m_1\mathbb{Z}$ has p elements of order p and similarly, $\mathbb{Z}/m_2\mathbb{Z}$ has p elements of order p, so G has at least p^2 elements of order p. However, all elements of order p in G are roots of the polynomial $X^p - 1$ which over the field F cannot have more than p roots, a contradiction. Thus, $r = 1$. □

1.2.5. THEOREM. *Let F be a field of characteristic zero or a finite field. Let L be a finite field extension of F. Then there exists an element $a \in L$ such that $L = F(a) = F[a]$.*

Proof. If F is of characteristic 0, then F is infinite. By 1.2.3 there are $n = |L : F|$ distinct F-homomorphisms $\sigma_i \colon L \longrightarrow C$. Put $V_{ij} = \{a \in L : \sigma_i(a) = \sigma_j(a)\}$. Then V_{ij} are proper F-vector subspaces of L for $i \neq j$ of dimension $< n$, and since F is infinite, there union $\cup_{i \neq j} V_{ij}$ is different from L. Then there is $a \in L \setminus (\cup V_{ij})$. Since the set $\{\sigma_i(a)\}$ is of cardinality n, the minimal polynomial of a over F has at least n distinct roots. Then $| F(a) : F | \geqslant n = | L : F |$ and hence $L = F(a)$.

If F is finite, then $L^\times$ is cyclic by 1.2.4. Let a be any of its generators. Then $L = F(a)$. □

1.2.6. DEFINITION. An algebraic extension L of F (inside C) is called the *splitting field of polynomials f_i* if $L = F(\{a_{ij}\})$ where a_{ij} are all the roots of f_i.

An algebraic extension L of F is called *a Galois extension* if L is the splitting field of some separable polynomials f_i over F.

EXAMPLE. Let $L = F(a)$ be a finite extension of F. Then L/F is a Galois extension if the polynomial $f_a(X)$ of a over F has $\deg f_a$ distinct roots in L.

So quadratic extensions of $\mathbb{Q}$ and cyclotomic extensions of $\mathbb{Q}$ are its Galois extensions.

1.2.7. LEMMA. *Let L be the splitting field of an irreducible polynomial $f(X)$ in $F[X]$. Then $\sigma(L) = L$ for every $\sigma \in \mathrm{Hom}_F(L,C)$.*

Proof. σ permutes the roots of $f(X)$. Thus, $\sigma(L) = F(\sigma(a_1),\dots,\sigma(a_n)) = L$. □

1.2.8. THEOREM. *A finite extension L of F is a Galois extension if and only if $\sigma(L) = L$ for every $\sigma \in \mathrm{Hom}_F(L,C)$ and $\mid \mathrm{Hom}_F(L,L) \mid = \mid L : F \mid$. The set $\mathrm{Hom}_F(L,L)$ equals to the set $\mathrm{Iso}_F(L,L)$ which is a finite group with respect to the composite of field isomorphisms. This group is called the Galois group $\mathrm{Gal}(L/F)$ of the extension L/F.*

Proof. Sketch. Let L be a Galois extension of F. The right arrow follows from the previous Proposition and properties of separable extensions. On the other hand, if $L = F(\{b_i\})$ and $\sigma(L) = L$ for every $\sigma \in \mathrm{Hom}_F(L,C)$ then $\sigma(b_i)$ belong to L and L is the splitting field of polynomials $f_{b_i}(X)$. If $\mid \mathrm{Hom}_F(L,L) \mid = \mid L : F \mid$ then one can show by induction that each of $f_{b_i}(X)$ is separable.

Now suppose we are in the situation of 1.2.5. Then $L = F(a)$ for some $a \in L$. L is the splitting field of some polynomials f_i over F, and hence L is the splitting field of their product. By 1.2.7 and induction we have $\sigma L = L$. Then $L = F(a_i)$ for any root a_i of f_a, and elements of $\mathrm{Hom}_F(L,L)$ correspond to $a \mapsto a_i$. Therefore $\mathrm{Hom}_F(L,L) = \mathrm{Iso}_F(L,L)$. Its elements correspond to some permutations of the set $\{a_i\}$ of all roots of $f_a(X)$. □

1.2.9. THEOREM. (without proof) *Let L/F be a finite Galois extension and M be an intermediate field between F and L. Then L/M is a Galois extension with the Galois group*

$$\mathrm{Gal}(L/M) = \{\sigma \in \mathrm{Gal}(L/F) : \sigma|_M = \mathrm{id}_M\}.$$

For a subgroup H of $\mathrm{Gal}(L/F)$ denote

$$L^H = \{x \in L : \sigma(x) = x \quad \text{for all } \sigma \in H\}.$$

This set is an intermediate field between L and F.

1.2.10. THEOREM. (without proof) *Main Theorem of Galois theory*

Let L/F be a finite Galois extension with Galois group $G = \mathrm{Gal}(L/F)$. Then

$$H \longmapsto L^H$$

is a one-to-one correspondence between subgroups H of G and subfields of L which contain F. The inverse map is given by

$$M \longmapsto \mathrm{Gal}(L/M) = H.$$

Normal subgroups H of G correspond to Galois extensions M/F and

$$\mathrm{Gal}(M/F) \simeq G/H.$$

1.3. *Finite fields*

Every finite field F has positive characteristic, since the homomorphism

$$\mathbb{Z} \longrightarrow F,$$

sending 1 to the identity element of F with respect to multiplication, is not injective. Let F be of prime characteristic p. Then the image of $\mathbb{Z}$ in F can be identified with the finite field $\mathbb{F}_p$ consisting of p elements. If the degree of $F/\mathbb{F}_p$ is n, then the number of elements in F is p^n. By 1.2.4 the group $F^\times$ is cyclic of order $p^n - 1$, so every non-zero element of F is a root of the polynomial $X^{p^n-1} - 1$. Therefore, all p^n elements of F are all p^n roots of the polynomial $f_n(X) = X^{p^n} - X$. The polynomial f_n is separable, since its derivative in characteristic p is equal to $p^n X^{p^n-1} - 1 = -1$. Thus, F is the splitting field of f_n over $\mathbb{F}_p$. We conclude that $F/\mathbb{F}_p$ is a Galois extension of degree $n = | F : \mathbb{F}_p |$.

LEMMA. *The Galois group of $F/\mathbb{F}_p$ is cyclic of order n: it is generated by an automorphism ϕ of F called the Frobenius automorphism:*

$$\phi(x) = x^p \quad \textit{for all } x \in F.$$

Proof. $\phi^m(x) = x^{p^m} = x$ for all $x \in F$ if and only if $n \mid m$. □

On the other hand, for every $n \geqslant 1$ the splitting field of f_n over $\mathbb{F}_p$ is a finite field consisting of p^n elements. Thus,

THEOREM. *For every n there is a unique (up to isomorphism) finite field $\mathbb{F}_{p^n}$ consisting of p^n elements; it is the splitting field of the polynomial $f_n(X) = X^{p^n} - X$. The finite extension $\mathbb{F}_{p^{nm}}/\mathbb{F}_{p^n}$ is a Galois extension with cyclic group of degree m generated by the Frobenius automorphism $\phi_n : x \mapsto x^{p^n}$.*

LEMMA. *Let $g(X)$ be an irreducible polynomial of degree m over a finite field $\mathbb{F}_{p^n}$. Then $g(X)$ divides $f_{nm}(X)$ and therefore is a separable polynomial.*

Proof. Let a be a root of $g(X)$. Then $\mathbb{F}_{p^n}(a)/\mathbb{F}_{p^n}$ is of degree m, so $\mathbb{F}_{p^n}(a) = \mathbb{F}_{p^{nm}}$. Since a is a root of $f_{nm}(X)$, g divides f_{nm}. The latter is separable and so is g. □

2. Integrality

Rings of algebraic integers generalise the ring of integers. Their study proves many properties of usual integers, not seen at the level of usual integers.

2.1. *Integrality over rings*

2.1.1. DEFINITION–PROPOSITION. Let B be a ring and A its subring.

An element $b \in B$ is called *integral over* A if it satisfies one of the following equivalent conditions:

(i) there exist $a_i \in A$ such that $f(b) = 0$ where $f(X) = X^n + a_{n-1}X^{n-1} + \cdots + a_0$;

(ii) the subring of B generated by A and b is an A-module of finite type;

(iii) there exists a subring C of B which contains A and b and which is an A-module of finite type.

Proof. (i)$\Rightarrow$(ii): note that the subring $A[b]$ of B generated by A and b coincides with the A-module M generated by $1, \ldots, b^{n-1}$. Indeed,

$$b^{n+j} = -a_0 b^j - \cdots - b^{n+j-1}$$

and by induction $b^j \in M$.

(ii)$\Rightarrow$ (iii): obvious.

(iii)$\Rightarrow$(i): let $C = c_1 A + \cdots + c_m A$. Then $bc_i = \sum_j a_{ij} c_j$, so $\sum_j (\delta_{ij} b - a_{ij})$ $c_j = 0$. Denote by d the determinant of $M = (\delta_{ij} b - a_{ij})$. Note that $d = f(b)$ where $f(X) \in A[X]$ is a monic polynomial. From linear algebra we know that $dE = M^* M$ where M^* is the adjugate matrix to M and E is the identity matrix of the same order of that of M. Denote by $\mathscr{C}$ the column consisting of c_j. Now we get $M\mathscr{C} = 0$ implies $M^* M \mathscr{C} = 0$ implies $dE\mathscr{C} = 0$ implies $d\mathscr{C} = 0$. Thus $dc_j = 0$ for all $1 \leqslant j \leqslant m$. Every $c \in C$ is a linear combination of c_j. Hence $dc = 0$ for all $c \in C$. In particular, $d1 = 0$, so $f(b) = d = 0$. □

EXAMPLES.

1. Every element of A is integral over A.

2. If A, B are fields, then an element $b \in B$ is integral over A if and only if b is algebraic over A.

3. Let $A = \mathbb{Z}$, $B = \mathbb{Q}$. A rational number r/s with relatively prime r and s is integral over $\mathbb{Z}$ if and only if $(r/s)^n + a_{n-1}(r/s)^{n-1} + \cdots + a_0 = 0$ for some integer a_i. Multiplying by s^n we deduce that s divides r^n, hence $s = \pm 1$ and $r/s \in \mathbb{Z}$. Hence integral in $\mathbb{Q}$ elements over $\mathbb{Z}$ are just all integers.

4. If B is a field, then it contains the field of fractions F of A. Let $\sigma \in \mathrm{Hom}_F(B,C)$ where C is an algebraically closed field containing B. If $b \in B$ is integral over A, then $\sigma(b) \in \sigma(B)$ is integral over A.

5. If $b \in B$ is a root of a non-zero polynomial $f(X) = a_nX^n + \cdots \in A[X]$, then $a_n^{n-1}f(b) = 0$ and $g(a_nb) = 0$ for $g(X) = X^n + a_{n-1}X^{n-1} + \cdots + a_n^{n-1}a_0$, $g(a_nX) = a_n^{n-1}f(X)$. Hence a_nb is integral over A. Thus, for every algebraic over A element b of B there is a non-zero $a \in A$ such that ab is integral over A.

2.1.2. COROLLARY. *Let A be a subring of an integral domain B. Let I be a non-zero A-module of finite type, $I \subset B$. Let $b \in B$ satisfy the property $bI \subset I$. Then b is integral over A.*

Proof. Indeed, as in the proof of $(iii) \Rightarrow (i)$ we deduce that $dc = 0$ for all $c \in I$. Since B is an integral domain, we deduce that $d = 0$, so $d = f(b) = 0$. □

2.1.3. PROPOSITION. *Let A be a subring of a ring B, and let $b_i \in B$ be such that b_i is integral over $A[b_1,\dots,b_{i-1}]$ for all i. Then $A[b_1,\dots,b_n]$ is an A-module of finite type.*

Proof. Induction on n. The case of $n = 1$ is the previous Proposition. If $C = A[b_1,\dots,b_{n-1}]$ is an A-module of finite type, then $C = \sum_{i=1}^m c_iA$. Now by the previous Proposition $C[b_n]$ is a C-module of finite type, so $C[b_n] = \sum_{j=1}^l d_jC$. Thus, $C[b_n] = \sum_{i,j} d_jc_iA$ is an A-module of finite type. □

2.1.4. COROLLARY 1. *If $b_1, b_2 \in B$ are integral over A, then $b_1 + b_2$, $b_1 - b_2, b_1b_2$ are integral over A.*

COROLLARY 2. *The set B' of elements of B which are integral over A is a subring of B containing A.*

DEFINITION. B' is called the *integral closure of A in B*. If A is an integral domain and B is its field of fractions, B' is called the *integral closure of A*.

A ring A is called *integrally closed* if A is an integral domain and A coincides with its integral closure in its field of fractions.

A ring B is said to be *integral over A* if every element of B is integral over A. If B is of characteristic zero, its elements integral over $\mathbb{Z}$ are called *integral elements* of B.

Let F be an algebraic number field. The integral closure of $\mathbb{Z}$ in F is called the *ring $\mathcal{O}_F$ of (algebraic) integers of F*.

From Example 5 in 2.1.1 it follows that the fraction field of $\mathcal{O}_F$ is F.

EXAMPLES.

1. A UFD is integrally closed. Indeed, if $x = a/b$ with relatively prime $a, b \in A$ is a root of polynomial $f(X) = X^n + \cdots + a_0 \in A[X]$, then b divides a^n, so b is a unit of A and $x \in A$.

In particular, the integral closure of $\mathbb{Z}$ in $\mathbb{Q}$ is $\mathbb{Z}$.

2. $\mathscr{O}_F$ is integrally closed (see below in 2.1.6).

2.1.5. LEMMA. *Let A be integrally closed and F be its fraction field. Let B be a field. Let $b \in B$ be algebraic over F. Then b is integral over A if and only if the monic irreducible polynomial $f_b(X) \in F[X]$ over F has coefficients in A.*

Proof. Let L be a finite extension of F which contains B and all $\sigma(b)$ for all F-homomorphisms from B to an algebraically closed field C. Since $b \in L$ is integral over A, $\sigma(b) \in L$ is integral over A for every σ. Then $f_b(X) = \prod(X - \sigma(b))$ has coefficients in F which belong to the ring generated by A and all $\sigma(b)$ and therefore are integral over A. Since A is integrally closed, $f_b(X) \in A[X]$.

If $f_b(X) \in A[X]$ then b is integral over A by 2.1.1. □

EXAMPLES.

1. Let F be an algebraic number field. Then an element $b \in F$ is integral if and only if its monic irreducible polynomial has integer coefficients.

For example, $\sqrt{d}$ for integer d is integral.

If $d \equiv 1 \mod 4$ then the monic irreducible polynomial of $(1+\sqrt{d})/2$ over $\mathbb{Q}$ is $X^2 - X + (1-d)/4 \in \mathbb{Z}[X]$, so $(1+\sqrt{d})/2$ is integral. Note that $\sqrt{d}$ belongs to $\mathbb{Z}[(1+\sqrt{d})/2]$, and hence $\mathbb{Z}[\sqrt{d}]$ is a subring of $\mathbb{Z}[(1+\sqrt{d})/2]$.

Thus, the integral closure of $\mathbb{Z}$ in $\mathbb{Q}(\sqrt{d})$ contains the subring $\mathbb{Z}[\sqrt{d}]$ and the subring $\mathbb{Z}[(1+\sqrt{d})/2]$ if $d \equiv 1 \mod 4$. We show that there are no other integral elements.

An element $a + b\sqrt{d}$ with rational a and $b \neq 0$ is integral if and only if its monic irreducible polynomial $X^2 - 2aX + (a^2 - db^2)$ belongs to $\mathbb{Z}[X]$. Therefore $2a, 2b$ are integers. If $a = (2k+1)/2$ for an integer k, then it is easy to see that $a^2 - db^2 \in \mathbb{Z}$ if and only if $b = (2l+1)/2$ with integer l and $(2k+1)^2 - d(2l+1)^2$ is divisible by 4. The latter implies that d is a quadratic residue mod 4, i.e. $d \equiv 1 \mod 4$. In turn, if $d \equiv 1 \mod 4$ then every element $(2k+1)/2 + (2l+1)\sqrt{d}/2$ is integral.

Thus, integral elements of $\mathbb{Q}(\sqrt{d})$ are equal to

$$\begin{cases} \mathbb{Z}[\sqrt{d}] & \text{if } d \not\equiv 1 \mod 4 \\ \mathbb{Z}[(1+\sqrt{d})/2] & \text{if } d \equiv 1 \mod 4 \end{cases}$$

2. $\mathscr{O}_{\mathbb{Q}^m}$ is equal to $\mathbb{Z}[\zeta_m]$ (see 2.4).

2.1.6. LEMMA. If B is integral over A and C is integral over B, then C is integral over A.

Proof. Let $c \in C$ be a root of the polynomial $f(X) = X^n + b_{n-1}X^{n-1} + \cdots + b_0$ with $b_i \in B$. Then c is integral over $A[b_0, \ldots, b_{n-1}]$. Since $b_i \in B$ are integral over A, Proposition 2.1.3 implies that $A[b_0, \ldots, b_{n-1}, c]$ is an A-module of finite type. From 2.1.1 we conclude that c is integral over A. □

COROLLARY. *$\mathscr{O}_F$ is integrally closed.*

Proof. An element of F integral over $\mathscr{O}_F$ is integral over $\mathbb{Z}$ due to the previous lemma. □

2.1.7. PROPOSITION. *Let B be an integral domain and A be its subring such that B is integral over A. Then B is a field if and only if A is a field.*

Proof. If A is a field, then $A[b]$ for $b \in B \setminus 0$ is a vector space of finite dimension over A, and the A-linear map $\varphi\colon A[b] \longrightarrow A[b], \varphi(c) = bc$ is injective, therefore surjective, so b is invertible in B.

If B is a field and $a \in A \setminus 0$, then the inverse $a^{-1} \in B$ satisfies an equation $a^{-n} + a_{n-1}a^{-n+1} + \cdots + a_0 = 0$ with some $a_i \in A$. Then $a^{-1} = -a_{n-1} - \cdots - a_0 a^{n-1}$, so $a^{-1} \in A$. □

2.2. *Norms and traces*

2.2.1. DEFINITION. Let A be a subring of a ring B such that B is a free A-module of finite rank n. In this situation, similarly to the situation of finite dimensional vector spaces over fields, for a $b \in B$ one has the operator m_b of multiplication by $b \in B$, $m_b\colon B \longrightarrow B$, $m_b(c) = bc$. One can work with its matrix M_b with respect to a specific basis of B over A, its characteristic polynomials $g_b(X) = \det(XE - M_b)$, trace $\mathrm{Tr}_{B/A}(b) = \mathrm{Tr}\, M_b$ and norm $N_{B/A}(b) = \det M_b$.

If $g_b(X) = X^n + a_{n-1}X^{n-1} + \cdots + a_0$ then

$$a_{n-1} = -\mathrm{Tr}_{B/A}(b), \quad a_0 = (-1)^n N_{B/A}(b).$$

2.2.2. We have

$$\mathrm{Tr}(b+b') = \mathrm{Tr}(b) + \mathrm{Tr}(b'), \mathrm{Tr}(ab) = a\,\mathrm{Tr}(b), \mathrm{Tr}(a) = na,$$
$$N(bb') = N(b)N(b'), N(ab) = a^n N(b), N(a) = a^n$$

for $a \in A$.

2.2.3. Everywhere below in this section F is either a finite field of a field of characteristic zero. Then every finite extension of F is separable.

PROPOSITION. *Let L be an algebraic extension of F of degree n. Let $b \in L$ and $b_1, \dots, b_n$ be roots of the monic irreducible polynomial of b over F each one repeated $|L : F(b)|$ times. Then the characteristic polynomial $g_b(X)$ of b with respect to L/F is $\prod (X - b_i)$, and $\mathrm{Tr}_{L/F}(b) = \sum b_i, N_{L/F}(b) = \prod b_i$.*

Proof. If $L = F(b)$, then use the basis $1, b, \dots, b^{n-1}$ to calculate g_b. Let $f_b(X) = X^n + c_{n-1}X^{n-1} + \dots + c_0$ be the monic irreducible polynomial of b over F, then the matrix of m_b is

$$M_b = \begin{pmatrix} 0 & 1 & 0 & \dots & 0 \\ 0 & 0 & 1 & \dots & 0 \\ \vdots & \vdots & \vdots & \vdots & \vdots \\ -c_0 & -c_1 & -c_2 & \dots & -c_{n-1} \end{pmatrix}.$$

Hence $g_b(X) = \det(XE - M_b) = f_b(X)$ and $\det M_b = \prod b_i$, $\mathrm{Tr}\, M_b = \sum b_i$.

In the general case when $|F(b) : F| = m < n$ choose a basis $\omega_1, \dots, \omega_{n/m}$ of L over $F(b)$ and take $\omega_1, \dots, \omega_1 b^{m-1}, \omega_2, \dots, \omega_2 b^{m-1}, \dots$ as a basis of L over F. The matrix M_b is a block matrix with the same block repeated n/m times on the diagonal and everything else being zero. Therefore, $g_b(X) = f_b(X)^{|L:F(b)|}$ where $f_b(X)$ is the monic irreducible polynomial of b over F. □

EXAMPLE. Let $F = \mathbb{Q}$, $L = \mathbb{Q}(\sqrt{d})$ with square-free integer d. Then

$$g_{a+b\sqrt{d}}(X) = (X - a - b\sqrt{d})(X - a + b\sqrt{d}) = X^2 - 2aX + (a^2 - db^2),$$

so

$$\mathrm{Tr}_{\mathbb{Q}(\sqrt{d})/\mathbb{Q}}(a + b\sqrt{d}) = 2a, \quad N_{\mathbb{Q}(\sqrt{d})/\mathbb{Q}}(a + b\sqrt{d}) = a^2 - db^2.$$

In particular, an integer number c is a sum of two squares if and only if $c \in N_{\mathbb{Q}(\sqrt{-1})/\mathbb{Q}} \mathcal{O}_{\mathbb{Q}(\sqrt{-1})}$.

More generally, c is in the form $a^2 - db^2$ with integer a, b and square-free d not congruent to $1 \mod 4$ if and only if

$$c \in N_{\mathbb{Q}(\sqrt{d})/\mathbb{Q}} \mathbb{Z}[\sqrt{d}].$$

2.2.4. COROLLARY 1. *Let σ_i be distinct F-homomorphisms of L into C. Then $\mathrm{Tr}_{L/F}(b) = \sum \sigma_i b$, $N_{L/F}(b) = \prod \sigma_i(b)$.*

Proof. In the previous Proposition $b_i = \sigma_i(b)$. □

COROLLARY 2. *Let A be an integral domain, and let F be its field of fractions. Let L be an extension of F of finite degree. Let A' be the integral closure of A in F. Then for an integral element $b \in L$ over A the polynomial $g_b(X)$ is in $A'[X]$, and $\mathrm{Tr}_{L/F}(b), N_{L/F}(b)$ belong to A'.*

Proof. All b_i are integral over A. □

COROLLARY 3. *If, in addition, A is integrally closed, then $\mathrm{Tr}_{L/F}(b), N_{L/F}(b)$ are in A.*

Proof. Since A is integrally closed, $A' \cap F = A$. □

2.2.5. LEMMA. *Let F be a finite field of a field of characteristic zero. If L is a finite extension of F and M/F is a subextension of L/F, then the following transitivity property holds*

$$\mathrm{Tr}_{L/F} = \mathrm{Tr}_{M/F} \circ \mathrm{Tr}_{L/M}, \qquad N_{L/F} = N_{M/F} \circ N_{L/M}.$$

Proof. Let $\sigma_1, \dots, \sigma_m$ be all distinct F-homomorphisms of M into C ($m = |M : F|$). Let $\tau_1, \dots, \tau_{n/m}$ be all distinct M-homomorphisms of L into C ($n/m = |L : M|$). The field $\tau_j(L)$ is a finite extension of F, and by 1.2.5 there is an element $a_j \in C$ such that $\tau_j(L) = F(a_j)$. Let E be the minimal subfield of C containing M and all a_j. Using 1.2.3 extend σ_i to $\sigma_i' : E \longrightarrow C$. Then the composition $\sigma_i' \circ \tau_j : L \longrightarrow C$ is defined. Note that $\sigma_i' \circ \tau_j = \sigma_{i_1}' \circ \tau_{j_1}$ implies $\sigma_i = \sigma_i' \circ \tau_j|_M = \sigma_{i_1}' \circ \tau_{j_1}|_M = \sigma_{i_1}$, so $i = i_1$, and then $j = j_1$. Hence $\sigma_i' \circ \tau_j$ for $1 \leqslant i \leqslant m, 1 \leqslant j \leqslant n/m$ are all n distinct F-homomorphisms of L into C. By Corollary 1 in 2.2.4

$$\begin{aligned} N_{M/F}(N_{L/M}(b)) &= N_{M/F}\left(\prod \tau_j(b)\right) = \prod \sigma_i'\left(\prod \tau_j(b)\right) \\ &= \prod (\sigma_i' \circ \tau_j)(b) = N_{L/F}(b). \end{aligned}$$

Similar arguments work for the trace. □

2.3. *Integral basis*

2.3.1. DEFINITION. Let A be a subring of a ring B such that B is a free A-module of rank n. Let $b_1, \dots, b_n \in B$. Then the *discriminant* $D(b_1, \dots, b_n)$ is defined as $\det(\mathrm{Tr}_{B/A}(b_i b_j))$.

2.3.2. PROPOSITION. *If* $c_i \in B$ *and* $c_i = \sum a_{ij} b_j$, $a_{ij} \in A$, *then*

$$D(c_1, \ldots, c_n) = (\det(a_{ij}))^2 D(b_1, \ldots, b_n).$$

Proof. $(c_i)^t = (a_{ij})(b_j)^t$, $(c_k c_l) = (c_k)^t (c_l) = (a_{ki})(b_i b_j)(a_{lj})^t$, $(\mathrm{Tr}(c_k c_l)) = (a_{ki})(\mathrm{Tr}(b_i b_j))(a_{lj})^t$. □

2.3.3. DEFINITION. The *discriminant* $\mathscr{D}_{B/A}$ of B over A is the principal ideal of A generated by the discriminant of any basis of B over A.

By Proposition 2.3.2 every basis of B over A generates the same principal ideal of A, since $(\det(a_{ij}))^2$ is invertible in A for the matrix (a_{ij}) relating two bases.

For the discriminant in finite extensions of fields of characteristic zero see 2.3.5.

2.3.4. PROPOSITION. *Let* $\mathscr{D}_{B/A} \neq 0$. *Let B be an integral domain. Then a set* $b_1, \ldots, b_n$ *is a basis of B over A if and only if* $D(b_1, \ldots, b_n)A = \mathscr{D}_{B/A}$.

Proof. Let $D(b_1, \ldots, b_n)A = \mathscr{D}_{B/A}$. Let $c_1, \ldots, c_n$ be a basis of B over A and let $b_i = \sum_j a_{ij} c_j$. Then $D(b_1, \ldots, b_n) = \det(a_{ij})^2 D(c_1, \ldots, c_n)$. Denote $d = D(c_1, \ldots, c_n)$.

Since $D(b_1, \ldots, b_n)A = D(c_1, \ldots, c_n)A$, we get $aD(b_1, \ldots, b_n) = d$ for some $a \in A$. Therefore $d(1 - a\det(a_{ij})^2) = 0$ and $\det(a_{ij})$ is invertible in A, so the matrix (a_{ij}) is invertible in the ring of matrices over A. Thus $b_1, \ldots, b_n$ is a basis of B over A. □

2.3.5. PROPOSITION. *Let F be a finite field or a field of characteristic zero. Let L be an extension of F of degree n and let* $\sigma_1, \ldots, \sigma_n$ *be distinct F-homomorphisms of L into C. Let* $b_1, \ldots, b_n$ *be a basis of L over F. Then*

$$D(b_1, \ldots, b_n) = \det(\sigma_i(b_j))^2 \neq 0.$$

Proof. We have

$$\det(\mathrm{Tr}(b_i b_j)) = \det(\sum_k \sigma_k(b_i)\sigma_k(b_j)) = \det((\sigma_k(b_i))^t(\sigma_k(b_j))) = \det(\sigma_i(b_j))^2.$$

To prove the rest we use Artin's trick. If $\det(\sigma_i(b_j)) = 0$, then there exist $a_i \in L$ not all zero such that $\sum_i a_i \sigma_i(b_j) = 0$ for all j. Then $\sum_i a_i \sigma_i(b) = 0$ for every $b \in L$.

Let $\sum a_i' \sigma_i(b) = 0$ for all $b \in L$ with the minimal number > 1 of non-zero $a_i' \in A$. Assume $a_1' \neq 0$.

Let $c \in L$ be such that $L = F(c)$ (see 1.2.5), then $\sigma_1(c) \neq \sigma_i(c)$ for $i > 1$.

We now have $\sum a_i'\sigma_i(bc) = \sum a_i'\sigma_i(b)\sigma_i(c) = 0$. Hence

$$\sigma_1(c)(\sum a_i'\sigma_i(b)) - \sum a_i'\sigma_i(b)\sigma_i(c) = \sum_{i>1} a_i'(\sigma_1(c) - \sigma_i(c))\sigma_i(b) = 0.$$

Put $a_i'' = a_i'(\sigma_1(c) - \sigma_i(c))$, so $\sum a_i''\sigma_i(b) = 0$ with smaller number of non-zero a_i'' than in a_i', a contradiction. □

Thus, for fields the discriminant measures the behaviour of elements of a basis with respect to Galois automorphisms action.

COROLLARY. *Under the assumptions of the Proposition the linear map*

$$L \longrightarrow \mathrm{Hom}_F(L,F): \quad b \mapsto (c \mapsto \mathrm{Tr}_{L/F}(bc))$$

between n-dimensional F-vector spaces is injective, and hence bijective.

Therefore for a basis $b_1,\dots,b_n$ of L/F there is a dual basis $c_1,\dots,c_n$ of L/F such that $\mathrm{Tr}_{L/F}(b_ic_j) = \delta_{ij}$.

Proof. If $b = \sum a_ib_i$, $a_i \in F$ and $\mathrm{Tr}_{L/F}(bc) = 0$ for all $c \in L$, then we get equations $\sum a_i \mathrm{Tr}_{L/F}(b_ib_j) = 0$. This is a system of linear equations in a_i with nondegenerate matrix $\mathrm{Tr}_{L/F}(b_ib_j)$, so the only solution is $a_i = 0$. Elements of the dual basis c_j correspond to $f_j \in \mathrm{Hom}_F(L,F)$, $f_j(b_i) = \delta_{ij}$. □

2.3.6. THEOREM. *Let A be an integrally closed ring and F be its field of fractions. Let L be an extension of F of degree n and A' be the integral closure of A in L. Let F be of characteristic 0. Then A' is an A-submodule of a free A-module of rank n.*

Proof. Let $e_1,\dots,e_n$ be a basis of F-vector space L. Then due to Example 5 in 2.1.1 there is $0 \neq a_i \in A$ such that $a_ie_i \in A'$. Then for $a = \prod a_i$ we get $b_i = ae_i \in A'$ form a basis of L/F.

Let $c_1,\dots,c_n$ be the dual basis for $b_1,\dots,b_n$. Claim: $A' \subset \sum c_iA$. Indeed, let $c = \sum a_ic_i \in A'$. Then

$$\mathrm{Tr}_{L/F}(cb_i) = \sum_j a_j \mathrm{Tr}_{L/F}(c_jb_i) = a_i \in A$$

by 2.2.5. Now $\sum c_iA = \oplus c_iA$, since $\{c_i\}$ is a basis of L/F. □

2.3.7. THEOREM. *Let A be a principal ideal ring and F be its field of fractions of characteristic 0. Let L be an extension of F of degree n. Then the integral closure A' of A in L is a free A-module of rank n.*

In particular, the ring of integers $\mathcal{O}_F$ of a number field F is a free $\mathbb{Z}$-module of rank equal to the degree of F.

Proof. The description of modules of finite type over PID and the previous Theorem imply that A' is a free A-module of rank $m \leqslant n$. On the other hand, by the first part of the proof of the previous Theorem A' contains n A-linear independent elements over A. Thus, $m = n$. □

DEFINITION. The discriminant d_F of any integral basis of $\mathcal{O}_F$ is called *the discriminant of* F. This is a non-zero integer.

Since every two integral bases are related via an invertible matrix with integer coefficients (whose determinant is therefore ± 1), 2.3.2 implies that d_F is uniquely determined.

2.3.8. EXAMPLES.

1. Let d be a square-free integer. By 2.1.5 the ring of integers of $\mathbb{Q}(\sqrt{d})$ has an integral basis $1, \alpha$ where $\alpha = \sqrt{d}$ if $d \not\equiv 1 \mod 4$ and $\alpha = (1+\sqrt{d})/2$ if $d \equiv 1 \mod 4$.

The discriminant of $\mathbb{Q}(\sqrt{d})$ is equal to

$$4d \quad \text{if } d \not\equiv 1 \mod 4, \qquad \text{and } d \quad \text{if } d \equiv 1 \mod 4\,.$$

To prove this calculate directly $D(1,\alpha)$ using the definitions, or use 2.3.9.

2. Let F be an algebraic number field of degree n and let $a \in F$ be an integral element over $\mathbb{Z}$. Assume that $D(1,a,\dots,a^{n-1})$ is a square free integer. Then $1,a,\dots,a^{n-1}$ is a basis of $\mathcal{O}_F$ over $\mathbb{Z}$, so $\mathcal{O}_F = \mathbb{Z}[a]$. Indeed: choose a basis $b_1,\dots,b_n$ of $\mathcal{O}_F$ over $\mathbb{Z}$ and let $\{c_1,\dots,c_n\} = \{1,a,\dots,a^{n-1}\}$. Let $c_i = \sum a_{ij} b_j$. By 2.3.2 we have $D(1,a,\dots,a^{n-1}) = (\det(a_{ij})^2 D(b_1,\dots,b_n)$. Since $D(1,a,\dots,a^{n-1})$ is a square free integer, we get $\det(a_{ij}) = \pm 1$, so (a_{ij}) is invertible in $M_n(\mathbb{Z})$, and hence $1,a,\dots,a^{n-1}$ is a basis of $\mathcal{O}_F$ over $\mathbb{Z}$.

2.3.9. EXAMPLE. Let F be of characteristic zero and $L = F(b)$ be an extension of degree n over F. Let $f(X)$ be the minimal polynomial of b over F whose roots are b_i. Then

$$f(X) = \prod (X - b_j), \quad f'(b_i) = \prod_{j \neq i} (b_i - b_j),$$

$$N_{L/F} f'(b) = \prod_i f'(\sigma_i b) = \prod_i f'(b_i).$$

Then

$$\begin{aligned} D(1,b,\dots,b^{n-1}) &= \det(b_i^j)^2 \\ &= (-1)^{n(n-1)/2} \prod_{i \neq j} (b_i - b_j) = (-1)^{n(n-1)/2} N_{L/F}(f'(b)). \end{aligned}$$

Let $f(X) = X^n + aX + c$. Then

$$b^n = -ab - c, \quad b^{n-1} = -a - cb^{-1}$$

and

$$e = f'(b) = nb^{n-1} + a = n(-a - cb^{-1}) + a,$$

so

$$b = -nc(e+(n-1)a)^{-1}.$$

The minimal polynomial $g(Y)$ of e over F corresponds to the minimal polynomial $f(X)$ of b; it is $(Y+(n-1)a)^n$ times $c^{-1}f(-nc(Y+(n-1)a)^{-1})$, i.e.

$$g(Y) = (Y+(n-1)a)^n - na(Y+(n-1)a)^{n-1} + (-1)^n n^n c^{n-1}.$$

Hence

$$\begin{aligned} N_{L/F}(f'(b)) &= g(0)(-1)^n \\ &= n^n c^{n-1} + (-1)^{n-1}(n-1)^{n-1}a^n, \end{aligned}$$

so

$$\begin{aligned} &D(1,b,\dots,b^{n-1}) \\ &= (-1)^{n(n-1)/2}(n^n c^{n-1} + (-1)^{n-1}(n-1)^{n-1}a^n). \end{aligned}$$

For $n=2$ one has a^2-4c, for $n=3$ one has $-27c^2-4a^3$.

For example, let $f(X) = X^3+X+1$. It is irreducible over $\mathbb{Q}$. Its discriminant is equal to (-31), so according to example 2.5.3 $\mathcal{O}_F = \mathbb{Z}[a]$ where a is a root of $f(X)$ and $F = \mathbb{Q}[a]$.

2.4. A little about cyclotomic fields.

2.4.1. DEFINITION. Let ζ_n be a primitive nth root of unity. The field $\mathbb{Q}(\zeta_n)$ is called the (nth) cyclotomic field.

2.4.2. THEOREM. *Let p be a prime number. The cyclotomic field $\mathbb{Q}(\zeta_p)$ is of degree $p-1$ over $\mathbb{Q}$. Its ring of integers coincides with $\mathbb{Z}[\zeta_p]$.*

Proof. Denote $z=\zeta_p$. Let $f(X) = (X^p-1)/(X-1) = X^{p-1}+\dots+1$. Recall that $z-1$ is a root of the polynomial $g(Y) = f(1+Y) = Y^{p-1}+\dots+p$ is a p-Eisenstein polynomial, so $f(X)$ is irreducible over $\mathbb{Q}$, $|\mathbb{Q}(z):\mathbb{Q}| = p-1$ and $1,z,\dots,z^{p-2}$ is a basis of the $\mathbb{Q}$-vector space $\mathbb{Q}(z)$.

Let O be the ring of integers of $\mathbb{Q}(z)$. Since the monic irreducible polynomial of z over $\mathbb{Q}$ has integer coefficients, $z \in O$. Since z^{-1} is a primitive root of unity, $z^{-1} \in O$. Thus, z is a unit of O.

Then $z^i \in O$ for all $i \in \mathbb{Z}$ ($z^{-1} = z^{p-1}$). We have $1-z^i = (1-z)(1+\dots+z^{i-1}) \in (1-z)O$.

Denote by Tr and N the trace and norm for $\mathbb{Q}(z)/\mathbb{Q}$. Note that $\mathrm{Tr}(z) = -1$ and since z^i for $1 \leqslant i \leqslant p-1$ are primitive pth roots of unity, $\mathrm{Tr}(z^i) = -1$; $\mathrm{Tr}(1) = p-1$. Hence

$$\mathrm{Tr}(1-z^i) = p \quad \text{for } 1 \leqslant i \leqslant p-1.$$

Furthermore, $N(z-1)$ is equal to the free term of $g(Y)$ times $(-1)^{p-1}$, hence $N(z-1) = (-1)^{p-1}p$ and

$$N(1-z) = \prod_{1 \leqslant i \leqslant p-1} (1-z^i) = p,$$

since $1-z^i$ are Galois conjugate to $1-z$ over $\mathbb{Q}$. Therefore $p\mathbb{Z}$ is contained in the ideal $I = (1-z)O \cap \mathbb{Z}$.

If $I = \mathbb{Z}$, then $1-z$ would be a unit of O and so would be its Galois conjugates $1-z^i$, which then implies that p as their product would be a unit of O. Then $p^{-1} \in O \cap \mathbb{Q} = \mathbb{Z}$, a contradiction. Thus,

$$I = (1-z)O \cap \mathbb{Z} = p\mathbb{Z}.$$

Now we prove another auxiliary result:

$$\mathrm{Tr}((1-z)O) \subset p\mathbb{Z}.$$

Indeed, every Galois conjugate of $y(1-z)$ for $y \in O$ is of the type $y_i(1-z^i)$ with appropriate $y_i \in O$, so $\mathrm{Tr}(y(1-z)) = \sum y_i(1-z^i) \in I = p\mathbb{Z}$.

Now let $x = \sum_{0 \leqslant i \leqslant p-2} a_i z^i \in O$ with $a_i \in \mathbb{Q}$. We aim to show that all a_i belong to $\mathbb{Z}$. From the calculation of the traces of z^i it follows that

$$\mathrm{Tr}((1-z)x) = a_0 \,\mathrm{Tr}(1-z) + \sum_{0 < i \leqslant p-2} a_i \,\mathrm{Tr}(z^i - z^{i+1}) = a_0 p$$

and so $a_0 p \in \mathrm{Tr}((1-z)O) \subset p\mathbb{Z}$; therefore, $a_0 \in \mathbb{Z}$. Since z is a unit of O, we deduce that $x_1 = z^{-1}(x - a_0) = a_1 + a_2 z + \cdots + a_{p-2} z^{p-3} \in O$. Applying the same arguments, we then deduce that $a_1 \in \mathbb{Z}$. Looking at $x_i = z^{-1}(x_{i-1} - a_{i-1}) \in O$ we then conclude $a_i \in \mathbb{Z}$ for all i. Thus $O = \mathbb{Z}[z]$. □

2.4.3. The discriminant of $\mathbb{Q}(\zeta_p)$ is $D(1, z, \ldots, z^{p-2})$.
By 2.3.9 it is equal $(-1)^{(p-1)(p-2)/2} N(f'(z))$. We have

$$f'(z) = pz^{p-1}/(z-1) = pz^{-1}/(z-1)$$

and

$$N(f'(z)) = N(p)N(z)^{-1}/N(z-1) = p^{p-1}(-1)^{p-1}/((-1)^{p-1}p) = p^{p-2}.$$

Thus, the discriminant of $\mathbb{Q}(\zeta_p)$ is $(-1)^{(p-1)(p-2)/2}\, p^{p-2}$.

2.4.4. In general, the extension $\mathbb{Q}(\zeta_m)/\mathbb{Q}$ is a Galois extension and elements of the Galois group $\mathrm{Gal}(\mathbb{Q}(\zeta_m)/\mathbb{Q})$ are determined by their action on the primitive mth root ζ_m of unity:

$$\sigma \mapsto i : \sigma(\zeta_m) = \zeta_m^i, \quad (i,m) = 1.$$

This map induces a group isomorphism

$$\mathrm{Gal}(\mathbb{Q}(\zeta_m)/\mathbb{Q}) \longrightarrow (\mathbb{Z}/m\mathbb{Z})^{\times}.$$

One can prove that the ring of integers of $\mathbb{Q}(\zeta_m)$ is $\mathbb{Z}(\zeta_m)$.

3. Dedekind Rings

3.1. *Noetherian rings in brief*

3.1.1. Recall (see the commutative algebra course linked to at the beginning of this text) that a module M over a ring is called a Noetherian module if one of the following equivalent properties is satisfied:

(i) every submodule of M is of finite type;

(ii) every increasing sequence of submodules stabilises;

(iii) every nonempty family of submodules contains a maximal element with respect to inclusion.

A ring A is called *Noetherian* if it is a Noetherian A-module.

EXAMPLE. A PID is a Noetherian ring, since every ideal of it is generated by one element.

LEMMA. *Let M be an A-module and N is a submodule of M. Then M is a Noetherian A-module if and only if N and M/N are.*

COROLLARY 1. *If N_i are Noetherian A-modules, so is $\oplus_{i=1}^n N_i$.*

COROLLARY 2. *Let A be a Noetherian ring and let M be an A-module of finite type. Then M is a Noetherian A-module.*

3.1.2. PROPOSITION. *Let A be a Noetherian integrally closed ring. Let K be its field of fractions and let L be a finite extension of K. Let A' be the integral closure of A in L. Suppose that K is of characteristic 0. Then A' is a Noetherian ring.*

Proof. According to 2.3.6 A' is a submodule of a free A-module of finite rank. Hence A' is a Noetherian A-module. Every ideal of A' is in particular an A-submodule of A'. Hence every increasing sequence ideals of A' stabilises and A' is a Noetherian ring. □

3.1.3. EXAMPLE. The ring of integers $\mathcal{O}_F$ of a number field F is a Noetherian ring. It is a free $\mathbb{Z}$-module of rank n where n is the degree of F.

LEMMA. *Every nonzero element of* $\mathcal{O}_F \setminus \{0\}$ *is either a unit or factorises into a product of prime elements and units (not uniquely in general).*

Proof. Indeed, assume the family of proper principal ideals (a) where a cannot be factorised into a product of prime elements is nonempty. Choose a maximal element (a) in this family. The element a is not a unit, and a is not prime. Hence there is a factorisation $a = bc$ with both $b, c \notin \mathcal{O}_F^*$. Then $(b), (c)$ are strictly larger than (a), so b and c are products of prime elements. Then a is, a contradiction. □

3.2. *Definition of Dedekind rings*

3.2.1. DEFINITION. An integral domain A is called a *Dedekind ring* if
(i) A is a Noetherian ring;
(ii) A is integrally closed;
(iii) every non-zero prime ideal of A is maximal.

LEMMA. *Every principal ideal domain A is a Dedekind ring.*

Proof. For (i) see 3.1.1 and for (ii) see 2.1.4. If (a) is a non-zero prime ideal and $(a) \subset (b) \neq A$, $(a) \neq (b)$. Then b isn't a unit of A, b divides a and a does not divide b. Write $a = bc$. Since (a) is prime, either b or c belongs to (a). If b does then $(a) = (b)$. If b doesn't, then c must belong to (a), so $c = ad$ for some $d \in A$, and $a = bc = bda$ which means that b is a unit of A, a contradiction. Thus, property (iii) is satisfied as well. □

3.2.2. LEMMA. *Let A be an integral domain. Let K be its field of fractions and let L be a finite extension of K. Let B be the integral closure of A in L. Let P be a non-zero prime ideal of B. Then $P \cap A$ is a non-zero prime ideal of A.*

Proof. Let P be a non-zero prime ideal of B. Then $P \cap A \neq A$, since otherwise $1 \in P \cap A$ and hence $P = B$.

If $c,d \in A$ and $cd \in P \cap A$, then either $c \in P \cap A$ or $d \in P \cap A$. Hence $P \cap A$ is a prime ideal of A.

Let $b \in P$, $b \neq 0$. Then b satisfies a polynomial relation $b^n + a_{n-1}b^{n-1} + \cdots + a_0 = 0$ with $a_i \in A$. We can assume that $a_0 \neq 0$. Then $a_0 = -(b^n + \cdots + a_1 b) \in A \cap P$, so $P \cap A$ is a non-zero prime ideal of A. □

3.2.3. THEOREM. *Let A be a Dedekind ring. Let K be its field of fractions and let L be a finite extension of K. Let B be the integral closure of A in L. Suppose that K is of characteristic 0. Then B is a Dedekind ring.*

Proof. B is Noetherian by 3.1.2. It is integrally closed due to 2.1.6. By 3.2.2 if P is a non-zero proper prime ideal of B, then $P \cap A$ is a non-zero prime ideal of A. Since A is a Dedeking ring, it is a maximal ideal of A. The quotient ring B/P is integral over the field $A/(P \cap A)$. Hence by 2.1.7 B/P is a field and P is a maximal ideal of B. □

3.2.4. EXAMPLE. The ring of integers $\mathcal{O}_F$ of a number field F is a Dedekind ring.

3.3. *Factorisation in Dedekind rings*

3.3.1. LEMMA. *Every non-zero ideal in a Dedekind ring A contains some product of maximal ideals.*

Proof. If not, then the set of non-zero ideals which do not contain products of maximal ideals is non-empty. Let I be a maximal element with this property. The ideal I is not A and is not a maximal ideal, since it doesn't contain a product of maximal ideals. Hence I is not a prime ideal. Therefore there are $a,b \in A$ such that $ab \in I$ and $a,b \notin I$. Since $I + aA$ and $I + bA$ are strictly greater than I, there are maximal ideals P_i and Q_j such that $\prod P_i \subset I + aA$ and $\prod Q_j \subset I + bA$. Then $\prod P_i \prod Q_j \subset (I + aA)(I + bA) \subset I$, a contradiction. □

3.3.2. LEMMA. *Let a prime ideal P of A contain $I_1 \dots I_m$, where I_j are ideals of A. Then P contains one of I_j.*

Proof. If $I_k \not\subset P$ for all $1 \leqslant k \leqslant m$, then take $a_k \in I_k \setminus P$ and consider the product $a_1 \dots a_m$. It belongs to P, therefore one of a_i belongs to P, a contradiction. □

3.3.3. The next Proposition shows that for every non-zero ideal I of a Dedekind ring A there is an ideal J such that IJ is a principal non-zero ideal of A. Moreover, the Proposition gives an explicit description of J.

PROPOSITION. *Let I be a non-zero ideal of a Dedekind ring A and b be a non-zero element of I. Let K be the field of fractions of A. Define*

$$J = \{a \in K : aI \subset bA\}.$$

Then J is an ideal of A and $IJ = bA$.

Proof. Since $b \in I$, we get $bA \subset I$.

If $a \in J$ then $aI \subset bA \subset I$, so $aI \subset I$. Now we use the Noetherian and integrality property of Dedekind rings: Since I is an A-module of finite type, by Remark in 2.1.1 a is integral over A. Since A is integrally closed, $a \in A$. Thus, $J \subset A$.

The set J is closed with respect to addition and multiplication by elements of A, so J is an ideal of A. It is clear that $IJ \subset bA$. Assume that $IJ \neq bA$ and get a contradiction.

The ideal $b^{-1}IJ$ is a proper ideal of A, and hence it is contained in a maximal ideal P. Note that $b \in J$, since $bI \subset bA$. So $b^2 \in IJ$ and $b \in b^{-1}IJ$, $bA \subset b^{-1}IJ$. By 3.3.1 there are non-zero prime ideals P_i such that $P_1 \dots P_m \subset bA$. Let m be the minimal number with this property.

We have

$$P_1 \dots P_m \subset bA \subset b^{-1}IJ \subset P.$$

By 3.3.2 P contains one of P_i. Without loss of generality we can assume that $P_1 \subset P$. Since P_1 is maximal, $P_1 = P$.

If $m = 1$, then $P \subset bA \subset b^{-1}IJ \subset P$, so $P = bA$. Since $bA \subset I$ we get $P \subset I$. Since P is maximal, either $I = P$ or $I = A$. The definition of J implies in the first case $J = \{a \in K : aI = aP \subset bA = P\} = A$ and $IJ = bA$ and in the second case $b \in J$ implies $bA \subset J = \{a \in K : aA \subset bA\} \subset \{a \in K : a \in bA\} = bA$ and so $J = bA$ and $IJ = bA$.

Let $m > 1$. Note that $P_2 \dots P_m \not\subset bA$ due to the definition of m. Therefore, there is $d \in P_2 \dots P_m$ such that $d \notin bA$. Since $b^{-1}IJ \subset P$, $db^{-1}IJ \subset dP \subset PP_2 \dots P_m \subset bA$. So $(db^{-1}J)I \subset bA$, and the defining property of J implies that $db^{-1}J \subset J$. Since J is an A-module of finite type, by 2.1.1 db^{-1} belongs to A, i.e. $d \in bA$, a contradiction. □

3.3.4. COROLLARY 1. (*Cancellation property*)
Let I, J, H be non-zero ideals of A, then $IH = JH$ implies $I = J$.

Proof. Let H' be an ideal such that $HH' = aA$ is a principal ideal. Then $aI = aJ$ and $I = J$. □

3.3.5. COROLLARY 2. (*Factorisation property*)
Let I and J be ideals of A. Then $I \subset J$ if and only if $I = JH$ for an ideal H.

Proof. If $I \subset J$ and J is non-zero, then let J' be an ideal of A such that $JJ' = aA$ is a principal ideal. Then $IJ' \subset aA$, so $H = a^{-1}IJ'$ is an ideal of A. Now

$$JH = Ja^{-1}IJ' = a^{-1}IJJ' = a^{-1}aI = I.$$

□

3.3.6. THEOREM. *Every proper ideal of a Dedekind ring factorises into a product of maximal ideals whose collection is uniquely determined.*

Proof. Let I be a non-zero ideal of A. There is a maximal ideal P_1 which contains I. Then by the factorisation property 3.3.5 $I = P_1Q_1$ for some ideal Q_1. Note that $I \subset Q_1$ is a proper inclusion, since otherwise $AQ_1 = Q_1 = I = P_1Q_1$ and by the cancellation property 3.3.4 $P_1 = A$, a contradiction. If $Q_1 \neq A$, then there is a maximal ideal P_2 such that $Q_1 = P_2Q_2$. Continue the same argument: eventually we have $I = P_1 \dots P_nQ_n$ and $I \subset Q_1 \subset \dots \subset Q_n$ are all proper inclusions. Since A is Noetherian, $Q_m = A$ for some m and then $I = P_1 \dots P_m$.

If $P_1 \dots P_m = Q_1 \dots Q_n$, then $P_1 \supset Q_1 \dots Q_n$ and by 3.3.2 P_1 being a prime ideal contains one of Q_i, so $P_1 = Q_i$. Using 3.3.4 cancel P_1 on both sides and use induction. □

3.3.7. REMARK. A maximal ideal P of A is involved in the factorisation of I if and only if $I \subset P$. Indeed, if $I \subset P$, then $I = PQ$ by 3.3.5.

THEOREM. *Let $I = \prod P_i^{r_i}, J = \prod P_i^{s_i}$ be factorisations of non-zero ideals I, J of a Dedekind ring, with non-negative integer r_i, s_i such that $r_i + s_i > 0$ for all i. Then*

$$I \cap J = \prod P_i^{\max\{r_i, s_i\}}, \quad I + J = \prod P_i^{\min\{r_i, s_i\}}$$

and

$$IJ = (I \cap J)(I + J).$$

Proof. For the first equality, $P_i \supset$ the LHS, so the previous Remark implies that the LHS $= \prod P_i^{m_i}$ with non-negative integer m_i. Then $m_i \geqslant \max\{r_i, s_i\}$, so the LHS $\subset$ the RHS. The opposite inclusion is obvious.

For the second equality, if the LHS $= \prod Q_j^{n_j}$ with positive n_j then the previous Remark implies Q_j equals to one of P_i. Then the LHS $= \prod P_i^{n_i}$, $n_i \leqslant \min\{r_i, s_i\}$, so the LHS $\supset$ the RHS. The opposite inclusion is obvious.

The last equality follows immediately from the first two. □

3.3.8. EXAMPLE. Let $A = \mathbb{Z}[\sqrt{-5}]$. This is a Dedekind ring, since $-5 \not\equiv 1 \mod 4$, and A is the ring of integers of $\mathbb{Q}(\sqrt{-5})$.

We have the norm map $N(a+b\sqrt{-5}) = a^2+5b^2$. If an element u is a unit of A then $uv = 1$ for some $v \in A$, and the product of two integers $N(u)$ and $N(v)$ is 1, thus $N(u) = 1$. Conversely, if $N(u) = 1$ then u times its conjugate u' is one, and so u is a unit of A. Thus, $u \in A^\times$ if and only if $N(u) \in \mathbb{Z}^\times$.

The norms of $2, 3, 1 \pm \sqrt{-5}$ are $4, 9, 6$. It is easy to see that $2, 3$ are not in the image $N(A)$.

If, say, 2 were not a prime element in A, then $2 = \pi_1\pi_2$ and $4 = N(\pi_1)N(\pi_2)$ with both norms being proper divisors of 4, a contradiction. Hence 2 is a prime element of A, and similarly $3, 1 \pm \sqrt{-5}$ are.

Now $2, 3, 1 \pm \sqrt{-5}$ are prime elements of A and

$$6 = 2 \cdot 3 = (1+\sqrt{-5})(1-\sqrt{-5}).$$

Note that $2, 3, 1 \pm \sqrt{-5}$ are not associated with each other (the quotient is not a unit) since their norms differ not by a unit of $\mathbb{Z}$. Thus A isn't a UFD.

The ideals

$$(2, 1+\sqrt{-5}), (3, 1+\sqrt{-5}), (3, 1-\sqrt{-5})$$

are maximal.

For instance, $|A/(2)| = 4$, and it is easy to show that $A \neq (2, 1+\sqrt{-5}) \neq (2)$, so $|A/(2, 1+\sqrt{-5})| = 2$, therefore $A/(2, 1+\sqrt{-5})$ is isomorphic to $\mathbb{Z}/2\mathbb{Z}$, i.e. is a field.

We get factorisation of ideals

$$\begin{aligned}
(2) &= (2, 1+\sqrt{-5})^2, \\
(3) &= (3, 1+\sqrt{-5})(3, 1-\sqrt{-5}), \\
(1+\sqrt{-5}) &= (2, 1+\sqrt{-5})(3, 1+\sqrt{-5}), \\
(1-\sqrt{-5}) &= (2, 1+\sqrt{-5})(3, 1-\sqrt{-5}).
\end{aligned}$$

To prove the first equality note that $(1+\sqrt{-5})^2 = -4+2\sqrt{-5} \in (2)$, so the RHS$\subset$ the LHS; we also have $2 = 2(1+\sqrt{-5}) - 2^2 - (1+\sqrt{-5})^2 \in$ the RHS, so the LHS= the RHS.

For the second equality use $(1+\sqrt{-5})(1-\sqrt{-5}) = 6 \in (3)$, $3 = 3^2 - (1+\sqrt{-5})(1-\sqrt{-5}) \in$ the RHS.

For the third equality use $6 \in (1+\sqrt{-5})$, $1+\sqrt{-5} = 3(1+\sqrt{-5}) - 2(1+\sqrt{-5}) \in$ the RHS.

For the fourth equality use conjugate the third equality and use $(2, 1+\sqrt{-5}) = (2, 1-\sqrt{-5})$.

Thus

$$\begin{aligned}(2)\cdot(3) &= (2,1+\sqrt{-5})^2(3,1+\sqrt{-5})(3,1-\sqrt{-5})\\ &= (2,1+\sqrt{-5})(3,1+\sqrt{-5})(2,1+\sqrt{-5})(3,1-\sqrt{-5})\\ &= (1+\sqrt{-5})(1-\sqrt{-5}).\end{aligned}$$

3.3.9. LEMMA. *Let $I+J=A$. Then $I^n+J^m=A$ for every $n,m\geqslant 1$.*

Proof. We have $A=(I+J)\dots(I+J)=I(\dots)+J^m\subset I+J^m$, so $I+J^m=A$. Similarly $I^n+J^m=A$. □

PROPOSITION. *Let P be a maximal ideal of A. Then there is an element $\pi\in P$ such that*

$$P=\pi A+P^n$$

for every $n\geqslant 2$.

Hence the ideal P/P^n is a principal ideal of the quotient ring A/P^n. Moreover, it is the only maximal ideal of that ring.

Every ideal of the ring A/P^n is principal of the form $P^m/P^n=(\pi^m A+P^n)/P^n$ for some $m\leqslant n$.

Proof. If $P=P^2$, then $P=A$ by cancellation property, a contradiction. Let $\pi\in P\setminus P^2$. Since $\pi A+P^n\subset P$, factorisation property implies that $\pi A+P^n=PQ$ for an ideal Q.

Note that $Q\not\subset P$, since otherwise $\pi\in P^2$, a contradiction.

Therefore, $P+Q=A$. The Lemma implies $P^{n-1}+Q=A$. Then

$$P=P(Q+P^{n-1})\subset PQ+P^n=\pi A+P^n\subset P,$$

so $P=\pi A+P^n$.

For $m\leqslant n$ we deduce $P^m\subset\pi^m A+P^n\subset P^m$, so $P^m=\pi^m A+P^n$.

Let I be a proper ideal of A containing P^n. Then by factorisation property $P^n=IK$ with some ideal K. Hence the factorisation of I involves powers of P only, so $I=P^m$, $0<m\leqslant n$. Hence ideals of A/P^n are P^m/P^n with $m\leqslant n$. □

3.3.10. COROLLARY. *Every ideal in a Dedekind ring is generated by 2 elements.*

Proof. Let I be a non-zero ideal, and let a be a non-zero element of I. Then $aA=P_1^{n_1}\dots P_m^{n_m}$ with distinct maximal ideals P_i.

By Lemma 3.3.9 we have $P_1^{n_1}+P_k^{n_k}=A$ if $l\neq k$, so we can apply the Chinese Remainder Theorem which gives

$$A/aA\simeq A/P_1^{n_1}\times\dots\times A/P_m^{n_m}.$$

For the ideal I/aA of A/aA we get

$$I/aA \simeq (I+P_1^{n_1})/P_1^{n_1} \times \cdots \times (I+P_m^{n_m})/P_m^{n_m}.$$

Each of ideals $(I+P_i^{n_i})/P_i^{n_i}$ is of the form $(\pi_i^{l_i}A+P_i^{n_i})/P_i^{n_i}$ by 3.3.9. Hence I/aA is isomorphic to $\prod(\pi_i^{l_i}A+P_i^{n_i})/P_i^{n_i}$. Using the Chinese Remainder Theorem find $b \in A$ such that $b-\pi_i^{l_i}$ belongs to $P_i^{n_i}$ for all i. Then $I/aA = (aA+bA)/aA$ and $I = aA+bA$. □

3.3.11. THEOREM. *A Dedekind ring A is a UFD if and only if A is a PID.*

Proof. Let A be not a PID. Since every proper ideal is a product of maximal ideals, there is a maximal ideal P which isn't principal. Consider the family $\mathscr{F}$ of non-zero ideals I such that PI is principal. It is nonempty by 3.3.3. Let I be a maximal element of this family and $PI = aA$, $a \neq 0$.

Note that I isn't principal, because otherwise $I = xA$ and $PI = xP = aA$, so a is divisible by x. Put $y = ax^{-1}$, then $(x)P = (x)(y)$ and by 3.3.4 $P = (y)$, a contradiction.

Claim: a is a prime element of A. First, a is not a unit of A: otherwise $P \supset PI = aA = A$, a contradiction. Now, if $a = bc$, then $bc \in P$, so either $b \in P$ or $c \in P$. By 3.3.5 then either $bA = PJ$ or $cA = PJ$ for an appropriate ideal J of A. Since $PI \subset PJ$, we get $aI = IPI \subset IPJ = aJ$ and $I \subset J$. Note that $J \in \mathscr{F}$. Due to maximality of I we deduce that $I = J$, and hence either bA or cA is equal to aA. Then one of b, c is associated to a, so a is a prime element.

$P \not\subset aA$, since otherwise $aA = PI \subset aI$, so $A = I$, a contradiction.

$I \not\subset aA$, since otherwise $aA \subset I$ implies $aA = I$, I is principal, a contradiction.

Thus, there are $d \in P$ and $e \in I$ not divisible by a. We also have $ed \in PI = aA$ is divisible by the prime element a. This can never happen in UFD. Thus, A isn't a UFD. □

Using this Theorem, to establish that the ring $\mathbb{Z}[\sqrt{-5}]$ of 3.3.8 is not a unique factorisation domain it is sufficient to indicate a non-principal ideal of it.

3.4. *The norm of an ideal*

In this subsection F is a number field of degree n, $\mathscr{O}_F$ is the ring of integers of F.

3.4.1. PROPOSITION. *For a non-zero element $a \in \mathscr{O}_F$*

$$|\mathscr{O}_F : a\mathscr{O}_F| = |N_{F/\mathbb{Q}}(a)|.$$

Proof. We know that $\mathcal{O}_F$ is a free $\mathbb{Z}$-module of rank n. The ideal $a\mathcal{O}_F$ is a free submodule of $\mathcal{O}_F$ of rank n, since if $x_1,\dots,x_m$ are generators of $a\mathcal{O}_F$, then $a^{-1}x_1,\dots,a^{-1}x_m$ are generators of $\mathcal{O}_F$, so $m=n$. By the Theorem on the structure of modules over principal ideal domains, there is a basis $a_1,\dots,a_n$ of $\mathcal{O}_F$ such that $e_1a_1,\dots,e_na_n$ is a basis of $a\mathcal{O}_F$ with appropriate $e_1|\dots|e_n$. Then $\mathcal{O}_F/a\mathcal{O}_F$ is isomorphic to $\prod\mathbb{Z}/e_i\mathbb{Z}$, so $|\mathcal{O}_F : a\mathcal{O}_F| = \prod|e_i|$. By the definition $N_{F/\mathbb{Q}}(a)$ is equal to the determinant of the matrix of the linear operator $f:\mathcal{O}_F \longrightarrow \mathcal{O}_F$, $b\mapsto ab$. Note that $a\mathcal{O}_F$ has another basis: $aa_1,\dots,aa_n$, so $(aa_1,\dots,aa_n) = (e_1a_1,\dots,e_na_n)M$ with an invertible matrix M with integer entries. Thus, the determinant of M is ±1 and $N_{F/\mathbb{Q}}(a)$ is equal to $\pm\prod e_i$. □

3.4.2. COROLLARY. *$|\mathcal{O}_F : a\mathcal{O}_F| = |a|^n$ for every non-zero $a\in\mathbb{Z}$.*

Proof. $N_{F/\mathbb{Q}}(a) = a^n$. □

3.4.3. DEFINITION. *The norm* $N(I)$ of a non-zero ideal I of $\mathcal{O}_F$ is its index $|\mathcal{O}_F : I|$.

Note that if $I\neq 0$ then $N(I)$ is a finite number.

Indeed, by 3.4.1 $N(a\mathcal{O}_F) = |N_{F/\mathbb{Q}}(a)|$ for a non-zero a which belongs to I. Then $a\mathcal{O}_F \subset I$ and $N(I) \leqslant N(a\mathcal{O}_F) = |N_{F/\mathbb{Q}}(a)|$.

3.4.4. PROPOSITION. *If I,J are non-zero ideals of $\mathcal{O}_F$, then*

$$N(IJ) = N(I)N(J).$$

Proof. Since every ideal factors into a product of maximal ideals by 3.3.6, it is sufficient to show that $N(IP) = N(I)N(P)$ for a maximal ideal P of $\mathcal{O}_F$.

The LHS = $|\mathcal{O}_F : IP| = |\mathcal{O}_F : I||I : IP|$. Recall that P is a maximal ideal of $\mathcal{O}_F$, so $\mathcal{O}_F/P$ is a field.

The quotient I/IP can be viewed as a vector space over $\mathcal{O}_F/P$. Its subspaces correspond to ideals between IP and I according to the description of ideals of the quotient ring. If $IP\subset J\subset I$, then by 3.3.5 $J = IQ$ for an ideal Q of $\mathcal{O}_F$.

By 3.3.3 there is a non-zero ideal I' such that II' is a principal non-zero ideal $a\mathcal{O}_F$. Then $IP\subset IQ$ implies $aP\subset aQ$ implies $P\subset Q$. Therefore either $Q=P$ and then $J=IP$ or $Q=\mathcal{O}_F$ and then $J=I$. Thus, the only subspaces of the vector space I/IP are itself and the zero subspace IP/IP. Hence I/IP is of dimension one over $\mathcal{O}_F/P$ and therefore $|I : IP| = |\mathcal{O}_F : P|$. □

REMARK. If I is a non-zero ideal of $\mathcal{O}_F$ and $N(I)$ is prime, then I is a maximal ideal. Indeed, $\mathcal{O}_F/I$ is a finite commutative ring with a prime number of elements, hence a field.

3.5. *Splitting of prime ideals in field extensions*

In this subsection F is a number field and L is a finite extension of F. Let $\mathcal{O}_F$ and $\mathcal{O}_L$ be their rings of integers.

3.5.1. PROPOSITION–DEFINITION. Let P be a maximal ideal of $\mathcal{O}_F$ and Q a maximal ideal of $\mathcal{O}_L$. Denote by $P\mathcal{O}_L$ the ideal of $\mathcal{O}_L$ generated by its subset P.

Then Q is said to lie over P and P is said to lie under Q if one of the following equivalent conditions is satisfied:

(i) $P\mathcal{O}_L \subset Q$;

(ii) $P \subset Q$;

(iii) $Q \cap \mathcal{O}_F = P$.

Proof. (i) is equivalent to (ii), since $1 \in \mathcal{O}_L$. (ii) implies $Q \cap \mathcal{O}_F$ contains P, so either $Q \cap \mathcal{O}_F = P$ or $Q \cap \mathcal{O}_F = \mathcal{O}_F$, the latter is impossible since $1 \notin Q$. (iii) implies (ii). □

3.5.2. PROPOSITION. *Every maximal ideal of $\mathcal{O}_L$ lies over a unique maximal ideal P of $\mathcal{O}_F$. For a maximal ideal P of $\mathcal{O}_F$ the ideal $P\mathcal{O}_L$ is a proper non-zero ideal of $\mathcal{O}_L$. Let $P\mathcal{O}_L = \prod Q_i$ be the factorisation into a product of prime ideals of $\mathcal{O}_L$. Then Q_i are exactly those maximal ideals of $\mathcal{O}_L$ which lie over P.*

Proof. The first assertion follows from 3.2.2.

Choose a $b \in P \setminus P^2$, it exists by 3.3.9. By 3.3.3 for $b \in P \setminus P^2$ there is an ideal J of $\mathcal{O}_F$ such that $PJ = b\mathcal{O}_F$. Then $J \not\subset P$, since otherwise $b \in P^2$, a contradiction. Take an element $c \in J \setminus P$. Then $cP \subset b\mathcal{O}_F$.

If $P\mathcal{O}_L = \mathcal{O}_L$, then $c\mathcal{O}_L = cP\mathcal{O}_L \subset b\mathcal{O}_L$, so $cb^{-1} \in \mathcal{O}_L \cap F = \mathcal{O}_F$ and $c \in b\mathcal{O}_F \subset P$, a contradiction. Thus, $P\mathcal{O}_L$ is a proper ideal of $\mathcal{O}_L$.

According to 3.5.1 a prime ideal Q of $\mathcal{O}_L$ lies over P if and only if $P\mathcal{O}_L \subset Q$ which is equivalent by 3.3.7 to the fact that Q is involved in the factorisation of $P\mathcal{O}_L$. □

3.5.3. LEMMA. *Let P be a maximal ideal of $\mathcal{O}_F$ which lie under a maximal ideal Q of $\mathcal{O}_L$. Then the finite field $\mathcal{O}_F/P$ is a subfield of the finite field $\mathcal{O}_L/Q$.*

Proof. $\mathcal{O}_L/Q$ is finite by 3.4.3. The kernel of the homomorphism $\mathcal{O}_F \longrightarrow \mathcal{O}_L/Q$ is equal to $Q \cap \mathcal{O}_F = P$, so $\mathcal{O}_F/P$ can be identified with a subfield of $\mathcal{O}_L/Q$. □

3.5.4. COROLLARY. *Let P be a maximal ideal of $\mathcal{O}_F$. Then $P \cap \mathbb{Z} = p\mathbb{Z}$ for a prime number p and $N(P)$ is a positive power of p.*

Proof. $P \cap \mathbb{Z} = p\mathbb{Z}$ for a prime number p by 3.2.2. Then $\mathcal{O}_F/P$ is a vector space over $\mathbb{Z}/p\mathbb{Z}$ of finite positive dimension, therefore $|\mathcal{O}_F : P|$ is a power of p. □

3.5.5. DEFINITION. Let a maximal ideal P of $\mathcal{O}_F$ lie under a maximal ideal Q of $\mathcal{O}_L$. The degree of $\mathcal{O}_L/Q$ over $\mathcal{O}_F/P$ is called the *inertia degree* $f(Q|P)$.

If $P\mathcal{O}_L = \prod Q_i^{e_i}$ is the factorisation of $P\mathcal{O}_L$ with distinct prime ideals Q_i of $\mathcal{O}_L$, then e_i is called the *ramification index* $e(Q_i|P)$.

3.5.6. LEMMA. *Let M be a finite extension of L and $P \subset Q \subset R$ be maximal ideals of $\mathcal{O}_F$, $\mathcal{O}_L$ and $\mathcal{O}_M$ correspondingly. Then $f(R|P) = f(Q|P)f(R|Q)$ and $e(R|P) = e(Q|P)e(R|Q)$.*

Proof. The first assertion follows from 1.1.1. From $P\mathcal{O}_L = Q^{e(Q|P)} \dots$ we get $P\mathcal{O}_M = Q^{e(Q|P)}\mathcal{O}_M \dots = (Q\mathcal{O}_M)^{e(Q|P)} \dots = (R^{e(R|Q)})^{e(Q|P)} \dots$, so the second assertion follows. □

3.5.7. THEOREM. *Let $Q_1, \dots Q_m$ be different maximal ideals of $\mathcal{O}_L$ which lie over a maximal ideal P of $\mathcal{O}_F$. Let $n = |L : F|$. Then*

$$\sum_{i=1}^{m} e(Q_i|P)f(Q_i|P) = n.$$

Proof. We consider only the case $F = \mathbb{Q}$. Apply the norm to the equality $p\mathcal{O}_L = \prod Q_i^{e_i}$. Then by 3.4.2, 3.4.4

$$p^n = N(p\mathcal{O}_L) = \prod N(Q_i)^{e_i} = \prod p^{f(Q_i|P)e(Q_i|P)}.$$

□

3.5.8. EXAMPLE. One can describe in certain situations how a prime ideal (p) factorises in finite extensions of $\mathbb{Q}$, provided the factorisation of the monic irreducible polynomial of an integral generator (if it exists) modulo p is known.

Let the ring of integers $\mathcal{O}_F$ of an algebraic number field F be generated by one element α: $\mathcal{O}_F = \mathbb{Z}[\alpha]$, and $f(X) \in \mathbb{Z}[X]$ be the monic irreducible polynomial of α over $\mathbb{Q}$.

Let $f_i(X) \in \mathbb{Z}[X]$ be monic polynomials such that

$$\overline{f}(X) = \prod_{i=1}^{m} \overline{f_i}(X)^{e_i} \in \mathbb{F}_p[X]$$

is the factorisation of $\overline{f}(X)$ where $\overline{f_i}(X)$ is an irreducible polynomial over $\mathbb{F}_p$.

Since $\mathcal{O}_F \simeq \mathbb{Z}[X]/(f(X))$, we have

$$\mathcal{O}_F/p\mathcal{O}_F \simeq \mathbb{Z}[X]/(p, f(X)) \simeq \mathbb{F}_p[X]/(\overline{f}(X)),$$

and

$$\mathcal{O}_F/(p, f_i(\alpha)) \simeq \mathbb{Z}[X]/(p, f(X), f_i(X)) \simeq \mathbb{F}_p[X]/(\overline{f_i}(X)).$$

Putting $P_i = (p, f_i(\alpha))$ we see that $\mathcal{O}_F/P_i$ is isomorphic to the field $\mathbb{F}_p[X]/(\overline{f_i}(X))$, hence P_i is a maximal ideal of $\mathcal{O}_F$ dividing (p). We also deduce that

$$N(P_i) = p^{|\mathbb{F}_p[X]/(\overline{f_i}(X)):\mathbb{F}_p|} = p^{\deg \overline{f_i}}.$$

Now $\prod P_i^{e_i} = \prod(p, f_i(\alpha))^{e_i} \subset p\mathcal{O}_F$, since $\prod f_i(\alpha)^{e_i} - f(\alpha) \in p\mathcal{O}_F$. We also get $N(\prod P_i^{e_i}) = p^{\sum e_i \deg \overline{f_i}} = p^n = N(p\mathcal{O}_F)$. Therefore from 3.5.7 we deduce that $p\mathcal{O}_F = \prod_{i=1}^m P_i^{e_i}$ is the factorisation of $p\mathcal{O}_F$.

So we have proved

THEOREM. *Let the ring of integers $\mathcal{O}_F$ of an algebraic number field F be generated by one element α: $\mathcal{O}_F = \mathbb{Z}[\alpha]$, and $f(X) \in \mathbb{Z}[X]$ be the monic irreducible polynomial of α over $\mathbb{Q}$. Let $f_i(X) \in \mathbb{Z}[X]$ be irreducible polynomials such that*

$$\overline{f}(X) = \prod_{i=1}^{m} \overline{f_i}(X)^{e_i} \in \mathbb{F}_p[X]$$

is the factorisation of $\overline{f}(X)$ where $\overline{f_i}(X)$ is an irreducible polynomial over $\mathbb{F}_p$.

Then in $\mathcal{O}_F$

$$p\mathcal{O}_F = \prod_{i=1}^{m} P_i^{e_i}$$

where $P_i = (p, f_i(\alpha))$ is a maximal ideal of $\mathcal{O}_F$ with norm $p^{\deg \overline{f_i}}$.

EXAMPLE. Let $F = \mathbb{Q}$ and $L = \mathbb{Q}(\sqrt{d})$ with a square free integer d.

Then one can take $\sqrt{d}$ for $d \not\equiv 1 \mod 4$ and $(1+\sqrt{d})/2$ for $d \equiv 1 \mod 4$ as α. Then $f(X) = X^2 - d$ and $f(X) = X^2 - X + (1-d)/4$ resp.

Let p be a prime in $\mathbb{Z}$ and let $p\mathcal{O}_L = \prod_{i=1}^m Q_i^{e_i}$. Then there are three cases:

(i) $m = 2$, $e_1 = e_2 = 1$, $f(Q_i|P) = 1$. Then $p\mathcal{O}_L = Q_1 Q_2$, $Q_1 \neq Q_2$. We say that p *splits in* L. From 3.5.8 we know that $Q_i = (p, f_i(\alpha))$.

(ii) $m = 1$, $e_1 = 2$, $f(Q_1|P) = 1$. Then $p\mathcal{O}_L = Q_1^2$. We say that p *ramifies in* L. From 3.5.8 we know that $Q_1 = (p, f_1(\alpha))$.

(iii) $m = 1$, $e_1 = 1$, $f(Q_1|P) = 2$. Then $p\mathcal{O}_L = Q_1$. We say that p *remains prime in* L. Here $Q_1 = (p)$ as ideal of $\mathcal{O}_L$.

Using the previous Theorem we see that p splits ($p\mathcal{O}_F = P_1 \ldots P_m$) if and only if $\overline{f}$ is separable and reducible, p ramifies ($p\mathcal{O}_F = P^e$) if and only if $\overline{f}$ is a power > 1 of an irreducible polynomial over $\mathbb{F}_p$, p remains prime in $\mathcal{O}_F$ if and only if $\overline{f}$ is irreducible over $\mathbb{F}_p$.

3.5.9. We have $X^2 - X + (1-d)/4 = 1/4(Y^2 - d)$ where $Y = 2X - 1$, so if p is odd (so the image of 2 is invertible in $\mathbb{F}_p$), the factorisation of $f(X)$ corresponds to the factorisation of $X^2 - d$ independently of what d is. The factorisation of

X^2-d certainly depends on whether d is a quadratic residue modulo p, or not. If $d \equiv c^2 \mod p$, then

$$X^2-d \equiv f_1 f_2 \mod p, \quad f_1 = X-c, f_2 = X+c,$$

$$X^2-X+(1-d)/4 \equiv f_1 f_2 \mod p, \quad f_1 = X-(1+c)/2, f_2 = X-(1-c)/2.$$

Let $p=2$. If $d \not\equiv 1 \mod 4$ then

$$f(X) = X^2-d \equiv X^2+d^2 \equiv (X-d)^2 \mod 2.$$

If $d \equiv 1 \mod 4$ then $f(X) = X^2+X+(1-d)/4$. So, if $d \equiv 1 \mod 8$ then

$$X^2+X+(1-d)/4 = X(X+1) \mod 2,$$

if $d \not\equiv 1 \mod 8, d \equiv 1 \mod 4$ then $X^2+X+(1-d)/4 = X^2+X+1 \mod 2$ is irreducible in $\mathbb{F}_2[X]$. Thus, we get

THEOREM. *If p is odd prime, then*

(1) *p splits in $L = \mathbb{Q}(\sqrt{d})$ if and only if d is a quadratic residue mod p. Then $f_i = X \pm c, \alpha = \sqrt{d}$ if $d \not\equiv 1 \mod 4$ and $f_i = X-(1\pm c)/2, \alpha = (1+\sqrt{d})/2$ if $d \equiv 1 \mod 4$.*

(2) *p ramifies in L if and only if d is divisible by p. Then $f_1 = X$ if $d \not\equiv 1 \mod 4$ and $f_1 = X-a$, $2a \equiv 1 \mod p$ if $d \equiv 1 \mod 4$.*

(3) *p remains prime in L if and only if d is a quadratic non-residue mod p.*

If $p=2$ then

(1) *if $d \equiv 1 \mod 8$, then 2 splits in $\mathbb{Q}(\sqrt{d})$. Then $f_1 = X, f_2 = X+1, \alpha = (1+\sqrt{d})/2$.*

(2) *if $d \not\equiv 1 \mod 4$ then 2 ramifies in $\mathbb{Q}(\sqrt{d})$. Then $f_1 = X-d, \alpha = \sqrt{d}$.*

(3) *if $d \equiv 1 \mod 4, d \not\equiv 1 \mod 8$ then 2 remains prime in $\mathbb{Q}(\sqrt{d})$.*

COROLLARY. *Only finitely many primes ramify in $\mathbb{Q}(\sqrt{d})$.*
The only quadratic extension of $\mathbb{Q}$ in which no primes ramify is $\mathbb{Q}(\sqrt{-1})$.

See Proposition 4.6 of Chapter 3 for a much more general property.

3.5.10. Let p be an odd prime. Recall from 2.4.2 that the ring of integers of the pth cyclotomic field $\mathbb{Q}(\zeta_p)$ is generated by ζ_p. Its irreducible monic polynomial is $f(X) = X^{p-1}+\dots+1 = (X^p-1)/(X-1)$. Since $X^p-1 \equiv (X-1)^p \mod p$ we deduce that $(f(X),p) = ((X-1)^{p-1},p)$.

Therefore by 3.5.8, $p\mathcal{O}_{\mathbb{Q}(\zeta_p)} = (\zeta_p-1)^{p-1}\mathcal{O}_{\mathbb{Q}(\zeta_p)}$ and p ramifies in $\mathbb{Q}(\zeta_p)/\mathbb{Q}$.

For any other prime l one can show that the polynomial $f(X)$ modulo l is the product of distinct irreducible polynomials over $\mathbb{F}_l$. Thus, no other prime ramifies in $\mathbb{Q}(\zeta_p)/\mathbb{Q}$.

3.6. *Finiteness of the ideal class group*

In this subsection $\mathcal{O}_F$ is the ring of integers of a number field F.

3.6.1. DEFINITION. For two non-zero ideals I and J of $\mathcal{O}_F$ define the equivalence relation $I \sim J$ if there are non-zero $a, b \in \mathcal{O}_F$ such that $aI = bJ$. In other words, I and J are proportional to each other. Classes of equivalence are called *ideal classes*. Define the product of two classes with representatives I and J as the class containing IJ. Then the class of $\mathcal{O}_F$ (consisting of all nonzero principal ideals) is the identity element. By 3.3.3 for every non-zero ideal I there is a non-zero ideal J such that IJ is a principal ideal, i.e. every ideal class is invertible. Thus ideal classes form an abelian group which is called the *ideal class group* Cl_F of the number field F.

The ideal class group shows how far from PID the ring $\mathcal{O}_F$ is. Note that Cl_F consists of one element if and only if $\mathcal{O}_F$ is a PID if and only if $\mathcal{O}_F$ is a UFD.

DEFINITION. One can also consider *fractional ideals* of F, i.e. $\mathcal{O}_F$-submodules of the $\mathcal{O}_F$-module F that are proportional to ideals of $\mathcal{O}_F$, i.e. such that aI is an ideal of $\mathcal{O}_F$ for some non-zero $a \in \mathcal{O}_F$. Principal fractional ideals are $b\mathcal{O}_F$ with $b \in F$.

Proposition 3.3.3 immediately implies that for every non-zero fractional ideal I there is a non-zero fractional ideal J such that $IJ = \mathcal{O}_F$ and $J = \{b \in F : bI \subset \mathcal{O}_F\}$. The fractional ideal J is called the inverse I^{-1} of the fractional ideal I. Theorem 3.3.6 implies that every non-zero fractional ideal is the product $\prod P_i^{n_i}$ of maximal ideals P_i with non-zero integers n_i, uniquely up to permutation. The quotient of the group of non-zero fractional ideals by its subgroup of non-zero principal fractional ideals is isomorphic to the class group of $\mathcal{O}_F$.

3.6.2. PROPOSITION. *There is a positive real number c such that every non-zero ideal I of $\mathcal{O}_F$ contains a non-zero element a with*

$$|N_{F/\mathbb{Q}}(a)| \leqslant cN(I).$$

Proof. Let $n = |F : \mathbb{Q}|$. According to 2.3.7 there is a basis $a_1, \dots, a_n$ of the $\mathbb{Z}$-module $\mathcal{O}_F$ which is also a basis of the $\mathbb{Q}$-vector space F. Let $\sigma_1, \dots, \sigma_n$ be all distinct $\mathbb{Q}$-homomorphisms of F into $\mathbb{C}$. Put

$$c = \prod_{i=1}^{n} \Big(\sum_{j=1}^{n} |\sigma_i a_j| \Big).$$

Then $c > 0$.

For a non-zero ideal I let m be the positive integer satisfying the inequality $m^n \leqslant N(I) < (m+1)^n$. In particular, $|\mathcal{O}_F : I| < (m+1)^n$. Consider $(m+1)^n$ elements $\sum_{j=1}^n m_j a_j$ with $0 \leqslant m_j \leqslant m$, $m_j \in \mathbb{Z}$. There are two of them which have the same image in $\mathcal{O}_F/I$. Their difference $0 \neq a = \sum_{j=1}^n n_j a_j$ belongs to I and satisfies $|n_j| \leqslant m$.

Now we deduce

$$\begin{aligned} |N_{F/\mathbb{Q}}(a)| &= \prod_{i=1}^n |\sigma_i a| = \prod_{i=1}^n |\sum_{j=1}^n n_j \sigma_i a_j| \\ &\leqslant \prod_{i=1}^n \left(\sum_{j=1}^n |n_j||\sigma_i a_j| \right) \leqslant m^n c \leqslant cN(I). \end{aligned}$$

□

Thus every non-zero ideal I of $\mathcal{O}_F$ contains a non-zero principal ideal $a\mathcal{O}_F$ whose index in I does not exceed c.

3.6.3. COROLLARY. *Every ideal class of $\mathcal{O}_F$ contains an ideal J with $N(J) \leqslant c$.*

Proof. Given ideal class, consider an ideal I of the inverse ideal class. Let $a \in I$ be as in the Theorem. By 3.3.3 there is an ideal J such that $IJ = a\mathcal{O}_F$, so $(I)(J) = (a\mathcal{O}_F) = 1$ in Cl_F. Then J belongs to the given ideal class. Using 3.4.1 and 3.4.4 we deduce that $N(I)N(J) = N(IJ) = N(a\mathcal{O}_F) = |N_{F/\mathbb{Q}}(a)| \leqslant cN(I)$. Thus, $N(J) \leqslant c$. □

3.6.4. THEOREM. *The ideal class group Cl_F is finite. The number $|\mathrm{Cl}_F|$ is called the class number of F.*

Proof. By 3.5.4 and 3.5.2 for each prime p there are finitely many maximal ideals P lying over (p), and $N(P) = p^m$ for $m \geqslant 1$. From $N(\prod P_i^{e_i}) \leqslant c$ we have bounds $e_i \leqslant \log_2 c$.

Hence there are finitely many ideals $\prod P_i^{e_i}$ satisfying $N(\prod P_i^{e_i}) \leqslant c$. □

EXAMPLE. The class number of $\mathbb{Q}(\sqrt{-19})$ is 1, i.e. every ideal of the ring of integers of $\mathbb{Q}(\sqrt{-19})$ is principal.

Indeed, by 2.3.8 we can take $a_1 = 1$, $a_2 = (1+\sqrt{-19})/2$ as an integral basis of the ring of integers of $\mathbb{Q}(\sqrt{-19})$. Then

$$c = \left(1 + |(1+\sqrt{-19})/2|\right)\left(1 + |(1-\sqrt{-19})/2|\right) = 10.4\ldots.$$

So every ideal class of $\mathcal{O}_{\mathbb{Q}(\sqrt{-19})}$ contains an ideal J with $N(J) \leqslant 10$.

Let $J = \prod P_i^{e_i}$ be the factorisation of J, then $N(P_i) \leqslant 10$ for every i.

By Corollary 3.5.4 we know that $N(P_i)$ is a positive power of a prime integer, say p_i, and so $p_i \leqslant 10$.

From 3.5.2 we know that P_i is a prime divisor of the ideal (p_i) of $\mathcal{O}_{\mathbb{Q}(\sqrt{-19})}$. So we need to look at prime integer numbers not greater than 7 and their prime ideal divisors as potential candidates for non-principal ideals. Now prime number 3 has the property that -19 is a quadratic non-residue modulo them, so by Theorem 3.5.9 it remains prime in $\mathcal{O}_{\mathbb{Q}(\sqrt{-19})}$.

Odd prime numbers 5, 7 have the property that -19 is a quadratic residue module them, so by Theorem 3.5.9 they split in $\mathcal{O}_{\mathbb{Q}(\sqrt{-19})}$. By 3.5.8 and 3.5.9 we have $-19 \equiv 1^2 \mod 5$, so $f_1 = X - 1, f_2 = X$, $-19 \equiv 3^2 \mod 7$, so $f_1 = X - 2, f_2 = X + 1$, and

$$\begin{aligned} 5O &= (5, (1+\sqrt{-19})/2 - 1)(5, (1+\sqrt{-19})/2) \\ &= (5, (1-\sqrt{-19})/2)(5, (1+\sqrt{-19})/2) \\ 7O &= (7, (1+\sqrt{-19})/2 - 2)(7, (1+\sqrt{-19})/2 + 1) \\ &= (7, (3-\sqrt{-19})/2)(7, (3+\sqrt{-19})/2). \end{aligned}$$

Now we have

$$5 = (1+\sqrt{-19})/2 \cdot (1-\sqrt{-19})/2, \quad 7 = (3+\sqrt{-19})/2 \cdot (3-\sqrt{-19})/2,$$

so

$$5O = ((1-\sqrt{-19})/2)((1+\sqrt{-19})/2), \; 7O = ((3-\sqrt{-19})/2)((3+\sqrt{-19})/2)$$

and the prime ideal factors of $5O, 7O$ are principal.

Finally, 2 remains prime in $\mathcal{O}_{\mathbb{Q}(\sqrt{-19})}$, as follows from 3.5.9.

Thus, $\mathcal{O}_{\mathbb{Q}(\sqrt{-19})}$ is a principal ideal domain.

REMARKS. 1. The bound given by c is not good in practical applications. A more refined estimation is given by Minkowski's Theorem 3.6.6.

2. For adelic proofs of the finiteness of the class number see Remark 1 of 4.7 and Remark 2 of 5.6 Chapter 3.

3.6.5. DEFINITION. Let F be of degree n over $\mathbb{Q}$. Let $\sigma_1, \dots, \sigma_n$ be all $\mathbb{Q}$-homomorphisms of F into $\mathbb{C}$. Let

$$\tau \colon \mathbb{C} \longrightarrow \mathbb{C}$$

be the complex conjugation. Then $\tau \circ \sigma_i$ is a $\mathbb{Q}$-homomorphism of F into $\mathbb{C}$, so it is equal to certain σ_j. Note that $\sigma_i = \tau \circ \sigma_i$ if and only if $\sigma_i(F) \subset \mathbb{R}$. Let r_1 be the number of $\mathbb{Q}$-homomorphisms of this type, say, after renumeration, $\sigma_1, \dots, \sigma_{r_1}$. For every $i > r_1$ we have $\tau \circ \sigma_j \neq \sigma_j$, so we can form couples $(\sigma_j, \tau \circ \sigma_j)$. Then $n - r_1$ is an even number $2r_2$, and $r_1 + 2r_2 = n$.

Reorder the σ_j's so that $\sigma_{i+r_2} = \tau \circ \sigma_i$ for $r_1 + 1 \leqslant i \leqslant r_1 + r_2$. Define the *canonical embedding* of F by

$$\sigma\colon a \mapsto (\sigma_1(a), \dots, \sigma_{r_1+r_2}(a)) \in \mathbb{R}^{r_1} \times \mathbb{C}^{r_2}, \qquad a \in F.$$

The field F is isomorphic to its image $\sigma(F) \subset \mathbb{R}^{r_1} \times \mathbb{C}^{r_2}$. The image $\sigma(F)$ is called the geometric image of F and it can be partially studied by geometric tools.

3.6.6. THEOREM. (*Minkowski's Bound Theorem*)

Let F be an algebraic number field of degree n with parameters r_1, r_2. Then every class of Cl_F contains an ideal I such that its norm $N(I)$ satisfies the inequality

$$N(I) \leqslant (4/\pi)^{r_2} n! \sqrt{|d_F|}/n^n$$

where d_F is the discriminant of F.

Proof. One of the proofs uses the geometric image of F and some geometric combinatorial considerations. In particular, one can use Minkowski's Lattice Point Theorem:

Let L be a free $\mathbb{Z}$-module of rank n in an n-dimensional vector space V over $\mathbb{R}$ (then L is called a complete lattice in V). Denote by $\mathrm{Vol}\,(L)$ the volume of the set

$$\{a_1e_1 + \cdots + a_ne_n : 0 \leqslant a_i \leqslant 1\},$$

where $e_1, \dots, e_n$ is a basis of L. Notice that $\mathrm{Vol}\,(L)$ does not depend on the choice of basis. Let X be a centrally symmetric convex subset of V. Suppose that $\mathrm{Vol}\,(X) > 2^n \mathrm{Vol}\,(L)$. Then X contains at least one nonzero point of L.

Details can be found in many textbooks. □

REMARK. In relation to an adelic proof of Minkowski's Bound Theorem see Remark 2 in 4.7 of Chapter 3.

3.6.7. EXAMPLES.

1. Let $F = \mathbb{Q}(\sqrt{5})$. Then $r_1 = 2$, $r_2 = 0$, $n = 2$, $|d_F| = 5$.

$$(4/\pi)^{r_2} n! \sqrt{|d_F|}/n^n = 2!\sqrt{5}/2^2 = 1.1...,$$

so $N(I) = 1$ and therefore $I = \mathcal{O}_F$. Thus, every ideal of $\mathcal{O}_F$ is principal and $\mathrm{Cl}_F = \{1\}$.

Similarly, the class groups of $\mathbb{Q}(\sqrt{-1})$, $\mathbb{Q}(\sqrt{-2})$, $\mathbb{Q}(\sqrt{-3})$, $\mathbb{Q}(\sqrt{-7})$ are trivial, since their discriminants are $-4, -8, -3, -7$, $r_2 = 1$, $r_1 = 0$, as $(2/\pi)\sqrt{8} < 2$.

2. Let $F = \mathbb{Q}(\sqrt{-5})$. Then $r_1 = 0$, $r_2 = 1$, $n = 2$, $|d_F| = 20$, $(2/\pi)\sqrt{|20|} < 3$. Hence, similar to Example in 3.6.4 we only need to look at prime numbers 2 (< 3)

and prime ideal divisors of the ideal (2) as potential candidates for non-principal ideals.

From 3.3.8 we know that $2O = (2, 1+\sqrt{-5})^2$ and $2 = N(2, 1+\sqrt{-5})$. So the ideal $(2, 1+\sqrt{-5})$ is maximal by 3.4.5.

Alternatively, from 3.5.9 we get $2O = (2, 5-\sqrt{-5})^2 = (2, 1+\sqrt{-5})^2$ and $(2, 1+\sqrt{-5})$ is maximal.

The ideal $(2, 1+\sqrt{-5})$ is not principal: Indeed, if $(2, 1+\sqrt{-5}) = a\mathscr{O}_L$ then

$$2 = N(2, 1+\sqrt{-5}) = N(a\mathscr{O}_L) = |N_{L/\mathbb{Q}}(a)|.$$

If $a = c + d\sqrt{-5}$ with $c, d \in \mathbb{Z}$ we deduce that $c^2 + 5d^2 = \pm 2$, a contradiction.

We conclude that $\mathrm{Cl}_{\mathbb{Q}(\sqrt{-5})}$ is a cyclic group of order 2.

3. Let $F = \mathbb{Q}(\sqrt{14})$. Then $r_1 = 2$, $r_2 = 0$, $n = 2$, $|d_F| = 56$ and $(1/2)\sqrt{56} = 3.7... < 4$. So we only need to inspect prime ideal divisors of (2) and of (3).

By 3.5.8 and 3.5.9 we get $2O = (2, \sqrt{14})^2$. Note that $(4+\sqrt{14}) \subset (2, \sqrt{14})$ and

$$2 = (4+\sqrt{14})(4-\sqrt{14}) \in (4+\sqrt{14}), \quad \sqrt{14} = 4+\sqrt{14}-4 \in (4+\sqrt{14}),$$

hence $(2, \sqrt{14}) = (4+\sqrt{14})$ is principal.

14 is quadratic non-residue modulo 3, so by Theorem 3.5.9 we deduce that 3 remains prime in $\mathscr{O}_F$. Thus, every ideal of the ring of integers of $\mathbb{Q}(\sqrt{14})$ is principal, $\mathrm{Cl}_{\mathbb{Q}(\sqrt{14})} = \{1\}$.

4. Let $F = \mathbb{Q}(\sqrt{-13})$.

The discriminant of F is -52. We have $4 < 2/\pi\sqrt{52} < 5$.

Hence we only need to look at primes 2 and 3 (< 5) and prime ideal divisors in $\mathscr{O}_F$ of the ideals (2) and (3) as potential candidates for non-principal ideals of $\mathscr{O}_F$.

By 3.5.9 the ideal (3) remains prime in F since -13 is quadratic non-residue modulo 3.

By 3.5.9 2 ramifies in F. By 3.5.8 we get the following factorisation into maximal ideals:

$$(2) = (2, \sqrt{-13}-13)^2 = (2, 1+\sqrt{-13})^2.$$

The ideal $(2, 1+\sqrt{-13})$ is not principal: indeed, if $(2, 1+\sqrt{-13}) = a\mathscr{O}_F$ then

$$2 = N(2, 1+\sqrt{-13}) = N(a\mathscr{O}_F) = |N_{F/\mathbb{Q}}(a)|.$$

If $a = c + d\sqrt{-13}$ with $c, d \in \mathbb{Z}$ we deduce that $c^2 + 13d^2 = \pm 2$, a contradiction.

Thus, the class group of F is cyclic of order 2 and is generated by the class of the ideal $(2, 1+\sqrt{-13})$.

5. It is known that for negative square-free d the only quadratic fields $\mathbb{Q}(\sqrt{d})$ with class number 1 are the following:

$$\mathbb{Q}(\sqrt{-1}),\quad \mathbb{Q}(\sqrt{-2}),\quad \mathbb{Q}(\sqrt{-3}),\quad \mathbb{Q}(\sqrt{-7}),\quad \mathbb{Q}(\sqrt{-11}),$$
$$\mathbb{Q}(\sqrt{-19}),\quad \mathbb{Q}(\sqrt{-43}),\quad \mathbb{Q}(\sqrt{-67}),\quad \mathbb{Q}(\sqrt{-163}).$$

For $d > 0$ there are many more quadratic fields with class number 1. Gauß conjectured that there are infinitely many such fields, but this is still unproved.

3.7. *On Fermat's Last Theorem*

3.7.1. Already Euler noticed that for an infinitely differentiable function $f(x)$ one has

$$f(x+1) = e^D f(x)$$

where D is the operator d/dx.

If we denote $g(x) = f(x+1) - f(x) = (1-e^D)f(x)$, then

$$f(x) = (1-e^D)^{-1}g(x) = (a_1D^{-1} + a_0 + a_1D + a_2D^2 + \ldots)g(x)$$

where the coefficients are of the Taylor expansion of $\frac{x}{1-e^x}$ at $x = 0$. This is how one comes for what Euler called (Jacob) Bernoulli numbers

$$\frac{t}{e^t - 1} = \sum_{i=0}^{\infty} \frac{b_i}{i!} t^i,$$

$b_0 = 1$, $b_1 = -1/2$, $b_2 = 1/6$, $b_i = 0$ for odd $i > 1$.

Now we can state one of the main achievements of Kummer.

THEOREM. (*Kummer's Theorem*)

Let p be an odd prime. Let $F = \mathbb{Q}(\zeta_p)$ be the pth cyclotomic field.

If p doesn't divide $|\mathrm{Cl}_F|$, or, equivalently, p does not divide numerators of (rational) Bernoulli numbers $b_2, b_4, \ldots, b_{p-3}$, then the Fermat equation

$$X^p + Y^p = Z^p$$

does not have positive integer solutions, i.e. Fermat's Last Theorem (FLT) holds in this case.

Among primes < 100 only 37, 59 and 67 don't satisfy the condition that p does not divide $|\mathrm{Cl}_F|$, so Kummer's Theorem implies that for any other prime number smaller 100 the Fermat equation does not have positive integer solutions.

3.7.2. *Full proofs of FLT.*

1. In 1995 Wiles and Taylor published a proof of modularity of elliptic curves over rational numbers with semi-stable reduction, as part of activity in the Langlands program. Using a previously established result of K. Ribet, their result implies FLT.

2. Another proof of FLT, entirely independent from 1, by Mochizuki, Fesenko, Minamide, Hoshi, Porowski, was published in their 2022 paper. It is based on the fundamental IUT theory of Mochizuki and its enhanced version developed in that paper, and the first proof of effective abc inequalities. FLT follows as one of first applications of the established effective abc inequalities. To deduce this application, one uses old classical results of Vandiver, old computer verifications of FLT, and new lower bounds for positive integer solutions of the Fermat's equation when their product is divisible by p obtained by Mihăilescu in 2021.

3.8. *On Dirichlet's Unit Theorem*

3.8.1. THEOREM. *Let F be a number field of degree n, $r_1+2r_2=n$. Let $\mathcal{O}_F$ be its ring of integers and U be the group of units of $\mathcal{O}_F$. Then U is the direct product of a finite cyclic group T consisting of all roots of unity in F and a free abelian group U_1 of rank r_1+r_2-1:*

$$U \simeq T \times U_1 \simeq T \times \mathbb{Z}^{r_1+r_2-1}.$$

A basis of the free abelian group U_1 is called a fundamental system of units in $\mathcal{O}_F$.

Proof. Consider the canonical embedding σ of F into $\mathbb{R}^{r_1} \times \mathbb{C}^{r_2}$. Define

$$f\colon \mathcal{O}_F \setminus \{0\} \longrightarrow \mathbb{R}^{r_1+r_2},$$

$$f(x) = \big(\log|\sigma_1(x)|,\dots,\log|\sigma_{r_1}(x)|,\log(|\sigma_{r_1+1}(x)|^2),\dots,\log(|\sigma_{r_1+r_2}(x)|^2)\big).$$

The map f induces a homomorphism $g\colon U \longrightarrow \mathbb{R}^{r_1+r_2}$.

Let Z be a bounded set of $\mathbb{R}^{r_1+r_2}$. If $u \in g^{-1}(Z)$ then there is c such that $|\sigma_i(u)| \leqslant c$ for all i. The coefficients of the characteristic polynomial $g_u(X) = \prod_{i=1}^n (X-\sigma_i(u))$ of u over F being functions of $\sigma_i(u)$ are integers bounded by $\max(c^n, nc^{n-1},\dots)$, so the number of different characteristic polynomials of elements of $g^{-1}(Z)$ is finite. So the sets $g^{-1}(Z)$ and $Z \cap g(U)$ is finite. Thus $g(U)$ is a discrete group.

Every finite subgroup of the multiplicative group of a field is cyclic by 1.2.4. Hence the kernel of g, being the preimage of 0, is a cyclic finite group. On the other hand, every root of unity belongs to the kernel of g, since

$mg(z) = g(z^m) = g(1) = 0$ implies $g(z) = 0$ for the vector $g(z)$. We conclude that the kernel of g consists of all roots of unity T in F.

Since for $u \in U$ the norm $N_{F/\mathbb{Q}}(u) = \prod \sigma_i(u)$, as the product of units, is a unit in $\mathbb{Z}$, it is equal to ± 1. Then $\prod |\sigma_i(u)| = 1$ and

$$\log|\sigma_1(u)| + \cdots + \log|\sigma_{r_1}(u)| + \log(|\sigma_{r_1+1}(u)|^2) + \cdots + \log(|\sigma_{r_1+r_2}(u)|^2) = 0.$$

We deduce that the image $g(U)$ is contained in the hyperplane $H \subset \mathbb{R}^{r_1+r_2}$ defined by the equation

$$y_1 + \cdots + y_{r_1+r_2} = 0.$$

Since $g(Z)$ is discrete, by 3.7.2 $g(U)$ has a $\mathbb{Z}$-basis $\{y_i\}$ consisting of $m \leqslant r_1 + r_2 - 1$ linearly independent vectors over $\mathbb{Z}$. Denote by U_1 the subgroup of U generated by z_i such that $g(z_i) = y_i$; it is a free abelian group, since there are no non-trivial relations among y_i. From the Main Theorem on Group Homomorphisms we deduce that $U/T \simeq g(U)$ and hence $U = TU_1$. Since U_1 has no non-trivial torsion, $T \cap U_1 = \{1\}$. Then U as a $\mathbb{Z}$-module is the direct product of the free abelian group U_1 of rank m and the cyclic group T of roots of unity.

It remains to show that $m = r_1 + r_2 - 1$, i.e. $g(U)$ contains $r_1 + r_2 - 1$ linearly independent vectors. Put $l = r_1 + r_2$. As an application of Minkowski's geometric method one can show that

for every integer k between 1 and l there is $c > 0$ such that for every non-zero $a \in \mathcal{O}_F \setminus \{0\}$ with $g(a) = (\alpha_1, \ldots, \alpha_l)$ there is a non-zero $b = h_k(a) \in \mathcal{O}_F \setminus \{0\}$ such that

$$|N_{F/\mathbb{Q}}(b)| \leqslant c \quad \text{and } g(b) = (\beta_1, \ldots, \beta_l) \text{ with } \beta_i < \alpha_i \text{ for } i \neq k.$$

(for the proof see [**27**], Theorem 38 of Chapter 5).

Fix k. Start with $a_1 = a$ and construct the sequence $a_j = h_k(a_{j-1}) \in \mathcal{O}_F$ for $j \geqslant 2$. Since $N(a_j\mathcal{O}_F) = |N_{F/\mathbb{Q}}(a_j)| \leqslant c$, in the same way as in the proof of 3.6.4 we deduce that there are only finitely many distinct ideals $a_j\mathcal{O}_F$. So $a_j\mathcal{O}_F = a_q\mathcal{O}_F$ for some $j < q \leqslant l$. Then $u_k = a_q a_j^{-1}$ is a unit and satisfies the property: the ith coordinate of $g(u_k) = f(a_q) - f(a_j) = (\alpha_1^{(k)}, \ldots, \alpha_l^{(k)})$ is negative for $i \neq k$. Then $\alpha_k^{(k)}$ is positive, since $\sum_i \alpha_i^{(k)} = 0$.

This way we get l units $u_1, \ldots, u_l$. We claim that there are $l-1$ linearly independent vectors among the images $g(u_i)$. To verify the claim it suffices to check that the first $l-1$ columns of the matrix $(\alpha_i^{(k)})$ are linearly independent.

If there were not, then there would be a non-zero vector $(t_1, \ldots, t_{l-1})$ such that $\sum_{i=1}^{l-1} t_i \alpha_i^{(k)} = 0$ for all $1 \leqslant k \leqslant l$. Without loss of generality one can assume that there is i_0 between 1 and $l-1$ such that $t_{i_0} = 1$ and $t_i \leqslant 1$ for $i \neq i_0$, $1 \leqslant i \leqslant l-1$.

Then $t_{i_0}\alpha_{i_0}^{(i_0)} = \alpha_{i_0}^{(i_0)}$ and for $i \neq i_0$ $t_i\alpha_i^{(i_0)} \geqslant \alpha_i^{(i_0)}$ since $t_i \leqslant 1$ and $\alpha_i^{(i_0)} < 0$. Now we would get

$$0 = \sum_{i=1}^{l-1} t_i\alpha_i^{(i_0)} \geqslant \sum_{k=1}^{l-1} \alpha_i^{(i_0)} > \sum_{i=1}^{l} \alpha_i^{(i_0)} = 0,$$

a contradiction.

Thus, $m = r_1 + r_2 - 1$. □

REMARK. For a full and very different proof of Dirichlet's Unit Theorem see 4.8 in Chapter 3.

3.8.2. EXAMPLE. Let $F = \mathbb{Q}(\sqrt{d})$ with a square free non-zero integer d.

If $d > 0$, then the group of roots of 1 in F is $\{\pm 1\}$, since $F \subset \mathbb{R}$ and there are only two roots of unity in $\mathbb{R}$.

Let $\mathscr{O}_F$ be the ring of integers of F. We have $n = 2$ and $r_1 = 2, r_2 = 0$ if $d > 0$; $r_1 = 0, r_2 = 1$ if $d < 0$. If $d < 0$, then

$$U(\mathscr{O}_F) = T$$

is a finite cyclic group consisting of all roots of unity in F. It has order 4 for $d = -1$, 6 for $d = -3$, and one can show it has order 2 for all other negative square free integers.

If $d > 0$, $U(\mathscr{O}_F)$ is the direct product of $\langle \pm 1 \rangle$ and the infinite group generated by a unit u (it is called a *fundamental unit* of $\mathscr{O}_F$):

$$U(\mathscr{O}_F) \simeq \langle \pm 1 \rangle \times \langle u \rangle = \{\pm u^k : k \in \mathbb{Z}\}.$$

Here is an algorithm how to find a fundamental unit if $d \not\equiv 1 \mod 4$ (there is a similar algorithm for an arbitrary square free positive d):

If $a + b\sqrt{d} > 1$ is a unit of $\mathscr{O}_F$ then $N_{F/\mathbb{Q}}(a + b\sqrt{d}) = a^2 - db^2 = \pm 1$. Let b be the minimal positive integer such that either $db^2 - 1$ or $db^2 + 1$ is a square of a positive integer, say, a.

Let $u = e + f\sqrt{d}$ be a fundamental unit. Changing the sign of e, f if necessary, we can assume that e, f are positive. Due to the definition of u there is an integer k such that $a + b\sqrt{d} = \pm u^k$. The sign is $+$, since the left hand side is positive; $k > 0$, since $u \geqslant 1$ and the left hand side is > 1. From $a + b\sqrt{d} = (e + f\sqrt{d})^k$ we deduce that if $k > 1$ then $b = f +$ some positive integer $> f$, a contradiction. Thus, $k = 1$ and $a + b\sqrt{d} > 1$ is a fundamental unit of $\mathscr{O}_F$.

For example, $1 + \sqrt{2}$ is a fundamental unit of $\mathbb{Q}(\sqrt{2})$ and $2 + \sqrt{3}$ is a fundamental unit of $\mathbb{Q}(\sqrt{3})$.

3.8.3. Now suppose that $d > 0$, and for simplicity, $d \not\equiv 1 \mod 4$. Let $u = e + f\sqrt{d}$ be a fundamental unit. From the previous we deduce that all integer solutions (a,b) of the equation

$$X^2 - dY^2 = \pm 1$$

satisfy $a + b\sqrt{d} = \pm(e + f\sqrt{d})^m$ for some integer m, which gives formulas for a and b as functions of e, f, m.

4. *p*-adic Numbers

This section introduces first features of p-adic numbers. Chapter 2 contains a more general presentation of local fields and its readers can essentially skip this section.

4.1. *p-adic valuation and p-adic norm*

4.1.1. Fix a prime p.

For a non-zero integer m let

$$k = v_p(m)$$

be the maximal integer such that p^k divides m, i.e. k is the power of p in the factorisation of m. Then $v_p(m_1 m_2) = v_p(m_1) + v_p(m_2)$.

Extend v_p to rational numbers putting $v_p(0) := \infty$ and

$$v_p(m/n) = v_p(m) - v_p(n),$$

this does not depend on the choice of a fractional representation: if $m/n = m'/n'$ then $mn' = m'n$, hence $v_p(m) + v_p(n') = v_p(m') + v_p(n)$ and $v_p(m) - v_p(n) = v_p(m') - v_p(n')$.

Thus we get the *p-adic valuation* $v_p\colon \mathbb{Q} \longrightarrow \mathbb{Z} \cup \{+\infty\}$. For non-zero rational numbers $a = m/n, b = m'/n'$ we get

$$\begin{aligned} v_p(ab) &= v_p(mm'/(nn')) = v_p(mm') - v_p(nn') \\ &= v_p(m) + v_p(m') - v_p(n) - v_p(n') \\ &= v_p(m) - v_p(n) + v_p(m') - v_p(n') \\ &= v_p(m/n) + v_p(m'/n') \\ &= v_p(a) + v_p(b). \end{aligned}$$

Thus v_p is a homomorphism from $\mathbb{Q}^\times$ to $\mathbb{Z}$.

4.1.2. *p-adic norm.* Define the *p-adic norm* of a rational number α by

$$|\alpha|_p = p^{-v_p(\alpha)}, \quad |0|_p = 0.$$

Then

$$|\alpha\beta|_p = |\alpha|_p|\beta|_p.$$

If $\alpha = m/n$ with integer m,n relatively prime to p, then $v_p(m) = v_p(n) = 0$ and $|\alpha|_p = 1$. In particular, $|-1|_p = |1|_p = 1$ and so $|-\alpha|_p = |\alpha|_p$ for every rational α.

4.1.3. *Ultrametric inequality.* For two integers m,n put

$$k = \min(v_p(m), v_p(n)).$$

Hence $m+n$ is divisible by p^k, thus

$$v_p(m+n) \geqslant \min(v_p(m), v_p(m)).$$

For two nonzero rational numbers $\alpha = m/n$, $\beta = m'/n'$

$$\begin{aligned} v_p(\alpha+\beta) &= v_p(mn'+m'n) - v_p(nn') \\ &\geqslant \min(v_p(m)+v_p(n'), v_p(m')+v_p(n)) - v_p(n) - v_p(n') \\ &\geqslant \min(v_p(m)-v_p(n), v_p(m')-v_p(n')) \\ &= \min(v_p(\alpha), v_p(\beta)). \end{aligned}$$

Hence for all rational α, β we get

$$v_p(\alpha+\beta) \geqslant \min(v_p(\alpha), v_p(\beta)).$$

This implies

$$|\alpha+\beta|_p \leqslant \max(|\alpha|_p, |\beta|_p).$$

This inequality is called *an ultrametric inequality.*

In particular, since $\max(|\alpha|_p, |\beta|_p) \leqslant |\alpha|_p + |\beta|_p$, we obtain

$$|\alpha+\beta|_p \leqslant |\alpha|_p + |\beta|_p,$$

so $|\ |_p$ is a metric (p-adic metric) on the set of rational numbers $\mathbb{Q}$ and

$$d_p(\alpha,\beta) = |\alpha-\beta|_p$$

gives *the p-adic distance* between rational α, β.

4.1.4. *All norms on* $\mathbb{Q}$. In general, for a field F a norm $|\ |: F \longrightarrow \mathbb{R}_{\geqslant 0}$ is a map which sends 0 to 0, which is a homomorphism from $F^\times$ to $\mathbb{R}^\times_{>0}$ and which satisfies the triangle inequality: $|\alpha+\beta| \leqslant |\alpha|+|\beta|$. In particular,

$$|1| = 1, 1 = |1| = |(-1)(-1)| = |-1|^2,$$

so $|-1| = 1$, and hence

$$|-a| = |-1||a| = |a|.$$

A norm is called nontrivial if there is a nonzero $a \in F$ such that $|a| \neq 1$.

In addition to p-adic norms on $\mathbb{Q}$ we get the usual absolute value on $\mathbb{Q}$ which we will denote by $|\ |_\infty$.

A complete description of norms on $\mathbb{Q}$ is supplied by the following result.

THEOREM. *(Ostrowski's Theorem) A non-trivial norm $|\ |$ on $\mathbb{Q}$ is either a power of the absolute value $|\ |_\infty^c$ with positive real c, or is a power of the p-adic norm $|\ |_p^c$ for some prime p with positive real c.*

Proof. For an integer $a > 1$ and an integer $b > 0$ write

$$b = b_n a^n + b_{n-1} a^{n-1} + \cdots + b_0$$

with $0 \leqslant b_i < a, a^n \leqslant b$. Then

$$|b| \leqslant (|b_n| + |b_{n-1}| + \cdots + |b_0|) \max(1, |a|^n)$$

and

$$|b| \leqslant (\log_a b + 1)\, d \max(1, |a|^{\log_a b}),$$

with $d = \max(|0|, |1|, \ldots, |a-1|)$.

Substituting b^s instead of b in the last inequality, we get

$$|b^s| \leqslant (s \log_a b + 1)\, d \max(1, |a|^{s \log_a b}),$$

hence

$$|b| \leqslant (s \log_a b + 1)^{1/s} d^{1/s} \max(1, |a|^{\log_a b}).$$

When $s \to +\infty$ we deduce

$$|b| \leqslant \max(1, |a|^{\log_a b}).$$

There are two cases to consider.

(1) Suppose there is an integer b such that $|b| > 1$. We can assume b is positive. Then

$$1 < |b| \leqslant \max(1, |a|^{\log_a b}),$$

and so $|a| > 1$, $|b| \leqslant |a|^{\log_a b}$ for every integer $a > 1$. Swapping a and b we get $|a| \leqslant |b|^{\log_b a}$, thus,

$$|a| = |b|^{\log_b a}$$

for every integer a and hence for every rational a.

Choose $c > 0$ such that $|b| = |b|_\infty^c$ then we obtain $|a| = |a|_\infty^c$ for every rational a.

(2) Suppose that $|a| \leqslant 1$ for all integer a. Since $|\ |$ is non-trivial, let a_0 be the minimal positive integer such that $|a_0| < 1$. If $a_0 = a_1 a_2$ with positive integers a_1, a_2, then $|a_1||a_2| < 1$ and either $a_1 = 1$ or $a_2 = 1$. This means that $a_0 = p$ is a prime. If $q \notin p\mathbb{Z}$, then $pp_1 + qq_1 = 1$ with some integers p_1, q_1 and hence $1 = |1| \leqslant |p||p_1| + |q||q_1| \leqslant |p| + |q|$. Writing q^s instead of q we get $|q|^s \geqslant 1 - |p| > 0$ and $|q| \geqslant (1 - |p|)^{1/s}$. The right hand side tends to 1 when s tends to infinity. So we obtain $|q| = 1$ for every q prime to p. Therefore, $|\alpha| = |p|^{v_p(\alpha)}$, and $|\ |$ is a power of the p-adic norm. □

4.1.5. LEMMA. *(Product formula) For every nonzero rational* α

$$\prod_{i \text{ prime or } \infty} |\alpha|_i = 1.$$

Proof. Due to the multiplicative property of the norms and factorisation of integers it is sufficient to consider the case when α a prime number p. Then $|p|_p = p^{-1}$, $|p|_\infty = p$ and $|p|_i = 1$ for all other i. □

4.2. *The field of p-adic numbers* $\mathbb{Q}_p$

4.2.1. DEFINITION. Similarly to the definition of real numbers as the completion of $\mathbb{Q}$ with respect to the absolute value $|\ |_\infty$ define $\mathbb{Q}_p$ as the completion of $\mathbb{Q}$ with respect to the p-adic norm $|\ |_p$. So $\mathbb{Q}_p$ consists of equivalences classes of all fundamental sequences (with respect to the p-adic norm) (a_n) of rational numbers a_n: two fundamental sequences (a_n), (b_n) are equivalent if and only if $|a_n - b_n|_p$ tends to 0.

The field $\mathbb{Q}_p$ is called the *field of p-adic numbers* and its elements are called p-adic numbers.

4.2.2. *p-adic series presentation of p-adic numbers.* As an analog of the decimal presentation of real numbers every element α of $\mathbb{Q}_p$ has a series representation: it can be written as an infinite convergent (with respect to the p-adic norm) series

$$\sum_{i=n}^{\infty} a_i p^i$$

with coefficients $a_i \in \{0, 1, \ldots, p-1\}$ and $a_n \neq 0$.

4.2.3. *The p-adic norm and p-adic distance.* We have an extension of the p-adic norm from $\mathbb{Q}$ to $\mathbb{Q}_p$ by continuity: if $\alpha \in \mathbb{Q}_p$ is the limit of a fundamental sequence (a_n) of rational numbers, then $|\alpha|_p := \lim |a_n|_p$. Since two fundamental sequences (a_n), (b_n) are equivalent if and only if $|a_n - b_n|_p$ tends to 0, the p-adic norm of α is well defined.

If we use the series representation $\alpha = \sum_{i=n}^{\infty} a_i p^i$ with $a_i \in \{0, 1, \dots, p-1\}$ and $a_n \neq 0$, then $|\alpha|_p = p^{-n}$.

The p-adic norm on $\mathbb{Q}_p$ satisfies the ultrametric inequality: let $\alpha = \lim a_n, \beta = \lim b_n$, (a_n), (b_n) are fundamental sequences of rational numbers, then $\alpha + \beta = \lim(a_n + b_n)$. Suppose that $|\alpha|_p \leqslant |\beta|_p$, then $|a_n|_p \leqslant |b_n|_p$ for all sufficiently large n, and so

$$\begin{aligned}|\alpha+\beta|_p &= \lim |a_n + b_n|_p \leqslant \lim \max(|a_n|_p, |b_n|_p) \\ &= \lim |b_n|_p = |\beta|_p = \max(|\alpha|_p, |\beta|_p).\end{aligned}$$

For α, β such that $|\alpha|_p < |\beta|_p$ we obtain $\beta = \gamma + \alpha$ where $\gamma = \beta - \alpha$. By the ultrametric inequality $|\beta|_p \leqslant \max(|\gamma|_p, |\alpha|_p)$, so $|\beta|_p \leqslant |\gamma|_p$ and by the ultrametric inequality $|\gamma|_p \leqslant \max(|\alpha|_p, |-\beta|_p) = \max(|\alpha|_p, |\beta|_p) = |\beta|_p$. Thus if $|\alpha|_p < |\beta|_p$ then $|\alpha - \beta|_p = |\beta|_p$.

Using the p-adic distance d_p we have shown that for every triangle with vertices in $0, \alpha, \beta$ if the p-adic length of its side connecting 0 and α is smaller than the p-adic length of its side connecting 0 and β then the p-adic length of the third side connecting α and β equals to the former. Thus, in every triangle two sides are of the same p-adic length!

4.2.4. *The ring of p-adic integers* $\mathbb{Z}_p$. Define the set $\mathbb{Z}_p$ of p-adic integers as those p-adic numbers whose p-adic norm does not exceed 1, i.e. whose p-adic series representation has $n_0 \geqslant 0$. For two elements $\alpha, \beta \in \mathbb{Z}_p$ we get $|\alpha\beta|_p \leqslant 1$, $|\alpha \pm \beta|_p \leqslant 1$. Hence $\mathbb{Z}_p$ is a subring of $\mathbb{Q}_p$.

The units $\mathbb{Z}_p^\times$ of the ring $\mathbb{Z}_p$ are those p-adic numbers u whose p-adic norm is 1.

Every nonzero p-adic number α can be uniquely written as $p^{v_p(\alpha)} u$ with $u \in \mathbb{Z}_p^\times$. Thus

$$\mathbb{Q}_p^\times \simeq \langle p \rangle \times \mathbb{Z}_p^\times$$

where $\langle p \rangle$ is the infinite cyclic group generated by p.

Let I be a non-zero ideal of $\mathbb{Z}_p$. Let $n = \min\{v_p(\alpha) : \alpha \in I\}$. Then $p^n u$ belongs to I for some unit u, and hence p^n belongs to I, so $p^n\mathbb{Z}_p \subset I \subset p^n\mathbb{Z}_p$, i.e. $I = p^n\mathbb{Z}_p$. Thus $\mathbb{Z}_p$ is a principal ideal domain and a Dedekind ring.

4.2.5. Note that $\mathbb{Z}_p$ is the closed ball of radius 1 in the p-adic norm.

Let α be its internal point, so $|\alpha|_p < 1$. Then for every β on the boundary of the open ball, i.e. $|\beta|_p = 1$ we obtain, applying 4.2.3, we obtain $|\alpha - \beta|_p = |\beta|_p = 1$. Thus, the p-adic distance from α to every point on the boundary of the ball is 1, i.e. every internal point of a p-adic ball is its centre.

4.3. *Henselian property*

Let $f(X) = \sum a_i X^i \in \mathbb{Z}_p[X]$, and let $a, b \in \mathbb{Z}_p$, $a - b \in p^n\mathbb{Z}_p$, $n > 0$. Then

$$f(a) - f(b) = \sum a_i(a^i - b^i) = \sum_{i>0} a_i(a-b)(a^{i-1} + \cdots + b^{i-1}) \in p^n\mathbb{Z}_p.$$

THEOREM. (*Henselian property*)

Let $f(X) \in \mathbb{Z}_p[X]$.

Let a be a p-adic integer such that $v_p(f'(a)) = r, v_p(f(a)) > 2r$ for a non-negative integer r.

Define a sequence $\alpha_n \in \mathbb{Q}_p$ as $\alpha_0 = a$,

$$\alpha_{n+1} = \alpha_n - \frac{f(\alpha_n)}{f'(\alpha_n)}, \quad n \geqslant 0.$$

Then this sequence converges to $\alpha \in \mathbb{Z}_p$ such that

$$f(\alpha) = 0, \quad v_p(\alpha - a) \geqslant r + 1.$$

Proof. By induction on $n \geqslant 0$ we prove that $\alpha_n \in \mathbb{Z}_p$, $f(\alpha_n) \in p^{2r+1+n}\mathbb{Z}_p$ for $n \geqslant 0$, $\alpha_n - \alpha_{n-1} \in p^{r+n}\mathbb{Z}_p$ for $n \geqslant 1$. Then the sequence α_n indeed converges, and passing to the limit we obtain that its limit $\alpha \in \mathbb{Z}_p$ satisfies $f(\alpha) = 0$ and $\alpha - a \in p^{r+1}\mathbb{Z}_p$.

Base of induction: $n = 0$ is clear. Induction step ($n \Longrightarrow n+1$): $\alpha_{n+1} - \alpha_n = -\frac{f(\alpha_n)}{f'(\alpha_n)}$. Since by the induction hypothesis $\alpha_n - \alpha_0 \in p^{r+1}\mathbb{Z}_p$ and $v_p(f'(\alpha_0)) = r$, using the property stated before the Lemma, we obtain $v_p(f'(\alpha_n)) = r$. Then by the induction hypothesis

$$\frac{f(\alpha_n)}{f'(\alpha_n)} \in p^{r+1+n}\mathbb{Z}_p \qquad (*)$$

so $\alpha_{n+1} - \alpha_n \in p^{r+n+1}\mathbb{Z}_p$ and α_{n+1} is in $\mathbb{Z}_p$.

Finally, represent $f(X)$ as a polynomial of $X - \alpha_n$:

$$f(X) = f(\alpha_n) + f'(\alpha_n)(X - \alpha_n) + (X - \alpha_n)^2 g(X)$$

for a polynomial $g(X) \in \mathbb{Z}_p[X]$. Substitute $X = \alpha_{n+1}$. Using the definition of $\alpha_{n+1} \in \mathbb{Z}_p$ we obtain

$$f(\alpha_{n+1}) = \left(\frac{f(\alpha_n)}{f'(\alpha_n)}\right)^2 g(\alpha_{n+1}),$$

hence by (*) we obtain $f(\alpha_{n+1}) \in p^{2(r+1+n)}\mathbb{Z}_p$. □

REMARK. Often, a different property which implies this Theorem is called Hensel Lemma: Let $f(X), g_0(X), h_0(X)$ be monic polynomials with coefficients in $\mathbb{Z}_p$ such that for their residue images in $\mathbb{F}_p[X]$ the equality $\overline{f}(X) = \overline{g}_0(X)\overline{h}_0(X)$ holds. Suppose that $\overline{g}_0(X), \overline{h}_0(X)$ are relatively prime in $\mathbb{F}_p[X]$. Then there exist monic polynomials $g(X), h(X)$ with coefficients in $\mathbb{Z}_p$, such that

$$f(X) = g(X)h(X),\ \overline{g}(X) = \overline{g}_0(X),\ \overline{h}(X) = \overline{h}_0(X).$$

COROLLARY 1. *Let $f(X) \in \mathbb{Z}_p[X]$, $a \in \mathbb{Z}_p$ such that $f(a) \in p\mathbb{Z}_p$ and $f'(a) \notin p\mathbb{Z}_p$. Then the polynomial f has a root $\alpha \in \mathbb{Z}_p$ such that $\alpha - a \in p\mathbb{Z}_p$.*

Proof. $r = 0$. □

COROLLARY 2. *The polynomial $X^{p-1} - 1$ has $p-1$ distinct roots in the field $\mathbb{Q}_p$, if $p > 2$.*

Proof. Choose any of $p-1$ elements of $\mathbb{F}_p^{\times}$, denote it by b. Let $a \in \mathbb{Z}_p$ whose image in $\mathbb{F}_p$ with respect to the surjective homomorphism $\mathbb{Z}_p \longrightarrow \mathbb{Z}_p/p\mathbb{Z}_p = \mathbb{F}_p$ is b. Then the image of $a^{p-1} - 1$ with respect to the same homomorphism is 0, i.e. $v_p(a^{p-1} - 1) \geqslant 1$. Since $(X^{p-1} - 1)' = (p-1)X^{p-2}$ and the image of $(p-1)a^{p-2}$ in $\mathbb{F}_p$ is not zero, we can apply Corollary 1 to deduce the existence of a root $\alpha \in \mathbb{Z}_p$ of $X^{p-1} - 1$, $\alpha - a \in p\mathbb{Z}_p$. □

COROLLARY 3. *If $p > 2$, the group $\mathbb{Z}_p^{\times}$ is the product of the cyclic group of order $p-1$ and the group $1 + p\mathbb{Z}_p$. The group $\mathbb{Z}_2^{\times}$ is the product of the cyclic group of order 2 and the group $1 + 4\mathbb{Z}_2$.*

Proof. If p is odd, let $\beta \in \mathbb{Z}_p^{\times}$, let $b \in \mathbb{F}_p^{\times}$ be its image with respect to the homomorphism of the previous proof and let $\alpha \in \mathbb{Z}_p$ be a root of $X^{p-1} - 1$ such that $\beta - \alpha \in p\mathbb{Z}_p$. Then $\gamma = \beta\alpha^{-1} \in 1 + p\mathbb{Z}_p$. The intersection of the group of roots of $X^{p-1} - 1$ and the group $1 + p\mathbb{Z}_p$ is $\{1\}$: indeed for $\delta \in p\mathbb{Z}_p$ we have $1 = (1+\delta)^{p-1} = 1 + (p-1)\delta +$ terms whose p-adic valuation is at least $\geqslant 2v_p(\delta) > v_p((p-1)\delta) = v_p(\delta)$, hence $\delta = 0$.

If $p = 2$ then ± 1 are roots in $\mathbb{Q}_2$. We can write $-1 = 1 + 2 + 2^2 + \dots$ in $\mathbb{Z}_2$. Hence, every element of $\mathbb{Z}_2^{\times} = 1 + 2\mathbb{Z}_2$ is the product of ± 1 and an element of $1 + 4\mathbb{Z}_2$. The intersection of the group $1 + 4\mathbb{Z}_2$ and the cyclic group of order 2 is $\{1\}$. □

COROLLARY 4. *The group* $\mathbb{Q}_p^\times$ *contains* $p-1$ *roots of unity if* $p>2$ *and* 2 *roots of unity if* $p=2$.

Proof. Let $\gamma \in \mathbb{Q}_p$ satisfy $\gamma^m = 1$, $m>0$. If $s = v_p(\gamma)$, then $ms = v_p(\gamma^m) = v_p(1) = 0$, so $s=0$ and $\gamma \in \mathbb{Z}_p^\times$. Using Corollary 3 we only need to show that $1+p\mathbb{Z}_p$ does not have non-trivial roots of unity if $p>2$ and $1+4\mathbb{Z}_2$ does not have non-trivial roots of unity.

Write an element of $1+p\mathbb{Z}_p$ as $1+p^r a$ with $a \in \mathbb{Z}_p^\times$, $r \geqslant 1$. If m is prime to p, then $(1+p^r a)^m = 1+mp^r a+\cdots+p^{rm}a^m \equiv 1+mp^r a \not\equiv 1 \mod p^{r+1}\mathbb{Z}_p$, so $(1+p^r a)^m \neq 1$. Hence we only need to look at elements of order p. If p is odd, we have $(1+p^r a)^p \equiv 1+p^{r+1}a \not\equiv 1 \mod p^{2r+1}\mathbb{Z}_p$, hence $(1+p^r a)^p \neq 1$ and $1+p\mathbb{Z}_p$ does not have elements of order p. If $p=2$ then $(1+2^r a)^2 = 1+2^{r+1}a+2^{2r}a^2 \equiv 1+2^{r+1}a \not\equiv 1 \mod 2^{2r}\mathbb{Z}_2$ and $(1+2^r a)^2 \neq 1$ if $r \geqslant 2$, $a \in \mathbb{Z}_2^\times$, hence $1+4\mathbb{Z}_2$ does not have elements of order 2. □

COROLLARY 5. $1+p\mathbb{Z}_p = (1+p\mathbb{Z}_p)^m$ *for every positive integer* m *prime to* p.

Proof. Let $\gamma \in 1+p\mathbb{Z}_p$. Put $f(X) = X^m - \gamma$, $a=1$ and apply the Hensel Lemma. □

COROLLARY 6. *The fields* $\mathbb{Q}_p$ *and* $\mathbb{Q}_q$, $p \neq q$, *are not isomorphic.*

Proof. Consider $1+pq \in 1+p\mathbb{Z}_p$. By the previous corollary $1+pq$ is a qth power in $\mathbb{Q}_p$. On the other hand, $1+pq \in 1+q\mathbb{Z}_q$ cannot be a qth power. Indeed, if $1+pq = (q^n\alpha)^q$ with $\alpha \in \mathbb{Z}_q^\times$, then comparing v_q on the LHS and RHS we deduce $n=0$. Looking at the images of the LHS and RHS in $\mathbb{Z}_q/q\mathbb{Z}_q \simeq \mathbb{F}_q$ we deduce $\alpha \in 1+q\mathbb{Z}_q$, so $\alpha = 1+q\gamma$ with $\gamma \in \mathbb{Z}_q$. Since $(1+q\gamma)^q \in 1+q^2\mathbb{Z}_q$ and $p \notin q\mathbb{Z}_q$, we get a contradiction. □

REMARK. For much more about p-adic fields see Chapter 2.

5. A Little about Class Field Theory

This section introduces first features of cyclotomic class field theory in a way quite different from the general presentation of class field theory in Chapter 3.

First, we need to talk a little about limits of algebraic objects.

5.1. *Algebraic limits*

5.1.1. *Direct limits.*

DEFINITION. Let A_n be a set of groups/rings, with group operation written additively in the case of groups. Suppose there are link group/ring homomorphisms $\varphi_{mn}\colon A_m \longrightarrow A_n$ for all $m \leqslant n$ such that $\varphi_{nn} = \mathrm{id}_{A_n}$, $\varphi_{rn} = \varphi_{mn} \circ \varphi_{rm}$ for all $r \leqslant m \leqslant n$. We call it an inductive system.

The *direct/inductive limit* $\varinjlim A_n$ of inductive system (A_n, φ_{mn}) is the equivalence classes of elements of the disjoint union of A_n with respect to the equivalence relation $a_n \in A_n$ is equivalent to $a_m \in A_m$ if $\varphi_{mn}(a_m) = a_n$, with the group/ring operation(s) induced by $(a_n) + (b_n) = (a_n + b_n)$ and $(a_n)(b_n) = (a_n b_n)$.

One can extend in the obvious way the definition of the inductive limit to the case when the maps φ_{nm} are defined for some specific pairs (n,m) and not necessarily for all $n \geqslant m$, and when the index set is a partially ordered set and not necessarily countable.

For every m one has a group/ring homomorphism $\varphi_m\colon A_m \longrightarrow \varinjlim A_n, a_m \mapsto (a_n)_{n\geqslant m}$ where $a_n = \varphi_{mn}(a_m)$ for $n \geqslant m$.

If A_n are also topological spaces, then $\varinjlim A_n$ is endowed with the strongest topology in which all φ_m are continuous.

EXAMPLES.

1. If $A_n = A$ for all n and $\varphi_{mn} = \mathrm{id}$ then $\varinjlim A_n = A$.

2. If a group/ring A is the union of its subgroups/subrings A_n, then $A \simeq \varinjlim A_n$. In particular, $\mathbb{Q}/\mathbb{Z} \simeq \varinjlim (1/n)\mathbb{Z}/\mathbb{Z}$ with respect to $\varphi_{mn}(a/m+\mathbb{Z}) = (an/m)/n+\mathbb{Z}$ for $m|n$. Since $(1/n)\mathbb{Z}/\mathbb{Z} \simeq \mathbb{Z}/n\mathbb{Z}$, $\mathbb{Q}/\mathbb{Z}$ is also isomorphic to the direct limit of $\mathbb{Z}/n\mathbb{Z}$ with respect to the appropriate link homomorphisms.

3. Let B_j, $j \geqslant 1$, be groups/rings. Denote $A_n = B_1 \oplus \cdots \oplus B_n$ and define homomorphisms $\varphi_{mn}\colon A_m \longrightarrow A_n$ for $m \leqslant n$ as $(b_1, \ldots, b_m) \mapsto (b_1, \ldots, b_m, 0, \ldots)$ (0s are at all places starting with the $(m+1)$st place. Then $\varinjlim A_n$ is called the *infinite direct sum* $\oplus B_j$ of B_j, its elements have zero components for almost all places.

5.1.2. *Inverse limits.*

DEFINITION. Let A_n be a set of groups/rings, with group operation written additively in the case of groups. Suppose there are link group/ring homomorphisms $\psi_{nm}\colon A_n \longrightarrow A_m$ for all $n \geqslant m$ such that $\psi_{nn} = \mathrm{id}_{A_n}$, $\psi_{nr} = \psi_{mr} \circ \psi_{nm}$ for all $n \geqslant m \geqslant r$. We call it a projective system.

The *inverse/projective limit* $\varprojlim A_n$ of projective system (A_n, φ_{nm}) is the set

$$\{(a_n) : a_n \in A_n, \psi_{nm}(a_n) = a_m \text{ for all } n \geqslant m\}$$

with the group/ring operation(s) $(a_n)+(b_n) = (a_n+b_n)$ and $(a_n)(b_n) = (a_n b_n)$.

One can extend in the obvious way the definition of the projective limit to the case when the maps ψ_{nm} are defined for some specific pairs (n,m) and not necessarily all $n \geqslant m$, and when the index set is a partially ordered set and not necessarily countable.

For every m one has a group/ring homomorphism $\psi_m \colon \varprojlim A_n \longrightarrow A_m$, $(a_n) \mapsto a_m$.

If A_n are also topological spaces, then $\varprojlim A_n$ is endowed with the weakest topology in which all ψ_m are continuous. Thus, open subsets of $\varprojlim A_n$ are intersections of $\psi_m^{-1}(U_m)$ for finitely many indices and open subsets U_m of A_m.

EXAMPLES.

1. If $A_n = A$ for all n and $\psi_{nm} = \text{id}$ then $\varprojlim A_n = A$.

2. If $A_n = \mathbb{Z}/p^n\mathbb{Z}$ and $\varphi_{nm}(a+p^n\mathbb{Z}) = a+p^m\mathbb{Z}$ then $(a_n) \in \varprojlim \mathbb{Z}/p^n\mathbb{Z}$ is equivalent to the condition that $p^{\min(n,m)}$ divides $a_n - a_m$ for all n,m.

The sequence (a_n) as above is a fundamental sequence with respect to the p-adic norm, and thus determines a p-adic number $a = \lim a_n \in \mathbb{Z}_p$. For its description, denote by r_m the integer between 0 and $p^m - 1$ such that $r_m \equiv a_m \mod p^m$. Then $r_m \equiv a_n \mod p^m$ for $n \geqslant m$ and $r_n \equiv r_m \mod p^m$ for $n \geqslant m$. Denote $c_0 = r_0$ and $c_m = (r_m - r_{m-1})p^{-m+1}$, so $c_m \in \{0,1,\ldots,p-1\}$. Then $a = \sum_{m \geqslant 0} c_m p^m = \lim r_m \in \mathbb{Z}_p$.

We have a group and ring homomorphism

$$f \colon \varprojlim \mathbb{Z}/p^n\mathbb{Z} \longrightarrow \mathbb{Z}_p, \quad (a_n) \mapsto a = \lim a_n \in \mathbb{Z}_p.$$

It is surjective: if $a = \sum_{m \geqslant 0} c_m p^m$ then define r_m by the inverse procedure to the above, then a is the image of $(r_n) \in \varprojlim \mathbb{Z}/p^n$; and its kernel is trivial, since $a = 0$ implies that for every k p^k divides a_n for all sufficiently large n, and so p^k divides a_k.

Thus, we get

$$\varprojlim \mathbb{Z}/p^n\mathbb{Z} \simeq \mathbb{Z}_p.$$

This can be used as another (algebraic) definition of the ring of p-adic integers. Moreover, this isomorphism is also a topological isomorphism.

In particular, we have a surjective homomorphism $\mathbb{Z}_p \longrightarrow \mathbb{Z}/p^n\mathbb{Z}$ whose kernel equals to $p^n\mathbb{Z}_p$.

From the above we immediately deduce that if $A_n = (\mathbb{Z}/p^n\mathbb{Z})^\times$ and $\psi_{nm}(a+p^n\mathbb{Z}) = a + p^m\mathbb{Z}$, $(a,p) = 1$, then similarly we have a homomorphism

$$f\colon \varprojlim (\mathbb{Z}/p^n\mathbb{Z})^\times \longrightarrow \mathbb{Z}_p^\times, \quad (a_n) \mapsto \lim r_m \in \mathbb{Z}_p^\times$$

(note that $(r_m, p) = 1$ and hence $\lim r_m \notin p\mathbb{Z}_p$). Thus, we get an algebraic and topological isomorphism

$$\varprojlim (\mathbb{Z}/p^n\mathbb{Z})^\times \xrightarrow{\sim} \mathbb{Z}_p^\times.$$

3. Let B_j, $j \geqslant 1$, be groups/rings. Let C_n be the direct product $\prod_{1\leqslant j\leqslant n} B_j$ and define homomorphisms $\psi_{nm}\colon C_n \longrightarrow C_m$ for $m \leqslant n$ as $(b_1,\ldots,b_n) \mapsto (b_1,\ldots,b_m)$. Then $\varprojlim C_n$ is called the *infinite (direct) product* $\prod B_j$ of B_j. If B_j are topological spaces then open subsets of $\prod B_j$ are $U_1 \times \cdots \times U_m \times B_{m+1} \times B_{m+2} \times \cdots$ with open subsets U_i of B_i and varying m.

Then the definitions imply

$$\varprojlim A_n = \{(a_n) \in \prod A_n : a_m = \varphi_{nm}(a_n) \quad \text{for all } m \leqslant n\}.$$

The projective limit $\varprojlim A_n$ is a closed subset of $\prod A_n$ with the induced topology.

4. Let $A_n = \mathbb{Z}/n\mathbb{Z}$ and let $\psi_{nm}\colon A_n \longrightarrow A_m$ be defined only if $m \mid n$ and then $\psi_{nm}(a+n\mathbb{Z}) = a + m\mathbb{Z}$. We have a special notation and name (*zet hat*) for the inverse limit

$$\widehat{\mathbb{Z}} = \varprojlim \mathbb{Z}/n\mathbb{Z}.$$

As an additive group, it is a procyclic group (topologically it is generated by its unity 1). This group is uncountable. We have a surjective homomorphism $\widehat{\mathbb{Z}} \longrightarrow \mathbb{Z}/n\mathbb{Z}$ whose kernel is $n\widehat{\mathbb{Z}}$.

By the Chinese Remainder Theorem $\mathbb{Z}/n\mathbb{Z} = \mathbb{Z}/p_1^{k_1}\mathbb{Z} \times \cdots \times \mathbb{Z}/p_r^{k_r}\mathbb{Z}$, where $n = p_1^{k_1} \ldots p_r^{k_r}$ is the factorisation of n. The maps ψ_{nm} induce the maps already defined in Example 2 for $\mathbb{Z}/p^r\mathbb{Z}$, and we deduce

$$\widehat{\mathbb{Z}} = \varprojlim \mathbb{Z}/n\mathbb{Z} = \varprojlim \mathbb{Z}/2^r\mathbb{Z} \times \varprojlim \mathbb{Z}/3^r\mathbb{Z} \times \cdots \simeq \mathbb{Z}_2 \times \mathbb{Z}_3 \times \cdots = \prod \mathbb{Z}_p.$$

5. Similarly we have

$$\widehat{\mathbb{Z}}^\times = \varprojlim (\mathbb{Z}/n\mathbb{Z})^\times = \varprojlim (\mathbb{Z}/2^r\mathbb{Z})^\times \times \varprojlim (\mathbb{Z}/3^r\mathbb{Z})^\times \times \cdots \simeq \prod \mathbb{Z}_p^\times.$$

6. Let A be the projective limit of groups/rings A_n with respect to group/ring homomorphisms ψ_{nm}. Let C be a group/ring. Denote by $C_n = \operatorname{Hom}(A_n, C)$ the group/ring of homomorphisms from A_n to C. Then we have group/ring homomorphisms $\varphi_{mn}\colon C_m \longrightarrow C_n$ for $m \leqslant n$ sending an element f of $\operatorname{Hom}(A_m, C)$ to composition $f \circ \psi_{nm}$. The definitions imply that

$$\operatorname{Hom}(\varprojlim A_n, C) \simeq \varinjlim \operatorname{Hom}(A_n, C).$$

Similarly, if we start with $\varinjlim C_n$ of groups/rings C_n, then

$$\mathrm{Hom}(\varinjlim C_n, C) \simeq \varprojlim \mathrm{Hom}(C_n, C).$$

5.1.3. *Profinite groups.* In this subsubsection we write group operation multiplicatively.

DEFINITION. A *profinite group* is the projective limit $\varprojlim A_n$ for a projective system (A_n, φ_{nm}) with all A_n finite groups. The topology of $\varprojlim A_n$ is the projective limit of discrete finite groups A_n.

LEMMA. *A profinite group G is a topological group which is Hausdorff, compact, and a base of neighbourhoods of its identity element can be chosen as open normal subgroups. Moreover, $G \simeq \varprojlim G/N$ algebraically and topologically, where N runs through all open normal subgroups of G and the projective link homomorphisms $G/N \longrightarrow G/M$ are defined when $M \supset N$.*

Proof. For compact Hausdorff finite groups A_n, their product $\prod A_n$ is a compact Hausdorff group. Since $G = \varprojlim A_n$ is a closed subgroup of $\prod A_n$, it is a topological group which is compact and Hausdorff. When N_i run through normal subgroups of A_i for finitely many indices $i \leqslant m$ and m varying, normal open subgroups $G \cap \prod_{1 \leqslant i \leqslant m} N_i \times \prod_{n>m} A_n$ is a base of open neighbourhoods of the identity element in G.

The following argument is applicable to every compact Hausdorff group G with open normal subgroups as a base of open neighbourhoods of the identity element. Since G is compact, its open normal subgroups N are of finite index. The kernel of homomorphism $f\colon G \longrightarrow \varprojlim G/N$ sending g to $(gN)_N$ is the intersection of those N, hence trivial since G is Hausdorff. The intersections U of $\varprojlim G/N$ with the subgroup of $\prod G/N$ (product is taken over all normal open subgroups N) which for finitely many N has component equal to the identity element of G/N, is a base of neighbourhoods of the identity element of $\varprojlim G/N$. The preimage with respect to f is the intersection of those finitely many N, hence open. Therefore, f is continuous. Then, since G is compact, $f(G)$ is closed in $\varprojlim G/N$. Also, the image of every closed subgroup, being compact, is closed. Therefore, f is an open map. For $g = (g_N N)_N \in \varprojlim G/N$ and its neighbourhood gU let $h \in G$ be such that $h \in g_{N'}N'$ where N' is the intersection of those finitely many N, then $f(h) \in gU$, so $f(G)$ is dense in $\varprojlim G/N$. Thus, f is an algebraic and topological isomorphism. □

EXAMPLES.

1. $\mathbb{Z}_p$, $\widehat{\mathbb{Z}}$, $\mathbb{Z}_p^\times$, $\widehat{\mathbb{Z}}$ are profinite groups.
2. Products and infinite products of profinite groups are profinite groups.
3. Galois groups are profinite groups, see below.

5.2. *Infinite Galois theory*

The following definition extends the definition of finite Galois extensions and their Galois groups.

DEFINITION. Let L/F be an extension of fields, finite or infinite. It is called a Galois extension if L is the composite of splitting fields of separable polynomials over F. Its Galois group is

$$\mathrm{Gal}(L/F) = \varprojlim \mathrm{Gal}(E/F),$$

where E runs through finite Galois subextensions of L/F and the projective link homomorphisms are taken with respect to $\mathrm{Gal}(E''/F) \longrightarrow \mathrm{Gal}(E'/F)$ for $E' \subset E'' \subset L$.

Thus, elements of $\mathrm{Gal}(L/F)$ are F-automorphisms of L to L that are distinguished/characterised by their action on finite Galois subextensions E/F of L/F. Of course, every element of L belongs to some finite Galois extension E/F, so this view to F-automorphisms of L is justified.

By Lemma 5.1.3 $\mathrm{Gal}(L/F)$ is a compact Hausdorff topological group. A base of neighbourhoods of $\sigma \in \mathrm{Gal}(L/F)$ are open normal subgroups $\sigma\,\mathrm{Gal}(L/E)$ with finite Galois extensions E/F.

By 5.1.2 open subgroups of $\mathrm{Gal}(L/F)$ are intersections of $\psi_E^{-1}(U_E)$ for finitely many indices and open subgroups U_E of finite $\mathrm{Gal}(E/F)$. Hence open subgroups of $\mathrm{Gal}(L/F)$ are all $\mathrm{Gal}(L/E)$ for finite subextensions E/F of L/F.

For every subextension M/F of L/F we have a homomorphism from the group $\mathrm{Gal}(L/M) = \varprojlim \mathrm{Gal}(R/M)$ to the group $\mathrm{Gal}(L/F) = \varprojlim \mathrm{Gal}(E/F)$, induced by embeddings $\mathrm{Gal}(ME/M) \simeq \mathrm{Gal}(E/M \cap E) \longrightarrow \mathrm{Gal}(E/F)$. Due to the description of open subgroups of Galois groups, this injective homomorphism is continuous. The image of compact $\mathrm{Gal}(L/M)$ with respect to it is a closed subgroup of $\mathrm{Gal}(L/F)$.

For a closed subgroup H of $G = \mathrm{Gal}(L/F)$ denote by L^H the subfield of fixed elements with respect to elements of H. The inclusion $\mathrm{Gal}(L/L^H) \supset H$ follows from the definitions. For any element $\sigma \in \mathrm{Gal}(L/L^H)$, its neighbourhood is $\sigma\,\mathrm{Gal}(L/M)$ for a finite extension M/L^H. The fixed field of the surjective image of H in $\mathrm{Gal}(M/L^H)$ is $M^H = L^H$, so by finite Galois theory the image of H is $\mathrm{Gal}(M/L^H)$. Take any element of H which surjects onto the restriction of σ on M, it belongs both to H and to $\sigma\,\mathrm{Gal}(L/M)$. Hence σ belongs to the closure of H, i.e. to H, proving the second inclusion. So $\mathrm{Gal}(L/L^H) = H$.

For a Galois extension L/M and an element $\alpha \in L \setminus M$, there is an element of a finite Galois extension E/M with $E \supset M(\alpha)$ acting non-trivially on α, by finite Galois theory. Hence, there is an element of $\mathrm{Gal}(L/M)$ acting non-trivially on α. Thus, $L^{\mathrm{Gal}(L/M)} = M$. We obtain

THEOREM. (*The Main Theorem of extended Galois theory*) Let L/F be a (possibly infinite) Galois extension, $G = \mathrm{Gal}(L/F)$.

There is a one-to-one correspondence between closed subgroups H of G and subfields M, $F \subset M \subset L$, given by $H \mapsto L^H$ and $M \mapsto \mathrm{Gal}(L/M)$. We have $\mathrm{Gal}(L/L^H) = H$ and $L^{\mathrm{Gal}(L/M)} = M$.

Open subgroups of G correspond to finite extensions M/F.

Normal closed subgroups H of G correspond to Galois extensions M/F and $\mathrm{Gal}(M/F) \simeq G/H$.

EXAMPLES.

1. As described in 1.3, $\mathrm{Gal}(\mathbb{F}_{q^m}/\mathbb{F}_q) \simeq \mathbb{Z}/m\mathbb{Z}$, where $q = p^n$ and the isomorphism is given by $\phi_n \mapsto 1 + m\mathbb{Z}$. An algebraic closure $\mathbb{F}_q^{\mathrm{alg}}$ of $\mathbb{F}_q$ is the compositum of all $\mathbb{F}_{q^m}$.

Since $\mathbb{F}_q$ is perfect, $\mathbb{F}_q^{\mathrm{alg}} = \mathbb{F}_q^{\mathrm{sep}}$.

Here and below for a field F an algebraic closure of F is denoted by F^{alg} and the separable closure of F in F^{alg} is denoted by F^{sep}. Separable and algebraic closures of fields are assumed suitably chosen where it is necessary to make such conventions.

The infinite Galois group

$$\mathrm{Gal}(\mathbb{F}_q^{\mathrm{sep}}/\mathbb{F}_q) = \varprojlim \mathrm{Gal}(\mathbb{F}_{q^m}/\mathbb{F}_q)$$

with respect to the surjective homomorphisms $\mathrm{Gal}(\mathbb{F}_{q^m}/\mathbb{F}_q) \longrightarrow \mathrm{Gal}(\mathbb{F}_{q^r}/\mathbb{F}_q)$, $r \mid m$. This corresponds to maps ψ_{mr} defined in Example 4 of 5.1.2.

Hence we get

$$\mathrm{Gal}(\mathbb{F}_q^{\mathrm{sep}}/\mathbb{F}_q) \simeq \varprojlim \mathbb{Z}/n\mathbb{Z} = \widehat{\mathbb{Z}}.$$

2. Similarly, for the maximal cyclotomic extension $\mathbb{Q}^{\mathrm{cycl}}$, the composite of all finite cyclotomic extensions $\mathbb{Q}(\zeta_m)$ of $\mathbb{Q}$, we have

$$\mathrm{Gal}(\mathbb{Q}^{\mathrm{cycl}}/\mathbb{Q}) \simeq \varprojlim \mathrm{Gal}(\mathbb{Q}(\zeta_m)/\mathbb{Q}) \simeq \varprojlim (\mathbb{Z}/m\mathbb{Z})^\times \simeq \widehat{\mathbb{Z}}^\times.$$

3. The Galois group $\mathrm{Gal}(F^{\mathrm{sep}}/F)$ of a fixed separable closure F^{sep} of F is called *the absolute Galois group G_F of F*.

5.3. *Cyclotomic extensions of* $\mathbb{Q}$

We have already seen the importance of cyclotomic fields in Kummer's Theorem 3.6.8.

Another very important property of cyclotomic fields is given by the following Theorem

THEOREM. (*Kronecker–Weber*) *Every finite abelian extension of* $\mathbb{Q}$ *is contained in some cyclotomic field* $\mathbb{Q}(\zeta_n)$. *Therefore the maximal abelian extension* $\mathbb{Q}^{\mathrm{ab}}$ *of* $\mathbb{Q}$ *coincides with the cyclotomic field* $\mathbb{Q}^{\mathrm{cycl}}$ *which is the compositum of all cyclotomic fields* $\mathbb{Q}(\zeta_n)$.

According to 2.4.4 we have an isomorphism

$$\mathrm{Gal}(\mathbb{Q}(\zeta_n)/\mathbb{Q}) \simeq (\mathbb{Z}/n\mathbb{Z})^\times.$$

These isomorphisms for different $m > n$ are compatible with respect to the natural surjective homomorphisms. Hence the infinite group $\mathrm{Gal}(\mathbb{Q}^{\mathrm{ab}}/\mathbb{Q})$ is isomorphic to the limit of $(\mathbb{Z}/n\mathbb{Z})^\times$ which by 5.1.2 coincides with the group of units of $\widehat{\mathbb{Z}} = \varprojlim \mathbb{Z}/n\mathbb{Z}$:

$$\mathrm{Gal}(\mathbb{Q}^{\mathrm{ab}}/\mathbb{Q}) \simeq \widehat{\mathbb{Z}}^\times.$$

The isomorphism

$$\Upsilon\colon \widehat{\mathbb{Z}}^\times \xrightarrow{\sim} \mathrm{Gal}(\mathbb{Q}^{\mathrm{ab}}/\mathbb{Q})$$

can be described as follows: if $a \in \widehat{\mathbb{Z}}^\times$ is congruent to integer m modulo positive integer n via the homomorphism

$$\widehat{\mathbb{Z}}/n\widehat{\mathbb{Z}} \longrightarrow \mathbb{Z}/n\mathbb{Z},$$

then $\Upsilon(a)(\zeta_n) = \zeta_n^m$.

Using 5.1 we have an isomorphism

$$\Psi\colon \prod \mathbb{Z}_p^\times \xrightarrow{\sim} \widehat{\mathbb{Z}}^\times \xrightarrow{\sim} \mathrm{Gal}(\mathbb{Q}^{\mathrm{ab}}/\mathbb{Q}).$$

On the left hand side we have an object $\widehat{\mathbb{Z}}^\times$ which is defined at the ground level of $\mathbb{Q}$, on the right hand side we have an object which incorporates information about all finite abelian extensions of $\mathbb{Q}$.

The restriction of the isomorphism to quadratic extensions of $\mathbb{Q}$ is related with the Gauß quadratic reciprocity law, see 6.6 in Chapter 3.

Abelian *class field theory* generalises the Kronecker–Weber Theorem for an arbitrary algebraic number field K to give a *reciprocity homomorphism* from an object called idele class group defined at the level of K to the Galois group of the maximal abelian extension K^{ab} over K. This theory is much more general than the cyclotomic theory and its follows different patterns and methods. It is presented in Chapter 3.

5.4. *Ideles and reciprocity map*

5.4.1. Recall (see 4.2.4) that $\mathbb{Q}_p^\times \simeq \langle p\rangle \times \mathbb{Z}_p^\times, \quad a \mapsto (n,u)$ where $n = v_p(a)$ and $u = ap^{-n}$,
v_p is the p-adic valuation.

Denote $\mathbb{Q}_\infty = \mathbb{R}$ and include ∞ in the set of "primes" of $\mathbb{Z}$. Form the so called *restricted product*

$$J_{\mathbb{Q}} = \prod{}' \mathbb{Q}_p^\times = \{(a_\infty, a_2, a_3, \dots) : a_p \in \mathbb{Q}_p^\times\}$$

of $\mathbb{R}^\times = \mathbb{Q}_\infty^\times$, $\mathbb{Q}_2^\times, \mathbb{Q}_3^\times, \dots$ such that almost all components a_p are p-adic units. Elements of $J_{\mathbb{Q}}$ are called *ideles of* $\mathbb{Q}$.

Define a homomorphism

$$f\colon J_{\mathbb{Q}} = \prod{}' \mathbb{Q}_p^\times \longrightarrow \mathbb{Q}^\times \times \mathbb{R}_+^\times \times \prod \mathbb{Z}_p^\times,$$
$$(a_\infty, a_2, a_3, \dots) \mapsto (a, a_\infty a^{-1}, a_2 a^{-1}, a_3 a^{-1}, \dots)$$

where $a = \operatorname{sign}(a_\infty)\prod p^{v_p(a_p)} \in \mathbb{Q}^\times$ and $\operatorname{sign}(a)$ is the sign of a.

It is easy to verify that f is an isomorphism.

5.4.2. Define a homomorphism

$$\Psi_{\mathbb{Q}}\colon J_{\mathbb{Q}} = \prod{}' \mathbb{Q}_p^\times \longrightarrow \operatorname{Gal}(\mathbb{Q}^{\mathrm{ab}}/\mathbb{Q})$$

by the following local-global formula:

$$\Psi_{\mathbb{Q}}(a_\infty, a_2, a_3, \dots) = \prod \Psi_{\mathbb{Q}_p}(a_p).$$

Here the *local reciprocity map* $\Psi_{\mathbb{Q}_p}$ is described as follows: if $a_p = p^n u$ where $n = v_p(a)$, then for a q^mth primitive root ζ with prime q and $q^m > 2$,

$$\Psi_{\mathbb{Q}_p}(a_p)(\zeta) = \begin{cases} \zeta^{p^n}, & \text{if } p \neq q \\ \zeta^{u^{-1}}, & \text{if } p = q. \end{cases}$$

In particular, if $p \neq q$, then $\Psi_{\mathbb{Q}_p}(p)$ sends ζ to ζ^p, the latter is kind of similar to the pth Frobenius automorphism defined in 1.3. So the local reciprocity map $\Psi_{\mathbb{Q}_p}(p)$ sends prime p to the pth Frobenius automorphism.

For $p = \infty$ put

$$\Psi_{\mathbb{Q}_\infty}(a_\infty)(\zeta) = \zeta^{\operatorname{sign}(a_\infty)}.$$

The homomorphism $\Psi_{\mathbb{Q}}$ is called the *reciprocity map for* $\mathbb{Q}$.

THEOREM.

1. Reciprocity Law: for every non-zero rational number a one has

$$\Psi_{\mathbb{Q}}(a,a,a,\dots)=1.$$

2. For units $u_p \in \mathbb{Z}_p^\times$ one has

$$\Psi_{\mathbb{Q}}(1,u_2,u_3,\dots)^{-1}=\Psi(u_2,u_3,\dots).$$

3. Using f define

$$g\colon J_{\mathbb{Q}} \longrightarrow \mathbb{Q}^\times \times \mathbb{R}_+^\times \times \prod \mathbb{Z}_p^\times \longrightarrow \prod \mathbb{Z}_p^\times,$$

$(a,b,u_2,u_3,\dots)\mapsto(u_2,u_3,\dots)$. *Then*

$$\Psi_{\mathbb{Q}}(\alpha)^{-1}=\Psi\circ g(\alpha).$$

4. The kernel of the reciprocity map $\Psi_{\mathbb{Q}}$ equals to $g^{-1}(1,1,1,\dots)=$ the product of the diagonal image of $\mathbb{Q}^\times$ in $J_{\mathbb{Q}}$ and of the image of $\mathbb{R}_+^\times$ in $J_{\mathbb{Q}}$ with respect to the homomorphism $\alpha\mapsto(\alpha,1,1,\dots)$. It induces an isomorphism

$$J_{\mathbb{Q}}/\mathbb{Q}^\times\mathbb{R}_+^\times \simeq \mathrm{Gal}(\mathbb{Q}^{\mathrm{ab}}/\mathbb{Q}).$$

Proof. To verify the first property, due to the multiplicativity of $\Psi_{\mathbb{Q}}$ it is sufficient to show that for a primitive q^mth root ζ, $q^m>2$,

$$\Psi_{\mathbb{Q}}(p,p,\dots)(\zeta)=\zeta \quad \text{for all positive prime numbers } p$$
$$\Psi_{\mathbb{Q}}(-1,-1,\dots)(\zeta)=\zeta.$$

From the definition of $\Psi_{\mathbb{Q}}$ we deduce that

$$\Psi_{\mathbb{Q}_l}(p)(\zeta)=\begin{cases}\zeta, & \text{if } l\neq q, l\neq p\\ \zeta^p, & \text{if } l\neq q, l=p\\ \zeta^{p^{-1}}, & \text{if } l=q,\ l\neq p\\ \zeta, & \text{if } l=q=p.\end{cases}$$

So $(\prod_l \Psi_{\mathbb{Q}_l}(p))(\zeta)=\zeta$ for $q\neq p$ and for $q=p$. Similarly one checks the second assertion.

The second property is easy: due to multiplicativity it suffices to show that

$$\Psi(1,\dots,u_p,1,\dots)^{-1}=\Phi_{\mathbb{Q}}(1,\dots,u_p,1,\dots)$$

and this follows immediately from the definition of Ψ, $\Psi_{\mathbb{Q}}$.

The third property follows from the definition of f and the first and second properties. The fourth property follows from the third. □

5.4.3. The previous description is part of *cyclotomic class field theory* for $\mathbb{Q}$, where one can use the Galois action on roots and roots generate the maximal abelian extension of $\mathbb{Q}$ (Kronecker–Weber Theorem).

For an algebraic number field F one can define, in a similar way, the idele group J_F as a restricted product of the multiplicative groups $F_P^\times$ of completions F_P of F with respect to non-zero prime ideals P of the ring of integers of F, and of real or complex completions of F with respect to real and complex embeddings of F into $\mathbb{C}$.

Except the case of $\mathbb{Q}$, imaginary quadratic fields and totally imaginary quadratic extensions of totally real fields, one does not have an explicit description of the maximal abelian extension by appropriate torsion elements, as in the Kronecker–Weber Theorem. Thus, one needs to directly define a global reciprocity map

$$\Psi_F : J_F \longrightarrow \mathrm{Gal}(F^{\mathrm{ab}}/F)$$

for all number fields F and study its properties.

This is done in general class field theory in a completely different way from cyclotomic class field theory. The Kronecker–Weber Theorem, unlike its version for completions F_P, plays no role in general class field theory. In this book it will be the last statement to include, as a corollary of general class field theory, at the end of Chapter 3.

The global reciprocity map is closely related to local reciprocity maps

$$F_P^\times \longrightarrow \mathrm{Gal}(F_P^{\mathrm{ab}}/F_P)$$

and homomorphisms $\mathrm{Gal}(F_P^{\mathrm{ab}}/F_P) \longrightarrow \mathrm{Gal}(F^{\mathrm{ab}}/F)$. The local reciprocity maps are defined and studied in local class field theory.

The local reciprocity maps and global reciprocity maps satisfy a number of important properties, including functorial properties which do not play any role in special class field theories such as cyclotomic class field theory.

The analog of the reciprocity law is that the kernel of Ψ_F contains the image of $F^\times$ in J_F.

One endows $J_F/F^\times$ with the quotient topology of an appropriate topology on J_F that takes into account topologies of appropriate topologies $F_P^\times$ which generalise the p-adic topology. The reciprocity maps are continuous homomorphisms.

A key part of class field theory is the *existence theorem*: every open subgroup N in $J_F/F^\times$ corresponds to its *class field* L, the unique finite abelian extension of F such that $N_{L/F}(J_L)F^\times = N$ and $N = \Psi_F^{-1}(\mathrm{Gal}(F^{\mathrm{ab}}/L))$.

See Chapter 3 for many more details and full proofs.

Chapter 2

Complete Discrete Valuation Fields

In Chapter 2, we will go relatively slow in Sections 1–13 in order to build a good understanding of and intuition about complete discrete valuation fields. This chapter includes less known but important topics about the group of principal units as a topological $\mathbb{Z}_p$-module, about the norm map behaviour in cyclic extensions of prime degree, Artin–Schreier extensions of local fields of mixed characteristic and positive characteristic. Section 15 includes an approach to the Hasse–Herbrand function based on the use of behaviour of the norm map. A more recent Fontaine–Wintenberger's theory of fields of norms is presented in Section 17. Studying the latter could be a good place to test the knowledge of local fields; on the other hand, the readers can skip this most difficult section of Chapter 2 when concentrating on the study of class field theory.

Galois theory of infinite extensions is summarised in Section 5.2 of Chapter 1.

The reader who is mostly interested in class field theory can start from Section 18.

1. Valuation Fields

1.1. DEFINITION. Let Γ be a totally ordered abelian group, we will write its group operation in the additive way. Add to Γ a formal element $+\infty$ with the properties $a \leqslant +\infty$, $+\infty \leqslant +\infty$, $a + (+\infty) = +\infty$, $(+\infty) + (+\infty) = +\infty$, for each $a \in \Gamma$; denote $\Gamma' = \Gamma \cup \{+\infty\}$.

For a field F a map $v\colon F \longrightarrow \Gamma'$ with the properties

$$v(\alpha) = +\infty \Leftrightarrow \alpha = 0$$
$$v(\alpha\beta) = v(\alpha) + v(\beta)$$
$$v(\alpha + \beta) \geqslant \min(v(\alpha), v(\beta))$$

is said to be a *valuation* on F.

The map v induces a homomorphism of $F^{\times}$ to Γ and its value group $v(F^{\times})$ is a totally ordered subgroup of Γ.

If $v(F^\times) = \{0\}$, then v is called the *trivial valuation.*

A field F which has a non-trivial valuation is said to be a valuation field.

It is immediate that if $v(\alpha) \neq v(\beta)$, then $v(\alpha+\beta) = \min(v(\alpha), v(\beta))$.

1.2. Denote $\mathcal{O}_v = \{\alpha \in F : v(\alpha) \geqslant 0\}$, $\mathcal{M}_v = \{\alpha \in F : v(\alpha) > 0\}$.

Then $\mathcal{M}_v$ coincides with the set of non-invertible elements of $\mathcal{O}_v$. Therefore, $\mathcal{O}_v$ is a local ring with the unique *maximal ideal* $\mathcal{M}_v$.

$\mathcal{O}_v$ is called the *ring of integers* (with respect to v), and the field $k(v) = \overline{F}_v = \mathcal{O}_v/\mathcal{M}_v$ is called the *residue field*, or residue class field.

The image of an element $\alpha \in \mathcal{O}_v$ in $\overline{F}_v$ is denoted by $\overline{\alpha}$, it is called the *residue* of α in $\overline{F}_v$.

The set of invertible elements of $\mathcal{O}_v$ is a multiplicative group $U_v = \mathcal{O}_v - \mathcal{M}_v$, it is called the *group of units*.

A valuation is called *discrete* if the totally ordered group $v(F^\times)$ is isomorphic to the naturally ordered group $\mathbb{Z}$.

1.3. Examples.

1. The p-adic valuation on $\mathbb{Q}$ and $\mathbb{Q}_p$.

2. Let K be a field. Let $p(X) \in K[X]$ be a monic irreducible polynomial over K. For a polynomial $f(X) \in K[X]$ denote by $v_{p(X)}(f(X))$ the largest integer k such that $p(X)^k$ divides polynomial $f(X)$. For two polynomials f, g put $v_{p(X)}(f/g) = v_{p(X)}(f) - v_{p(X)}(g)$. Put $v_{p(X)}(0) = +\infty$.

The map $v_{p(X)}$ is a discrete valuation of $K(X)$. Its the ring of integers

$$\mathcal{O}_{v_{p(X)}} = \left\{ \frac{f(X)}{g(X)} : f(X), g(X) \in K[X], g(X) \text{ is relatively prime to } p(X) \right\}$$

and the residue field is $K[X]/(p(X))$.

Another discrete valuation of $K(X)$ is $-\deg$ with the ring of integers $K[X^{-1}]$ and maximal ideal $X^{-1}K[X^{-1}]$.

3. Let $\Gamma_1, \dots \Gamma_n$ be totally ordered abelian groups. One can order the group $\Gamma_1 \times \cdots \times \Gamma_n$ lexicographically, namely setting $(a_1, \dots, a_n) < (b_1, \dots, b_n)$ if and only if $a_1 = b_1, \dots, a_{i-1} = b_{i-1}$, $a_i < b_i$ for some $1 \leqslant i \leqslant n$. A valuation v on F is said to be *discrete of rank n* if the value group $v(F^\times)$ is isomorphic to the lexicographically ordered group $(\mathbb{Z})^n = \underbrace{\mathbb{Z} \times \cdots \times \mathbb{Z}}_{n \text{ times}}$.

Note that the first component v_1 of a discrete valuation $v = (v_1, \dots, v_n)$ of rank n is a discrete valuation (of rank 1) on the field F.

4. Let F be a field with a valuation v. For $f(X) = \sum \alpha_i X^i \in F[X]$ put

$$v^*(f(X)) = \min_i (i, v(\alpha_i)) \in \mathbb{Z} \times v(F^\times).$$

One can naturally extend v^* to $F(X)$. If we order the group $\mathbb{Z} \times v(F^\times)$ lexicographically, we obtain the valuation v^* on $F(X)$ with the residue field $\overline{F}_v$.

Similarly, it is easy to define a valuation on $F(X_1)\dots(X_n)$ with the value group $(\mathbb{Z})^{n-1} \times v(F^\times)$ ordered lexicographically. In particular, for $F = \mathbb{Q}$, $v = v_p$ we get a discrete valuation of rank n on $\mathbb{Q}(X_1)\dots(X_{n-1})$ and for $F = K(X)$, $v = v_{p(X)}$ we get a discrete valuation of rank n on $K(X)(X_1)\dots(X_{n-1})$.

5. Let v be a discrete (surjective to $\mathbb{Z}$) valuation of F. Fix an integer c. For $f(X) = \sum \alpha_i X^i \in F[X]$ put

$$w_c(f(X)) = \min_i \{v(\alpha_i) + ic\}.$$

Extending w_c to $F(X)$ we obtain the discrete valuation w_c with residue field $\overline{F}_v(X)$ (make substitution $X = Y\beta$ with $v(\beta) = c$ to reduce to the case $c = 0$).

6. Let F, v be as in Example 4. For $f(X) = \sum \alpha_i X^i \in F[X]$ put

$$v_*(f(X)) = \min_i (v(\alpha_i), i) \in v(F^\times) \times \mathbb{Z}, \quad v_*(0) = (+\infty, +\infty)$$

for $v(F^\times) \times \mathbb{Z}$ ordered lexicographically. Extending v_* to $F(X)$, we obtain the valuation v_*. The residue field of v_* is $\overline{F}_v$.

2. Discrete Valuation Fields

2.1. A field F is said to be a *discrete valuation field* if it admits a non-trivial discrete valuation v. An element $\pi \in \mathcal{O}_v$ is said to be a *prime element* (*uniformising element, a uniformiser*) if $v(\pi) > 0$ generates the group $v(F^\times)$. Without loss of generality we shall often assume that the homomorphism

$$v \colon F^\times \longrightarrow \mathbb{Z}$$

is *surjective*.

We denote the characteristic of a field F by $\mathrm{char}(F)$.

2.2. LEMMA. *Assume that F and $\overline{F}_v$ have different characteristics. Then F is of characteristic 0 and $\overline{F}_v$ is of positive characteristic p.*

Proof. Suppose that $\mathrm{char}(F) = p \neq 0$. Then $p = 0$ in F and therefore in $\overline{F}_v$. Hence $p = \mathrm{char}(\overline{F}_v)$. □

2.3. LEMMA. *Let F be a discrete valuation field, and π be a prime element. Then the ring of integers $\mathcal{O}_v$ is a principal ideal ring, and every proper ideal of $\mathcal{O}_v$ can be written as $\pi^n\mathcal{O}_v$ for some $n>0$. In particular, $\mathcal{M}_v = \pi\mathcal{O}_v$. The intersection of all proper ideals of $\mathcal{O}_v$ is the zero ideal.*

Proof. Let I be a proper ideal of $\mathcal{O}_v$. Then there exists $n = \min\{v(\alpha) : \alpha \in I\}$ and hence $\pi^n\varepsilon \in I$ for some unit ε. It follows that $\pi^n\mathcal{O}_v \subset I \subset \pi^n\mathcal{O}_v$ and $I = \pi^n\mathcal{O}_v$. If α belongs to the intersection of all proper ideals $\pi^n\mathcal{O}_v$ in $\mathcal{O}_v$, then $v(\alpha) = +\infty$, i.e., $\alpha = 0$. □

2.4. LEMMA. *Any element $\alpha \in F^\times$ can be uniquely written as $\pi^n\varepsilon$ for some $n \in \mathbb{Z}$ and $\varepsilon \in U_v$.*

Proof. Let $n = v(\alpha)$. Then $\alpha\pi^{-n} \in U_v$ and $\alpha = \pi^n\varepsilon$ for $\varepsilon \in U_v$. If $\pi^n\varepsilon_1 = \pi^m\varepsilon_2$, then $n + v(\varepsilon_1) = m + v(\varepsilon_2)$. As $\varepsilon_1, \varepsilon_2 \in U_v$, we deduce $n = m$, $\varepsilon_1 = \varepsilon_2$. □

2.5. Let v be a discrete valuation on F, $0 < d < 1$. The mapping $d_v \colon F \times F \longrightarrow \mathbb{R}$ defined by $d_v(\alpha, \beta) = d^{v(\alpha-\beta)}$ is a metric on F. Therefore, it induces a Hausdorff topology on F. For every $\alpha \in F$ the sets $\alpha + \pi^n\mathcal{O}_v$, $n \in \mathbb{Z}$, form a basis of open neighbourhoods of α. This topology on F is called the *discrete valuation topology*.

2.6. LEMMA. *The field F with the above-defined topology is a topological field.*

Proof. As

$$\begin{aligned}
&v((\alpha-\beta)-(\alpha_0-\beta_0)) \geqslant \min(v(\alpha-\alpha_0), v(\beta-\beta_0)),\\
&v(\alpha\beta-\alpha_0\beta_0) \geqslant \min(v(\alpha-\alpha_0)+v(\beta), v(\beta-\beta_0)+v(\alpha_0)),\\
&v(\alpha^{-1}-\alpha_0^{-1}) = v(\alpha-\alpha_0)-v(\alpha)-v(\alpha_0),
\end{aligned}$$

we deduce the statement. □

2.7. LEMMA. *The topologies on F defined by two discrete valuations v_1, v_2 coincide if and only if $v_1 = v_2$ (recall that $v_1(F^\times) = v_2(F^\times) = \mathbb{Z}$).*

Proof. Let the topologies induced by v_1, v_2 coincide. Observe that α^n tends to 0 when n tends to $+\infty$ in the topology defined by a discrete valuation v if and only if $v(\alpha) > 0$. Therefore, $v_1(\alpha) > 0$ if and only if $v_2(\alpha) > 0$. Let π_1, π_2 be prime elements with respect to v_1 and v_2. Then we conclude that $v_2(\pi_1) \geqslant 1$ and $v_1(\pi_2) \geqslant 1$. If $v_2(\pi_1) > 1$ then $v_2(\pi_1\pi_2^{-1}) > 0$. Consequently, $v_1(\pi_1\pi_2^{-1}) > 0$, i.e., $v_1(\pi_2) < 1$, a contradiction. Thus, $v_2(\pi_1) = 1$ and this equality holds for all prime elements π_1 with respect to v_1. This shows the equality $v_1 = v_2$. □

2.8. PROPOSITION. *(Approximation Theorem) Let $v_1,\dots,v_n$ be distinct discrete valuations on F. Then for every $\alpha_1,\dots,\alpha_n \in F$, $c \in \mathbb{Z}$, there exists $\alpha \in F$ such that $v_i(\alpha_i - \alpha) > c$ for $1 \leqslant i \leqslant n$.*

Proof. If $v(\alpha) > 0$ then $v(\alpha^m(1+\alpha^m)^{-1}) \to +\infty$ as $m \to +\infty$, and if $v(\alpha) < 0$ then $v(\alpha^m(1+\alpha^m)^{-1} - 1) \to +\infty$ as $m \to +\infty$. We proceed by induction to deduce that there exists an element $\beta_1 \in F$ such that $v_1(\beta_1) < 0$, $v_i(\beta_1) > 0$ for $2 \leqslant i \leqslant n$.

Towards that aim for $n = 2$, one can first verify that there is an element $\gamma_1 \in F$ such that $v_1(\gamma_1) \geqslant 0$, $v_2(\gamma_1) < 0$. Using the proof of the previous Lemma, find elements $\pi_1, \pi_2 \in F$ with $v_2(\pi_1) \neq 1 = v_1(\pi_1)$, $v_1(\pi_2) \neq 1 = v_2(\pi_2)$. If $v_2(\pi_1) < 0$ put $\gamma_1 = \pi_1$. If $v_2(\pi_1) \geqslant 0$, then $v_2(\rho) \neq 0 = v_1(\rho)$ for $\rho = \pi_2 \pi_1^{-v_1(\pi_2)}$. Put $\gamma_1 = \rho$ or $\gamma_1 = \rho^{-1}$. Now let $\gamma_2 \in F$ be such that $v_2(\gamma_2) \geqslant 0$, $v_1(\gamma_2) < 0$. Then $\beta_1 = \gamma_1^{-1}\gamma_2$ is the desired element for $n = 2$.

Let $n > 2$. Then, by the induction assumption, there exists $\delta_1 \in F$ with $v_1(\delta_1) < 0$, $v_i(\delta_1) > 0$ for $2 \leqslant i \leqslant n-1$ and $\delta_2 \in F$ with $v_1(\delta_2) < 0$, $v_n(\delta_2) > 0$. One can put $\beta_2 = \delta_1$ if $v_n(\delta_1) > 0$, $\beta_2 = \delta_1^m \delta_2$ if $v_n(\delta_1) = 0$, and $\beta_2 = \delta_1 \delta_2^m (1 + \delta_2^m)^{-1}$ if $v_n(\delta_1) < 0$ for a sufficiently large m.

To complete the proof we take $\beta_1,\dots,\beta_n \in F$ with $v_i(\beta_i) < 0$, $v_i(\beta_j) > 0$ for $i \neq j$. Put $\alpha = \sum_{i=1}^n \alpha_i \beta_i^m (1 + \beta_i^m)^{-1}$. Then α is the desired element for a sufficiently large m. □

3. Completion

3.1. Let F be a field with a discrete valuation v (as usual, $v(F^\times) = \mathbb{Z}$). As F is a metric topological space one can introduce the notion of a fundamental (Cauchy) sequence. A sequence $(\alpha_n)_{n\geqslant 0}$ of elements of F is called a Cauchy sequence if for every real c there is $n_0 \geqslant 0$ such that $v(\alpha_n - \alpha_m) \geqslant c$ for $m, n \geqslant n_0$.

If (α_n) is a fundamental sequence then for every integer r there is n_r such that for all $n, m \geqslant n_r$ we have $v(\alpha_n - \alpha_m) \geqslant r$. We can assume $n_1 \leqslant n_2 \leqslant \dots$. If for every r there is $n'_r \geqslant n_r$ such that $v(\alpha_{n'_r}) \neq v(\alpha_{n'_r+1})$, then $\lim v(\alpha_n) = +\infty$. Thus, every fundamental sequence (α_n) has limit $\lim v(\alpha_n) \in \mathbb{Z}'$.

LEMMA. *The set A of all Cauchy sequences forms a ring with respect to component-wise addition and multiplication. The set of all Cauchy sequences $(\alpha_n)_{n\geqslant 0}$ with $\alpha_n \to 0$ as $n \to +\infty$ forms a maximal ideal M of A. The field A/M is a discrete valuation field with its discrete valuation $\widehat{v}$ defined by $\widehat{v}((\alpha_n)) = \lim v(\alpha_n)$ for a Cauchy sequence $(\alpha_n)_{n\geqslant 0}$.*

Proof. A sketch of the proof is as follows. It suffices to show that M is a maximal ideal of A. Let $(\alpha_n)_{n\geqslant 0}$ be a Cauchy sequence with $\alpha_n \nrightarrow 0$ as $n \to +\infty$. Hence, there is an $n_0 \geqslant 0$ such that $\alpha_n \neq 0$ for $n \geqslant n_0$. Put $\beta_n = 0$ for $n < n_0$ and $\beta_n = \alpha_n^{-1}$ for $n \geqslant n_0$. Then $(\beta_n)_{n\geqslant 0}$ is a Cauchy sequence and $(\alpha_n)(\beta_n) \in (1) + M$. Therefore, M is maximal. □

3.2. A discrete valuation field F is called a *complete discrete valuation field* if every Cauchy sequence $(\alpha_n)_{n\geqslant 0}$ is convergent, i.e., there exists $\alpha = \lim \alpha_n \in F$ with respect to v. A field $\widehat{F}$ with a discrete valuation $\widehat{v}$ is called a *completion* of F if it is complete, $\widehat{v}|_F = v$, and F is a dense subfield in $\widehat{F}$ with respect to $\widehat{v}$.

PROPOSITION. *Every discrete valuation field F has a completion which is unique up to an isomorphism over F.*

Proof. We verify that the field A/M with the valuation $\widehat{v}$ is a completion of F. F is embedded in A/M by the formula $\alpha \mapsto (\alpha) \mod M$. For a Cauchy sequence $(\alpha_n)_{n\geqslant 0}$ and real c, let $n_0 \geqslant 0$ be such that $v(\alpha_n - \alpha_m) \geqslant c$ for all $m, n \geqslant n_0$. Hence, for $\alpha_{n_0} \in F$ we have $\widehat{v}((\alpha_{n_0}) - (\alpha_n)_{n\geqslant 0}) \geqslant c$, which shows that F is dense in A/M. Let $((\alpha_n^{(m)})_n)_m$ be a Cauchy sequence in A/M with respect to $\widehat{v}$. Let $n(0)$, $n(1), \dots$ be an increasing sequence of integers such that $v(\alpha_{n_2}^{(m)} - \alpha_{n_1}^{(m)}) \geqslant m$ for n_1, $n_2 \geqslant n(m)$. Then $(\alpha_{n(m)}^{(m)})_m$ is a Cauchy sequence in F and the limit of $((\alpha_n^{(m)})_n)_m$ with respect to $\widehat{v}$ in A/M. Thus, we obtain the existence of the completion $A/M, \widehat{v}$.

If there are two completions $\widehat{F_1}$, $\widehat{v_1}$ and $\widehat{F_2}$, $\widehat{v_2}$, then we put $f(\alpha) = \alpha$ for $\alpha \in F$ and extend this homomorphism by continuity from F, as a dense subfield in $\widehat{F_1}$, to $\widehat{F_1}$. It is easy to verify that the extension $\widehat{f}\colon \widehat{F_1} \longrightarrow \widehat{F_2}$ is an isomorphism and $\widehat{v_2} \circ \widehat{f} = \widehat{v_1}$. □

We shall denote the completion of the field F with respect to v by $\widehat{F}_v$ or $\widehat{F}$.

3.3. LEMMA. *Let F be a field with a discrete valuation v and $\widehat{F}$ its completion with the discrete valuation $\widehat{v}$. Then the ring of integers $\mathcal{O}_v$ is dense in $\mathcal{O}_{\widehat{v}}$, the maximal ideal $\mathcal{M}_v$ is dense in $\mathcal{M}_{\widehat{v}}$, and the residue field $\overline{F}_v$ coincides with the residue field of $\widehat{F}$ with respect to $\widehat{v}$.*

Proof. It follows immediately from the construction of A/M in 3.1 and Proposition 3.2. □

3.4. Examples.

1. Embeddings of the field of rational numbers $\mathbb{Q}$ in the field of p-adic numbers $\mathbb{Q}_p$ (introduced in 4.2 of Chapter 1) for all prime p and in the field of real numbers $\mathbb{R}$ is sometimes a tool to solve various problems over $\mathbb{Q}$. An example is the *Minkowski–Hasse* Theorem: an equation $\sum a_{ij}X_iX_j = 0$ for $a_{ij} \in \mathbb{Q}$ has a non-trivial solution in $\mathbb{Q}$ if and only if it admits a non-trivial solution in $\mathbb{R}$ and in $\mathbb{Q}_p$ for all prime p. A generalisation of this result is the so-called *Hasse local-global principle* which is of great importance in algebraic number theory. It is interesting that, from the standpoint of model theory, the complex field $\mathbb{C}$ is locally equivalent to the algebraic closure of $\mathbb{Q}_p$ for each prime p.

2. The completion of $K(X)$ with respect to v_X is the formal power series field $K((X))$ of all formal series $\sum_{-\infty}^{+\infty} \alpha_n X^n$ with $\alpha_n \in K$ and $\alpha_n = 0$ for almost all negative n. The ring of integers with respect to v_X is $K[[X]]$, that is, the set of all formal series $\sum_0^{+\infty} \alpha_n X^n$, $\alpha_n \in K$. Its residue field may be identified with K.

3. Let F be a field with a discrete valuation v, and $\widehat{F}$ its completion. Then the valuation v^* on $F(X)$ defined in Example 4 of 1.3 can be naturally extended to $\widehat{F}((X))$. For $f(X) = \sum_{n \geqslant m} \alpha_n X^n$, $\alpha_n \in \widehat{F}$, $\alpha_m \neq 0$, put $v^*(f(X)) = (m, \widehat{v}(\alpha_m))$. The ring of integers of v^* on $\widehat{F}((X))$ is $\mathscr{O}_{\widehat{v}} + X\widehat{F}[[X]]$.

4. Let F be the same as in the previous Example. Then the valuation v_* on $F(X)$ defined in Example 6 of 1.3 can be naturally extended to the field

$$\widehat{F}\{\{X\}\} = \Big\{\sum_{-\infty}^{+\infty} \alpha_n X^n : \alpha_n \in \widehat{F}, \inf_n\{\widehat{v}(\alpha_n)\} > -\infty, \widehat{v}(\alpha_n) \to +\infty \text{ as } n \to -\infty\Big\}.$$

For $f(X) = \sum_{-\infty}^{+\infty} \alpha_n X^n \in \widehat{F}\{\{X\}\}$ put

$$v_*(f(X)) = \min_n(\widehat{v}(\alpha_n), n).$$

The ring of integers of v_* is $\mathscr{M}_{\widehat{v}}\{\{X\}\} + \mathscr{O}_{\widehat{v}}[[X]]$ and the maximal ideal is $\mathscr{M}_{\widehat{v}}\{\{X\}\} + X\mathscr{O}_{\widehat{v}}[[X]]$, where $\mathscr{M}_{\widehat{v}}\{\{X\}\} = \mathscr{M}_{\widehat{v}}\mathscr{O}_{\widehat{v}}\{\{X\}\}$, $\mathscr{O}_{\widehat{v}}\{\{X\}\} = \{\sum_{-\infty}^{+\infty} \alpha_n X^n : \alpha_n \in \mathscr{O}_{\widehat{v}}, \widehat{v}(\alpha_n) \to +\infty \text{ as } n \to -\infty\}$, and the residue field is $\overline{F}_v$.

3.5. DEFINITIONS.

1. A complete discrete valuation field with perfect residue field is called a *local field*.

For example, $\mathbb{Q}_p$ and $F((X))$ are local fields where F is a perfect field (of positive or zero characteristic). A local field with finite residue field is sometimes called a *local number field* if it is of characteristic zero and a *local functional field* if it is of positive characteristic.

2. Local fields are sometimes called 1-dimensional local fields. An *n-dimensional local field* ($n \geqslant 2$) is a complete discrete valuation field whose residue field

is an $(n-1)$-dimensional local field. For $n>1$ such fields are called *higher local fields*.

For example, $\mathbb{Q}_p((X_2))\dots((X_n))$, $F((X_1))\dots((X_n))$ (F is a perfect field), $K\{\{X_1\}\}\dots\{\{X_{n-1}\}\}$ (K is a 1-dimensional local field of characteristic zero) are n-dimensional local fields.

4. Filtrations of Discrete Valuation Fields

In this section we study natural filtrations on the multiplicative group of a discrete valuation field F; in particular, its behaviour with respect to raising to the pth power. For simplicity, we will often omit the index v in notations U_v, $\mathcal{O}_v$, $\mathcal{M}_v$, $\overline{F}_v$. We fix a prime element π of F.

4.1. A set R is said to be a *set of representatives* for a valuation field F if $R\subset\mathcal{O}$, $0\in R$ and R is mapped bijectively on $\overline{F}$ under the canonical map $\mathcal{O}\longrightarrow\mathcal{O}/\mathcal{M}=\overline{F}$. Denote by $\operatorname{rep}\colon\overline{F}\longrightarrow R$ the inverse bijective map. For a set S denote by $(S)_n^{+\infty}$ the set of all sequences $(a_i)_{i\geqslant n}$, $a_i\in S$. Let $(S)_{-\infty}^{+\infty}$ denote the union of increasing sets $(S)_n^{+\infty}$ where $n\longrightarrow-\infty$.

4.2. The additive group F has a natural filtration

$$\dots\supset\pi^i\mathcal{O}\supset\pi^{i+1}\mathcal{O}\supset\dots.$$

The factor filtration of this filtration is easy to calculate: $\pi^i\mathcal{O}/\pi^{i+1}\mathcal{O}\xrightarrow{\sim}\overline{F}$.

PROPOSITION. *Let F be a complete field with respect to a discrete valuation v. Let $\pi_i\in F$ for each $i\in\mathbb{Z}$ be an element of F with $v(\pi_i)=i$. Then the map*

$$\operatorname{Rep}\colon(\overline{F})_{-\infty}^{+\infty}\longrightarrow F,\qquad(a_i)_{i\in\mathbb{Z}}\mapsto\sum_{-\infty}^{+\infty}\operatorname{rep}(a_i)\pi_i$$

is a bijection. Moreover, if $(a_i)_{i\in\mathbb{Z}}\neq(0)_{i\in\mathbb{Z}}$ then $v(\operatorname{Rep}(a_i))=\min\{i:a_i\neq0\}$.

Proof. The map Rep is well defined, because for almost all $i<0$ we get $\operatorname{rep}(a_i)=0$ and the series $\sum\operatorname{rep}(a_i)\pi_i$ converges in F. If $(a_i)_{i\in\mathbb{Z}}\neq(b_i)_{i\in\mathbb{Z}}$ and

$$n=\min\{i\in\mathbb{Z}:a_i\neq b_i\},$$

then $v(a_n\pi_n-b_n\pi_n)=n$. Since $v(a_i\pi_i-b_i\pi_i)>n$ for $i>n$, we deduce that

$$v(\operatorname{Rep}(a_i)-\operatorname{Rep}(b_i))=n.$$

Therefore Rep is injective.

In particular, $v(\operatorname{Rep}(a_i))=\min\{i:a_i\neq0\}$. Further, let $\alpha\in F$. Then $\alpha=\pi^n\varepsilon$ with $n\in\mathbb{Z}$, $\varepsilon\in U$. We also get $\alpha=\pi_n\varepsilon'$ for some $\varepsilon'\in U$. Let a_n be the image of ε' in $\overline{F}$; then $a_n\neq0$ and $\alpha_1=\alpha-\operatorname{rep}(a_n)\pi_n\in\pi^{n+1}\mathcal{O}$. Continuing in this way for α_1, we obtain a convergent series $\alpha = \sum \mathrm{rep}(a_i)\pi_i$. Therefore, Rep is surjective. □

COROLLARY. *We often take* $\pi_n = \pi^n$. *Therefore, by the preceding Proposition, every element* $\alpha \in F$ *can be uniquely expanded as*

$$\alpha = \sum_{-\infty}^{+\infty} \theta_i \pi^i, \qquad \theta_i \in R \quad \text{and} \quad \theta_i = 0 \quad \text{for almost all } i < 0.$$

DEFINITION. *If* $\alpha - \beta \in \pi^n \mathcal{O}$, *we write* $\alpha \equiv \beta \mod \pi^n$.

4.3. DEFINITIONS. The group $1 + \pi\mathcal{O}$ is called the *group of principal units* U_1 and its elements are called *principal units*. Introduce also *higher groups of units* as follows: $U_i = 1 + \pi^i \mathcal{O}$ for $i \geqslant 1$.

4.4. The multiplicative group $F^\times$ has a natural filtration $F^\times \supset U \supset U_1 \supset U_2 \supset \dots$.

PROPOSITION. *Let F be a discrete valuation field. Then*

(1) *The choice of a prime element* π $(1 \in \mathbb{Z} \mapsto \pi \in F^\times)$ *splits the exact sequence*

$$1 \to U \longrightarrow F^\times \xrightarrow{v} \mathbb{Z} \longrightarrow 0.$$

The group $F^\times$ *is isomorphic to* $U \times \mathbb{Z}$.

(2) *The canonical map* $\mathcal{O} \longrightarrow \mathcal{O}/\mathcal{M} = \overline{F}$ *induces the surjective homomorphism*

$$\lambda_0 \colon U \longrightarrow \overline{F}^\times, \quad \varepsilon \mapsto \overline{\varepsilon};$$

λ_0 *maps* U/U_1 *isomorphically onto* $\overline{F}^\times$.

(3) *The map*

$$\lambda_i \colon U_i \longrightarrow \overline{F}, \qquad 1 + \alpha\pi^i \mapsto \overline{\alpha}$$

for $\alpha \in \mathcal{O}$ *induces the isomorphism* λ_i *of* U_i/U_{i+1} *onto* $\overline{F}$ *for* $i \geqslant 1$.

Proof. The statement (1) follows for example from Lemma 2.4.

(2) The kernel of λ_0 coincides with U_1 and λ_0 is surjective.

(3) The induced map $U_i/U_{i+1} \longrightarrow \overline{F}$ is a homomorphism, since

$$(1 + \alpha_1\pi^i)(1 + \alpha_2\pi^i) = 1 + (\alpha_1 + \alpha_2)\pi^i + \alpha_1\alpha_2\pi^{2i}.$$

This homomorphism is bijective, since $\lambda_i(1 + \mathrm{rep}(\overline{\alpha})\pi^i) = \overline{\alpha}$. □

4.5. COROLLARY. *Let l be not divisible by* $\operatorname{char}(\overline{F})$. *Raising to the lth power induces an automorphism of U_i/U_{i+1} for $i \geqslant 1$.*

If F is complete, then the group U_i for $i \geqslant 1$ is uniquely l-divisible.

Proof. If $\varepsilon = 1 + \alpha\pi^i$ with $\alpha \in \mathcal{O}$, then $\varepsilon^l \equiv 1 + l\alpha\pi^i \mod \pi^{i+1}$. Absence of non-trivial l-torsion in the additive group $\overline{F}$ implies the first property. It also shows that U_i has no non-trivial l-torsion.

For an element $\eta = 1 + \beta\pi^i$ with $\beta \in \mathcal{O}^\times$ we have $\eta = (1 + l^{-1}\beta\pi^i)^l\eta_1$ with $\eta_1 \in U_{i+1}$. Applying the same argument to η_1 and so on, we get an lth root of η in F in the case of complete F. □

4.6. Let $\operatorname{char}(\overline{F}) = p > 0$. Lemma 2.2 tells that either $\operatorname{char}(F) = p$ or $\operatorname{char}(F) = 0$. We shall study the operation of raising to the pth power. Denote this homomorphism by

$$\uparrow p \colon \alpha \mapsto \alpha^p.$$

The first and simplest case is $\operatorname{char}(F) = p$.

PROPOSITION. *Let* $\operatorname{char}(F) = \operatorname{char}(\overline{F}) = p > 0$. *Then the homomorphism $\uparrow p$ maps U_i injectively into U_{pi} for $i \geqslant 1$. For $i \geqslant 1$ it induces the commutative diagram*

$$\begin{array}{ccc} U_i/U_{i+1} & \xrightarrow{\uparrow p} & U_{pi}/U_{pi+1} \\ {\scriptstyle \lambda_i}\big\downarrow & & {\scriptstyle \lambda_{pi}}\big\downarrow \\ \overline{F} & \xrightarrow{\uparrow p} & \overline{F} \end{array}$$

Proof. Since $(1 + \varepsilon\pi^i)^p = 1 + \varepsilon^p\pi^{pi}$ and there is no non-trivial p-torsion in $\overline{F}^\times$ and $F^\times$, the assertion follows. □

COROLLARY. *Let F be a field of characteristic $p > 0$ and let $\overline{F}$ be perfect, i.e $\overline{F} = \overline{F}^p$. Then $\uparrow p$ maps the quotient group U_i/U_{i+1} isomorphically onto the quotient group U_{pi}/U_{pi+1} for $i \geqslant 1$.*

4.7. We now consider the case of $\operatorname{char}(F) = 0$, $\operatorname{char}(\overline{F}) = p > 0$. As $p = 0$ in the residue field $\overline{F}$, we conclude that $p \in \mathcal{M}$ and, therefore, for the surjective discrete valuation v of F we get $v(p) = e \geqslant 1$.

DEFINITION. *The number $e = e(F) = v(p)$ is called the absolute ramification index of F.*

Let π be a prime element in F. Let R be a set of representatives, and let $\overline{\theta}_0 \in \overline{F}$ be the element of $\overline{F}$ uniquely determined by the relation $p - \operatorname{rep}(\overline{\theta}_0)\pi^e \in \pi^{e+1}\mathcal{O}$.

Proposition. *Let F be a discrete valuation field of characteristic zero with residue field of positive characteristic p. Then the homomorphism $\uparrow p$ maps U_i to U_{pi} for $i \leqslant e/(p-1)$, and U_i to U_{i+e} for $i > e/(p-1)$. This homomorphism induces the following commutative diagrams*

(1) *if $i < e/(p-1)$,*

$$\begin{array}{ccc} U_i/U_{i+1} & \xrightarrow{\uparrow p} & U_{pi}/U_{pi+1} \\ \lambda_i \downarrow & & \lambda_{pi} \downarrow \\ \overline{F} & \xrightarrow{\overline{\alpha} \mapsto \overline{\alpha}^p} & \overline{F} \end{array}$$

(2) *if $i = e/(p-1)$ is an integer,*

$$\begin{array}{ccc} U_i/U_{i+1} & \xrightarrow{\uparrow p} & U_{pi}/U_{pi+1} \\ \lambda_i \downarrow & & \lambda_{pi} \downarrow \\ \overline{F} & \xrightarrow{\overline{\alpha} \mapsto \overline{\alpha}^p + \overline{\theta}_0 \overline{\alpha}} & \overline{F} \end{array}$$

(3) *if $i > e/(p-1)$,*

$$\begin{array}{ccc} U_i/U_{i+1} & \xrightarrow{\uparrow p} & U_{i+e}/U_{i+e+1} \\ \lambda_i \downarrow & & \lambda_{i+e} \downarrow \\ \overline{F} & \xrightarrow{\overline{\alpha} \mapsto \overline{\theta}_0 \overline{\alpha}} & \overline{F} \end{array}$$

The horizontal homomorphisms are injective in cases (1), (3) *and surjective in case* (3).

If a primitive pth root ζ_p of unity is contained in F, then $v(1-\zeta_p) = e/(p-1)$ and the kernel of the horizontal homomorphisms in case (2) *is of order p.*

If $e/(p-1) \in \mathbb{Z}$, $U_{pe/(p-1)+1} \subset U^p_{e/(p-1)+1}$ and there is no non-trivial p-torsion in $F^\times$, then the homomorphism is injective in case (2).

Proof. Let $v(\alpha) = i$. Writing

$$(1+\alpha)^p = 1 + p\alpha + \frac{p(p-1)}{2}\alpha^2 + \cdots + p\alpha^{p-1} + \alpha^p$$

and calculating $v(p\alpha) = e+i$, $v\left(\frac{p(p-1)}{2}\alpha^2\right) = e+2i, \ldots, v(p\alpha^{p-1}) = e+(p-1)i$, $v(\alpha^p) = pi$, we get

$$\begin{aligned} & v((1+\alpha)^p - 1) = v(\alpha^p + p\alpha), && \text{if} \quad v(\alpha^p) \neq v(p\alpha), \\ & v((1+\alpha)^p - 1) \geqslant v(\alpha^p + p\alpha), && \text{otherwise.} \end{aligned}$$

These formulas reveal the behaviour of $\uparrow p$ acting on the filtration in U_1, because $v(\alpha^p) \leqslant v(p\alpha)$ if and only if $i \leqslant e/(p-1)$. Moreover, for a unit α we obtain

$$
\begin{aligned}
&(1+\alpha\pi^i)^p \equiv 1+\alpha^p\pi^{pi} \mod \pi^{pi+1}, && \text{if } i < e/(p-1),\\
&(1+\alpha\pi^i)^p \equiv 1+\mathrm{rep}(\overline{\theta}_0)\alpha\pi^{i+e} \mod \pi^{i+e+1}, && \text{if } i > e/(p-1),\\
&(1+\alpha\pi^i)^p \equiv 1+(\alpha^p+\mathrm{rep}(\overline{\theta}_0)\alpha)\pi^{pi} \mod \pi^{pi+1}, && \text{if } i = e/(p-1) \in \mathbb{Z}.
\end{aligned}
$$

Thus, we conclude that the diagrams in the Proposition are commutative. Further, the homomorphism $\uparrow p$ is an isomorphism in case (3) and injective in case (1).

Assume that $\zeta_p \in F$. The assertions obtained above imply that $v(1-\zeta_p) = e/(p-1)$ and $e/(p-1) \in \mathbb{Z}$. Therefore, the homomorphism $\overline{\alpha} \mapsto \overline{\alpha}^p + \overline{\theta}_0\overline{\alpha}$ is not injective. Its kernel $\sqrt[p-1]{-\overline{\theta}_0}\,\mathbb{F}_p$ in this case is of order p.

Now let $e/(p-1)$ be an integer and let $U_{pe/(p-1)+1} \subset U^p_{e/(p-1)+1}$. Assume that the horizontal homomorphism in case (2) is not injective. Let $\overline{\alpha}_0 \in \overline{F}$ satisfy the equation $\overline{\alpha}_0^p + \overline{\theta}_0\overline{\alpha}_0 = 0$. Then $(1+\mathrm{rep}(\overline{\alpha}_0)\pi^{e/(p-1)})^p \in U_j$ for some $j > pe/(p-1)$. Therefore $(1+\mathrm{rep}(\overline{\alpha}_0)\pi^{e/(p-1)})^p = \varepsilon_1^p$ for some $\varepsilon_1 \in U_{e/(p-1)+1}$. Thus, $(1+\mathrm{rep}(\overline{\alpha}_0)\pi^{e/(p-1)})\varepsilon_1^{-1} \in U_{e/(p-1)}$ is a primitive pth root of unity. □

4.8. COROLLARY 1. *Let* $\mathrm{char}(F) = 0$ *and let* $\overline{F}$ *be a perfect field of characteristic* $p > 0$. *Then* $\uparrow p$ *maps the quotient group* U_i/U_{i+1} *isomorphically onto* U_{pi}/U_{pi+1} *for* $1 \leqslant i < e/(p-1)$ *and isomorphically onto* U_{i+e}/U_{i+e+1} *for* $i > e/(p-1)$.

COROLLARY 2. *Let F be a complete field. Let* $i > pe/(p-1)$. *Then* $U_i \subset U^p_{i-e}$. *Therefore, if* $F^\times$ *has no non-trivial p-torsion then the homomorphism is injective in case* (2).

In addition, if the residue field of F is finite and F contains no non-trivial pth roots of unity, then $U_i \subset U^p_{i-e}$ *for* $i \geqslant pe/(p-1)$.

Proof. Use the completeness of F. Due to the surjectivity of the homomorphisms in case (3) we get $U_i \subset U_{i+1}U^p_{i-e} \subset U_{i+2}U^p_{i-e} \subset \cdots \subset U^p_{i-e}$.

If the residue field of F is finite, then the injectivity of the homomorphism in case (2) implies its surjectivity. □

4.9. PROPOSITION. *Let F be a complete discrete valuation field.*

If $\mathrm{char}(F) = 0$, *then* $F^{\times n}$ *is an open subgroup in* $F^\times$ *for* $n \geqslant 1$.

If $\mathrm{char}(F) = p > 0$, *then* $F^{\times n}$ *is an open subgroup in* $F^\times$ *if and only if n is relatively prime to p.*

Proof. If $\mathrm{char}(\overline{F}) = 0$, then by Corollary 4.5 we get $U_1 \subset F^{\times n}$ for $n \geqslant 1$. It means that $F^{\times n}$ is open. If $\mathrm{char}(\overline{F}) = p$, then by Corollary 4.5 $U_1 \subset F^{\times n}$ for $(n,p) = 1$ and $F^{\times n}$ is open. In this case, if $\mathrm{char}(F) = p$, then by Proposition 4.6 $1+\pi^i \notin F^{\times p}$ for $(i,p) = 1$. Then $F^{\times p}$ is not open. If $\mathrm{char}(F) = 0$, then using Corollary 2 of 4.8 we obtain $U_i \subset F^{\times p^m}$ when $i > pe/(p-1)+(m-1)e$. Therefore $F^{\times n}$ is open for $n \geqslant 1$. □

This Proposition demonstrates that topological properties are closely connected with the algebraic ones for complete discrete valuation fields of characteristic 0 with residue field of characteristic p. This is not the case when $\mathrm{char}(F) = p$.

4.10. Finally, we deduce a multiplicative analog of the expansion in Proposition 4.2.

PROPOSITION. *Let F be a complete discrete valuation field. Let R be a set of representatives and let π_i be as in 4.2. Then for $\alpha \in F^\times$ there exist uniquely determined $n \in \mathbb{Z}$, $\theta_i \in R$, $\theta_0 \in R^\times$ for $i \geqslant 0$, such that α can be expanded in the convergent product*

$$\alpha = \pi^n \theta_0 \prod_{i \geqslant 1} (1 + \theta_i \pi_i).$$

Proof. The existence and uniqueness of n and θ_0 immediately follow from Proposition 4.4. Assume that $\varepsilon \in U_m$, then, using Proposition 4.2, one can find $\theta_m \in R$ with $\varepsilon(1+\theta_m\pi_m)^{-1} \in U_{m+1}$. Proceeding by induction, we obtain an expansion of α in a convergent product. If there are two such expansions $\prod(1+\theta_i\pi_i) = \prod(1+\theta_i'\pi_i)$, then the residues $\overline{\theta}_i$, $\overline{\theta_i'}$ coincide in $\overline{F}$. Thus, $\theta_i = \theta_i'$. □

5. Group of Principal Units as Topological $\mathbb{Z}_p$-Module

We study $\mathbb{Z}_p$-structure of the group of principal units of a complete discrete valuation field F with residue field $\overline{F}$ of characteristic $p > 0$ by using convergent series and results of the previous section. Everywhere in this section F is a complete discrete valuation field with residue field of positive characteristic p.

Let A be a $\mathbb{Z}_p$-module endowed with a topology compatible with the structure of the $\mathbb{Z}_p$-module and the p-adic topology of $\mathbb{Z}_p$. A set $\{a_i\}_{i \in I}$ of elements of A is called a set of topological generators of A if every element of A is a limit of a convergent sequence of elements of the $\mathbb{Z}_p$-submodule of A generated by this set. A set $\{a_i\}_{i \in I}$ of topological generators is called a topological basis if for every $j \in I$ and every non-zero $c \in \mathbb{Z}_p$ the element ca_j is not a limit of a convergent sequence of elements of the $\mathbb{Z}_p$-submodule of A generated by $\{a_i : i \neq j\}$.

5.1. Propositions 4.6, 4.7 imply that $\varepsilon^{p^n} \to 1$ as $n \to +\infty$ for $\varepsilon \in U_1$. This enables us to write

$$\varepsilon^a = \lim_{n\to\infty} \varepsilon^{a_n} \qquad \text{if} \quad \lim_{n\to\infty} a_n = a \in \mathbb{Z}_p, \quad a_n \in \mathbb{Z}.$$

LEMMA. *Let $\varepsilon \in U_1$, $a \in \mathbb{Z}_p$. Then $\varepsilon^a \in U_1$ is well defined and $\varepsilon^{a+b} = \varepsilon^a\varepsilon^b$, $\varepsilon^{ab} = (\varepsilon^a)^b$, $(\varepsilon\eta)^a = \varepsilon^a\eta^a$ for $\varepsilon, \eta \in U_1$, $a, b \in \mathbb{Z}_p$. The multiplicative group U_1 is a $\mathbb{Z}_p$-module under the operation of raising to a power. Moreover, the structure of the $\mathbb{Z}_p$-module U_1 is compatible with the topologies of $\mathbb{Z}_p$ and U_1.*

Proof. Assume that $\lim a_n = \lim b_n$; hence $a_n - b_n \to 0$ as $n \to +\infty$. Therefore, $\lim \varepsilon^{a_n - b_n} = 1$. Propositions 4.6, 4.7 show that a map $\mathbb{Z}_p \times U_1 \to U_1$ $((a, \varepsilon) \to \varepsilon^a)$ is continuous with respect to the p-adic topology on $\mathbb{Z}_p$ and the discrete valuation topology on U_1. This argument can be applied to verify the other assertions of the Lemma. □

5.2. PROPOSITION. *Let F be of characteristic p with perfect residue field. Let R be a set of representatives, and let R_0 be a subset of it such that the residues of its elements in $\overline{F}$ form a basis of $\overline{F}$ as a vector space over $\mathbb{F}_p$. Let an index-set J numerate the elements of R_0. Assume that π_i are as in 4.2. Let v_p be the p-adic valuation.*

Then every element $\alpha \in U_1$ can be uniquely represented as a convergent product

$$\alpha = \prod_{\substack{(i,p)=1 \\ i>0}} \prod_{j\in J} (1 + \theta_j \pi_i)^{a_{ij}}$$

where $\theta_j \in R_0$, $a_{ij} \in \mathbb{Z}_p$ and the sets $J_{i,c} = \{j \in J : v_p(a_{ij}) \leqslant c\}$ are finite for all $c \geqslant 0$, $(i,p) = 1$.

Proof. We first show that the element α can be written modulo U_n for $n \geqslant 1$ in the desired form with $a_{ij} \in \mathbb{Z}$. Proceeding by induction, it will suffice to consider an element $\varepsilon \in U_n$ modulo U_{n+1}. Let $\varepsilon \equiv 1 + \theta\pi_n \mod U_{n+1}$, $\theta \in R$. If $(n,p) = 1$, then one can find $\theta_1, \dots, \theta_m \in R_0$ and $b_1, \dots, b_m \in \mathbb{Z}$ such that $1 + \theta\pi_n \equiv \prod_{k=1}^m (1 + \theta_k\pi_n)^{b_k} \mod U_{n+1}$ for some m. If $n = p^s n'$ with an integer n', $(n', p) = 1$, then using the Corollary 4.6, one can find $\theta_1, \dots, \theta_m \in R_0$ and $b_1, \dots, b_m \in \mathbb{Z}$ such that $1 + \theta\pi_n \equiv \prod_{k=1}^m (1 + \theta_k\pi_{n'})^{p^s b_k} \mod U_{n+1}$ for some m. Now due to the continuity we get the desired expression for $\alpha \in U_1$ with the above conditions on the sets $J_{i,c}$.

Assume that there is a convergent product for 1 with θ_j, a_{ij}. Let $(i_0, p) = 1$ and $j_0 \in J$ be such that $n = p^{v_p(a_{i_0 j_0})} i_0 \leqslant p^{v_p(a_{ij})} i$ for all $(i,p) = 1$, $j \in J$. Then the choice of R_0 and 4.5, 4.6 imply $\prod (1 + \theta_j\pi_i)^{a_{ij}} \notin U_{n+1}$, which concludes the proof. □

COROLLARY. *The $\mathbb{Z}_p$-module group U_1 has a topological basis $1+\theta_j\pi_i$ where where $\theta_j \in R_0$, $(i,p)=1$.*

5.3. Let's have an additional look at the horizontal homomorphism

$$\psi\colon \overline{F} \longrightarrow \overline{F}, \quad \overline{\alpha} \mapsto \overline{\alpha}^p + \overline{\theta}_0\overline{\alpha}$$

of case (2) in Proposition 4.7.

Suppose that a primitive pth root of unity ζ_p belongs to F and

$$\zeta_p \equiv 1 + \operatorname{rep}(\overline{\theta}_1)\pi^{e/(p-1)} \mod \pi^{e/(p-1)+1},$$

$(v(\zeta_p - 1) = e/(p-1)$ according to Proposition 4.7. As $\overline{\theta}_1 \in \ker\psi$, we conclude that $\psi(\overline{\alpha}) = \overline{\theta}_1^p(\eta^p - \eta)$ where $\eta = \overline{\alpha}\overline{\theta}_1^{-1}$. The homomorphism $\eta \mapsto \eta^p - \eta$ is usually denoted by $\wp$. In this terminology we get $\psi(\overline{F}) = \overline{\theta}_1^p\wp(\overline{F})$.

The theory of Artin–Schreier extensions, it is briefly reviewed in 3.6 of Chapter 3, sets a correspondence between abelian extensions of exponent p of a field $\overline{F}$ of characteristic p and subgroups of $\overline{F}/\wp(\overline{F})$. In particular, if $\overline{F}$ is finite, then the cardinalities of the kernel of ψ and of the cokernel of ψ coincide. In this simple case $\psi(\overline{F}) = \overline{F}$ if and only if there is no non-trivial p-torsion in $F^\times$, and $\psi(\overline{F})$ is of index p if and only if $\zeta_p \in F^\times$ (see 4.7). The homomorphism $\wp$ plays an important role in class field theory.

5.4. PROPOSITION. *Let F be of characteristic 0 with perfect residue field of characteristic p. Let π_i be as in 4.2. If $e = v(p)$ is divisible by $p-1$, let $\psi\colon \overline{F} \longrightarrow \overline{F}$ be the map introduced in 5.3.*

Let R be a set of representatives and let R_0 (resp. R_0') be a subset of it such that the residues of its elements in $\overline{F}$ form a basis of $\overline{F}$ as a vector space over $\mathbb{F}_p$ (resp. form a basis of $\overline{F}/\psi(\overline{F})$ as a $\mathbb{F}_p$-module). Let the index-set J (resp. J') numerate the elements of R_0 (resp. R_0'). Let

$$I = \{i : i \in \mathbb{Z}, 1 \leqslant i < pe/(p-1), (i,p) = 1\}.$$

Let v_p be the p-adic valuation.

Then every element $\alpha \in U_1$ can be represented as a convergent product

$$\alpha = \prod_{i\in I}\prod_{j\in J}(1+\theta_j\pi_i)^{a_{ij}} \prod_{j\in J'}(1+\eta_j\pi_{pe/(p-1)})^{a_j}$$

where $\theta_j \in R_0$, $\eta_j \in R_0'$, $a_{ij}, a_j \in \mathbb{Z}_p$ (the second product occurs when $e/(p-1)$ is an integer) and the sets

$$J_{i,c} = \{j \in J : v_p(a_{ij}) \leqslant c\}, \quad J_c' = \{j \in J' : v_p(a_j) \leqslant c\}$$

are finite for all $c \geqslant 0$, $i \in I$.

Proof. We shall show how to obtain the required form for $\varepsilon \in U_n$ modulo U_{n+1}. Put $\pi_n = \pi^n$ for $n = pe/(p-1)$. Let $\varepsilon = 1 + \theta\pi_n \mod U_{n+1}$, $\theta \in R$. There are four cases to consider:

(1) $n \in I$. One can find $\theta_1, \dots, \theta_m \in R_0$ and $b_1, \dots, b_m \in \mathbb{Z}$ satisfying the congruence $1 + \theta\pi_n \equiv \prod_{k=1}^{m} (1 + \theta_k \pi_n)^{b_k} \mod U_{n+1}$ for some m.

(2) $n < pe/(p-1)$, $n = p^s n'$ with $n' \in I$. Corollary 1 in 4.8 and 4.5 show that there exist $\theta_1, \dots, \theta_m \in R_0$, $b_1, \dots, b_m \in \mathbb{Z}$ such that

$$1 + \theta\pi_n \equiv \prod_{k=1}^{m} (1 + \theta_k \pi_{n'})^{p^s b_k} \mod U_{n+1} \quad \text{for some } m.$$

(3) $e/(p-1) \in \mathbb{Z}$, $n = pe/(p-1)$. Proposition 4.7 and 4.5 and the definition of R_0' imply that if $n = p^s n'$ with $n' \in I$, then there exist $\theta_1, \dots, \theta_m \in R_0$, $\eta_1, \dots, \eta_r \in R_0'$, $b_1, \dots, b_m$, $c_1, \dots, c_r \in \mathbb{Z}$ such that

$$1 + \theta\pi_n \equiv \prod_{k=1}^{m} (1 + \theta_k \pi_{n'})^{p^s b_k} \prod_{l=1}^{r} (1 + \eta_l \pi_n)^{c_l} \mod U_{n+1} \quad \text{for some } m, r.$$

(4) $n > pe/(p-1)$. Proposition 4.7 and Corollary 1 in 4.8 imply that if $d = \min\{d : n - de \leqslant pe/(p-1)\}$ and $n' = n - de$, then

$$1 + \theta\pi_n \equiv (1 + \theta' \pi_{n'})^{p^d} \mod U_{n+1} \quad \text{for some } \theta' \in R.$$

Now applying the arguments of the preceding cases to $1 + \theta'\pi_{n'}$, we can write $1 + \theta\pi_n \mod U_{n+1}$ in the required form. □

5.5. From Proposition 4.7 we deduce that F contains finitely many roots of unity of order a power of p.

COROLLARY. *Let F be of characteristic 0 with perfect residue field of characteristic p.*

(1) *If F does not contain non-trivial pth roots of unity then the representation in Proposition 5.4 is unique. Therefore the elements $1 + \theta_j \pi_i$, $1 + \eta_j \pi_{pe/(p-1)}$ of Proposition 5.4 form a topological basis of $\mathbb{Z}_p$-module $U_{1,F}$.*

(2) *If F contains a non-trivial pth root of unity let r be the maximal integer such that F contains a primitive p^rth root of unity. Then the numbers a_{ij}, a_j of Proposition 5.4 are determined uniquely modulo p^r. Therefore the images of the elements $1 + \theta_j \pi_i$, $1 + \eta_j \pi_{pe/(p-1)}$ of Proposition 5.4 form a topological basis of $\mathbb{Z}/p^r\mathbb{Z}$-module $U_{1,F}/U_{1,F}^{p^r}$.*

(3) *If the residue field of F is finite then U_1 is isomorphic to the direct sum of a free $\mathbb{Z}_p$-module of rank ef and its torsion part, where f is the dimension of $\overline{F}$ over $\mathbb{F}_p$.*

Proof. (1) All horizontal homomorphisms of the diagrams in Proposition 4.7 are injective when $\zeta_p \notin F$. Repeating the arguments for uniqueness from the proof of Proposition 5.2, we get the first assertion of the Corollary.

(2) We can argue by induction on r and explain the induction step. Write a primitive p^rth root ζ_{p^r} in the form of Proposition 5.4

$$\zeta_{p^r} = \prod_{i\in I}\prod_{j\in J}(1+\theta_j\pi_i)^{c_{ij}} \prod_{j\in J'}(1+\eta_j\pi_{pe/(p-1)})^{c_j}$$

and raise the expression to the p^rth power which demonstrates the non-uniqueness of the expansion in Proposition 5.4.

Now if

$$1 = \prod_{i\in I}\prod_{j\in J}(1+\theta_j\pi_i)^{a_{ij}} \prod_{j\in J'}(1+\eta_j\pi_{pe/(p-1)})^{a_j}$$

then by the same argument as in the proof of Proposition 5.2 we deduce that $a_{ij} = pb_{ij}, a_j = pb_j$ with p-adic integers b_{ij}, b_j. Then

$$\prod_{i\in I}\prod_{j\in J}(1+\theta_j\pi_i)^{b_{ij}} \prod_{j\in J'}(1+\eta_j\pi_{pe/(p-1)})^{b_j}$$

is a pth root of unity, and so is equal to

$$\Big(\prod_{i\in I}\prod_{j\in J}(1+\theta_j\pi_i)^{c_{ij}} \prod_{j\in J'}(1+\eta_j\pi_{pe/(p-1)})^{c_j}\Big)^{p^{r-1}c}$$

for some integer c. Now by the induction assumption all $b_{ij} - p^{r-1}cc_{ij}, b_j - p^{r-1}cc_j$ are divisible by p^{r-1}. Thus, all a_{ij}, a_j are divisible by p^r.

(3) If the residue field of F is finite then U_1 is a module of finite type over the principal ideal domain $\mathbb{Z}_p$. Note that the group $\wp(\overline{F})$ is of index p in $\overline{F}$ because $\overline{F}$ is finite (see 5.3). If the p-torsion of $F^\times$ is of order p^r, we replace $1+\eta_1\pi_{pe/(p-1)}$ with a primitive p^rth root of unity. The cardinality of I is equal to $e = [pe/(p-1)] - [[pe/(p-1)]/p]$. □

6. Set of Multiplicative Representatives

We shall introduce a special set $\mathscr{R}$ of multiplicative representatives which is closed with respect to multiplication. We will describe coefficients of the sum and product of convergent power series with multiplicative representatives.

6.1. Assume that $\operatorname{char}(\overline{F}) = p > 0$.

Let $a \in \overline{F}$. An element $\alpha \in \mathscr{O}$ is said to be a *multiplicative representative* (*Teichmüller representative*) of a if $\overline{\alpha} = a$ and $\alpha \in \bigcap_{m\geqslant 0} F^{p^m}$. This definition is justified by the following Proposition.

PROPOSITION. *An element $a \in \overline{F}$ has a multiplicative representative if and only if $a \in \bigcap_{m \geqslant 0} \overline{F}^{p^m}$. A multiplicative representative for such a is unique. If a and b have the multiplicative representatives α and β, then $\alpha\beta$ is the multiplicative representative of ab.*

Proof. We need the following Lemma.

6.2. LEMMA. *Let $\alpha, \beta \in \mathcal{O}$ and $v(\alpha - \beta) \geqslant m$, $m > 0$. Then*

$$v(\alpha^{p^n} - \beta^{p^n}) \geqslant n + m.$$

Proof. Put $\alpha = \beta + \pi^m\gamma$; then $\alpha^p = \beta^p + p\beta^{p-1}\pi^m\gamma + \cdots + p\beta(\pi^m\gamma)^{p-1} + \pi^{pm}\gamma^p$, and as $v(p) \geqslant 1$ (recall $\mathrm{char}(\overline{F}) = p$), we have $v(p\beta^{p-1}\pi^m\gamma) \geqslant m + 1, \ldots, v(\pi^{pm}\gamma^p) \geqslant m + 1$, and $\alpha^p - \beta^p \in \pi^{m+1}\mathcal{O}$. Now the required assertion follows by induction. □

To prove the first assertion of the Proposition, suppose that $a \in \bigcap_{m \geqslant 0} \overline{F}^{p^m}$. Since $\overline{F}$ has no non-trivial p-torsion, there exist unique elements $a_m \in \overline{F}$ satisfying the equations $a_m^{p^m} = a$. Let $\beta_m \in \mathcal{O}$ be such that $\overline{\beta}_m = a_m$. Then $\overline{\beta_{m+1}^p} = \overline{\beta}_m$ and $v(\beta_{m+1}^p - \beta_m) \geqslant 1$. Lemma 6.2 implies $v(\beta_{m+1}^{p^{n+1}} - \beta_m^{p^n}) \geqslant n + 1$. Hence, the sequence $(\beta_m^{p^{m-n}})_{m \geqslant n}$ is Cauchy. It has the limit $\alpha_n = \lim \beta_m^{p^{m-n}} \in \mathcal{O}$. We see that $\alpha_n^{p^n} = \alpha_0$ for $n \geqslant 0$ and $\overline{\alpha}_0 = a$, i.e., α_0 is a multiplicative representative of a. Conversely, if $a \in \overline{F}$ has a multiplicative representative α, then $\overline{\alpha} \in \bigcap_{m \geqslant 0} \overline{F}^{p^m}$.

Furthermore, if α and β are multiplicative representatives of $a \in \overline{F}$, then writing $\alpha = \alpha_m^{p^m}$ and $\beta = \beta_m^{p^m}$ for some $\alpha_m, \beta_m \in \mathcal{O}$, we have $\overline{\alpha}_m^{p^m} = \overline{\beta}_m^{p^m}$ and $\overline{\alpha}_m = \overline{\beta}_m$, due the injectivity of $\uparrow p^m$ in $\overline{F}$. Now Lemma 6.2 implies $v(\alpha - \beta) \geqslant m + 1$, hence $\alpha = \beta$.

Finally, if α and β are the multiplicative representatives of a and b, then $\overline{\alpha\beta} = ab$ and $\alpha\beta \in \bigcap_{m \geqslant 0} F^{p^m}$. Therefore, $\alpha\beta$ is the multiplicative representative of ab. □

6.3. Denote the set of multiplicative representatives in $\mathcal{O}$ by $\mathcal{R}$.

COROLLARY 1. *If $\overline{F}$ is perfect (i.e. F is a local field) then every element of $\overline{F}$ has its multiplicative representative in $\mathcal{R}$. The map $r\colon \overline{F} \longrightarrow \mathcal{R}$ induces an isomorphism $\overline{F}^{\times} \xrightarrow{\sim} \mathcal{R} \setminus \{0\}$. The correspondence $r\colon \overline{F} \longrightarrow \mathcal{R}$ is called the Teichmüller map.*

If $\overline{F}$ is finite then $\mathcal{R} \setminus \{0\}$ is a cyclic group of order equal to $|\overline{F}| - 1$.

COROLLARY 2. *Let* $\mathrm{char}(F) = p$. *If* α, β *are the multiplicative representatives of* $a, b \in \overline{F}$, *then* $\alpha + \beta$ *is the multiplicative representative of* $a + b$.

Proof. Let $\alpha = \alpha_m^{p^m}, \beta = \beta_m^{p^m}$. Then $\alpha + \beta = (\alpha_m + \beta_m)^{p^m}$, hence $\alpha + \beta \in \bigcap_{m \geqslant 0} F^{p^m}$ and $\overline{\alpha + \beta} = a + b$. □

6.4. Consider the case where $\mathrm{char}(F) = 0$ and $\mathrm{char}(\overline{F}) = p$. Suppose that we have two elements $\alpha, \beta \in \mathcal{O}$, and ($\pi$ is a prime element)

$$\alpha = \sum_{i \geqslant 0} \theta_i \pi^i, \qquad \beta = \sum_{i \geqslant 0} \eta_i \pi^i,$$

with $\theta_i, \eta_i \in \mathcal{R}$. Suppose also that $\alpha + \beta$ and $\alpha\beta$ are written in the form

$$\alpha + \beta = \sum_{i \geqslant 0} \rho_i^{(+)} \pi^i, \qquad \alpha\beta = \sum_{i \geqslant 0} \rho_i^{(\times)} \pi^i,$$

and $\rho_i^{(+)}, \rho_i^{(\times)} \in \mathcal{R}$.

Corollary 4.2 implies that $\rho_i^{(+)}, \rho_i^{(\times)}$ are uniquely determined by θ_i, η_i. Let's find out the dependence of $\rho_n^{(+)}, \rho_n^{(\times)}$ on θ_i, η_i, $i \leqslant n$. In order to obtain a polynomial relation we introduce elements $\theta_i = \varepsilon_i^{p^{n-i}}$, $\eta_i = \xi_i^{p^{n-i}}$, $\rho_i^{(*)} = \lambda_i^{(*)p^{n-i}}$ for ε_i, ξ_i, $\lambda_i^{(*)} \in \mathcal{R}$ and $* = +$ or $* = \times$, $i \geqslant 0$.

Then we deduce that

$$\Big(\sum_{i=0}^{n} \pi^i \varepsilon_i^{p^{n-i}}\Big) * \Big(\sum_{i=0}^{n} \pi^i \xi_i^{p^{n-i}}\Big) \equiv \Big(\sum_{i=0}^{n} \pi^i \lambda_i^{(*)p^{n-i}}\Big) \mod \pi^{n+1}, \tag{$*$}$$

for $* = +$ or $* = \times$. We see that if the residues $\overline{\varepsilon}_i, \overline{\xi}_i$ for $0 \leqslant i \leqslant n$ and $\overline{\lambda_i^{(*)}}$ for $0 \leqslant i \leqslant n-1$ are known, then by using Lemma 6.2 we can calculate $\pi^i \varepsilon_i^{p^{n-i}}$, $\pi^i \xi_i^{p^{n-i}}$, $\pi^i \lambda_i^{p^{n-i}} \mod \pi^{n+1}$. Hence, $\overline{\lambda_n^{(*)}}$ are uniquely determined from ($*$).

6.5. Let $A = \mathbb{Z}[X_0, X_1, \dots, Y_0, Y_1, \dots]$ be the ring of polynomials in variables $X_0, X_1, \dots$ and $Y_0, Y_1, \dots$, with coefficients from $\mathbb{Z}$. Introduce polynomials

$$W_n(X_0, \dots, X_n) = \sum_{i=0}^{n} p^i X_i^{p^{n-i}}, \qquad n \geqslant 0.$$

In particular, $W_0(X_0) = X_0$, $W_1(X_0, X_1) = X_0^p + pX_1$, and

$$W_n(X_0, \dots, X_n) = p^n X_n + W_{n-1}(X_0^p, \dots, X_{n-1}^p).$$

PROPOSITION. *There exist unique polynomials*

$$\omega_n^{(*)}(X_0, \dots, X_n, Y_0, \dots, Y_n) \in A,\ n \geqslant 0$$

satisfying the equations

$$W_n(X_0, \dots, X_n) * W_n(Y_0, \dots, Y_n) = W_n(\omega_0^{(*)}, \dots, \omega_n^{(*)})$$

for $n \geqslant 0$, *where* $* = +$ *or* $* = \times$.

Moreover, the polynomial

$$\omega_n^{(*)}(X_0,\dots,X_n,Y_0,\dots,Y_n)^p - \omega_n^{(*)}(X_0^p,\dots,X_n^p,Y_0^p,\dots,Y_n^p)$$

belongs to pA.

Proof. We get

$$\omega_0^{(+)} = X_0+Y_0, \quad \omega_1^{(+)} = X_1+Y_1+(X_0^p+Y_0^p-(X_0+Y_0)^p)/p,$$
$$\omega_0^{(\times)} = X_0Y_0, \quad \omega_1^{(\times)} = X_1Y_0^p+Y_1X_0^p+pX_1Y_1,$$
$$\dots.$$

Assume now that $\omega_i^{(*)} \in A$ and the second assertion of the Proposition holds for $0 \leqslant i \leqslant n-1$, and proceed by induction.

For a suitable polynomial $f_n^* \in A$ we get

$$\begin{aligned} p^n\omega_n^{(*)} &= W_{n-1}(X_0^p,\dots,X_{n-1}^p) * W_{n-1}(Y_0^p,\dots,Y_{n-1}^p) \\ &\quad - W_{n-1}(\omega_0^{(*)p},\dots,\omega_{n-1}^{(*)\,p}) + p^n f_n^* \end{aligned} \tag{$**$}$$

For example, $f_n^+ = X_n + Y_n$.

For any $g \in A$ we get

$$g(X_0,Y_0,\dots)^p - g(X_0^p,Y_0^p,\dots) \in pA$$

and

$$g(X_0,Y_0,\dots)^{p^m} - g(X_0^p,Y_0^p,\dots)^{p^{m-1}} \in p^mA$$

for $m \geqslant 0$.

Using the second assertion of the Proposition for $i < n$ and Lemma 6.2 we now deduce that

$$W_{n-1}(\omega_0^{(*)p},\dots,\omega_{n-1}^{(*)p}) - W_{n-1}(\omega_0^{(*)}(X_0^p,Y_0^p),\dots,\omega_{n-1}^{(*)}(X_0^p,\dots,Y_0^p,\dots)) \in p^nA.$$

From it and from

$$\begin{aligned} &W_{n-1}(X_0^p,\dots,X_n^p) * W_{n-1}(Y_0^p,\dots,Y_{n-1}^p) \\ &= W_{n-1}(\omega_0^{(*)}(X_0^p,Y_0^p),\dots,\omega_{n-1}^{(*)}(X_0^p,\dots,Y_0^p)) \end{aligned}$$

using $(**)$ we conclude that $\omega_n^{(*)} \in A$.

The last assertion of the Proposition now follows from the first congruence for g above. □

6.6. We now return to the original problem to find an expression for $\rho_i^{(*)}$.

PROPOSITION. *Let* $(\sum \theta_i p^i) * (\sum \eta_i p^i) = \sum \rho_i^{(*)} p^i$ *with* $\theta_i, \eta_i, \rho_i^{(*)} \in \mathscr{R}$ *and* $* = +$ *or* $* = \times$. *Then*

$$\rho_i^{(*)} \equiv \omega_i^{(*)}(\theta_0^{p^{-i}}, \theta_1^{p^{-i+1}}, \dots, \theta_i, \eta_0^{p^{-i}}, \eta_1^{p^{-i+1}}, \dots, \eta_i) \mod p, \quad i \geqslant 0,$$

where $\omega_i^{(*)}$ *are defined in* 6.5.

Proof. Assume that the assertion of the Proposition holds for $i \leqslant n-1$. Using the notations of 6.4 this means that

$$\lambda_i^{(*)p^{n-i}} \equiv \omega_i^{(*)}(\varepsilon_0^{p^{n-i}}, \dots, \varepsilon_i^{p^{n-i}}, \xi_0^{p^{n-i}}, \dots, \xi_i^{p^{n-i}}) \mod p, \quad i \leqslant n-1.$$

From Proposition 6.5 we obtain that for $i \leqslant n-1$

$$\omega_i^{(*)}(\varepsilon_0^{p^{n-i}}, \dots, \varepsilon_i^{p^{n-i}}, \xi_0^{p^{n-i}}, \dots, \xi_i^{p^{n-i}}) \equiv \omega_i^{(*)}(\varepsilon_0, \dots, \varepsilon_i, \xi_0, \dots, \xi_i)^{p^{n-i}} \mod p.$$

Hence

$$\lambda_i^{(*)} \equiv \omega_i^{(*)}(\varepsilon_0, \dots, \varepsilon_i, \xi_0, \dots, \xi_i) \mod p, \quad i \leqslant n-1.$$

From (*) in 6.4 we know

$$W_n(\lambda_0^{(*)}, \dots, \lambda_n^{(*)}) \equiv W_n(\varepsilon_0, \dots, \varepsilon_n) * W_n(\xi_0, \dots, \xi_n) \mod p^{n+1}.$$

By Lemma 6.2 we have

$$p^i \lambda_i^{(*)p^{n-i}} \equiv p^i \omega_i^{(*)}(\varepsilon_0, \dots, \varepsilon_i, \xi_0, \dots, \xi_i)^{p^{n-i}} \mod p^{n+1}, \quad i \leqslant n-1.$$

Therefore

$$p^n \lambda_n^{(*)} \equiv p^n \omega_n^{(*)}(\varepsilon_0, \dots, \varepsilon_n, \xi_0, \dots, \xi_n) \mod p^{n+1}$$

which implies the assertion. □

COROLLARY 1. *Let* $(\sum \theta_i^{p^{-i}} p^i) * (\sum \eta_i^{p^{-i}} p^i) = \sum \rho_i^{(*)p^{-i}} p^i$ *with* $\theta_i, \eta_i, \rho_i^{(*)} \in \mathscr{R}$, $* = +$ *or* $* = \times$. *Then*

$$\rho_i^{(*)} \equiv \omega_i^{(*)}(\theta_0, \dots, \theta_i, \eta_0, \dots, \eta_i) \mod p.$$

Proof. In fact, this has already been shown in the proof of the Proposition. □

COROLLARY 2. *If* $(\sum \theta_i p^i) * (\sum \eta_i p^i) = \sum \rho_i^{(*)} p^i$ *then* $(\sum \theta_i^p p^i) * (\sum \eta_i^p p^i) = \sum \rho_i^{(*)p} p^i$.

Proof. This follows immediately from the Proposition and the last assertion of Proposition 6.5. □

7. Witt Ring

Witt vectors over a perfect field K of positive characteristic p form the ring of integers of a local field with prime element p and residue field K.

7.1. Let B be an arbitrary commutative ring with unity. Let the polynomials

$$W_n(X_0,\dots,X_n)=\sum_{i=0}^{n}p^iX_i^{p^{n-i}},\quad n\geqslant 0$$

over B be the images of the polynomials $W_n\in\mathbb{Z}[X_0,\dots,X_n]$ defined in 6.5 under the natural homomorphism $\mathbb{Z}\longrightarrow B$.

For $(a_i)_{i\geqslant 0}$, put

$$(a^{(i)})=(W_0(a_0),W_1(a_0,a_1),\dots)\in(B)_0^{+\infty}.$$

The sequences $(a_i)\in(B)_0^{+\infty}$ are called *Witt vectors* (or, more generally, p-Witt vectors), and the $a^{(i)}$ for $i\geqslant 0$ are called the ghost components of the Witt vector (a_i).

The map $(a_i)\mapsto(a^{(i)})$ is a bijection of $(B)_0^{+\infty}$ onto $(B)_0^{+\infty}$ if p is invertible in B.

Transfer the ring structure of $(a^{(i)})\in(B)_0^{+\infty}$ under the natural componentwise addition and multiplication on $(a_i)\in(B)_0^{+\infty}$. Then for $(a_i),(b_i)\in(B)_0^{+\infty}$ we get

$$(a_i)*(b_i)=(\omega_0^{(*)}(a_0,b_0),\omega_1^{(*)}(a_0,a_1,b_0,b_1),\dots)$$

for $*=+$ or $*=\times$, where the polynomial $\omega_i^{(*)}$ is the image of the polynomial

$$\omega_i^{(*)}\in\mathbb{Z}[X_0,X_1,\dots,Y_0,Y_1,\dots]$$

under the canonical homomorphism $\mathbb{Z}\longrightarrow B$.

If p is invertible in B, then the set of Witt vectors is clearly a commutative ring under the operations defined above. In the general case, when p is not invertible in B, the property of the set $(B)_0^{+\infty}$ of being a commutative ring under the operations $+,\times$ defined above can be expressed via certain equations for the coefficients of the polynomials $\omega_i^{(*)}\in B[X_0,X_1,\dots,Y_0,Y_1,\dots]$. This implies that if a ring B satisfies these conditions, then the same is true for a subring, quotient ring and the polynomial ring. Since every ring can be obtained in this way from a ring $\mathcal{B}$ in which p is invertible, one deduces that under the image in B of the above defined operations for $\mathcal{B}$ the set $(B)_0^{+\infty}$ is a commutative ring with the unity $(1,0,0,\dots)$. This ring is called the *Witt ring* of B and is denoted by $W(B)$. It is easy to verify that if B is an integral domain, then $W(B)$ is an integral domain as well.

7.2. Assume from now on that $p = 0$ in B.

LEMMA. *Define the maps*

$$r_0 \colon B \longrightarrow W(B),$$
$$\mathbf{V} \colon W(B) \longrightarrow W(B) \quad \text{(the "Verschiebung", i.e. "shift" map)},$$
$$\mathbf{F} \colon W(B) \longrightarrow W(B) \quad \text{(the "Frobenius" map)}$$

by the formulas

$$r_0(a) = (a, 0, 0, \dots) \in W(B),$$
$$\mathbf{V}(a_0, a_1, \dots) = (0, a_0, a_1, \dots),$$
$$\mathbf{F}(a_0, a_1, \dots) = (a_0^p, a_1^p, \dots).$$

Then

$$r_0(ab) = r_0(a) r_0(b),$$
$$\mathbf{F}(\alpha + \beta) = \mathbf{F}(\alpha) + \mathbf{F}(\beta), \mathbf{F}(\alpha\beta) = \mathbf{F}(\alpha)\mathbf{F}(\beta),$$
$$\mathbf{V}(\alpha + \beta) = \mathbf{V}(\alpha) + \mathbf{V}(\beta), \quad \mathbf{VF}(\alpha) = \mathbf{FV}(\alpha) = p\alpha$$

for $\alpha, \beta \in W(B)$.

Proof. All these properties can be deduced from properties of $\omega_i^{(*)}$. □

The map $\mathbf{F} - \mathrm{id}$ is often denoted by $\wp \colon W(B) \longrightarrow W(B)$.

Put $W_n(B) = W(B)/\mathbf{V}^n W(B)$. This is a ring consisting of finite sequences $(a_0, \dots, a_{n-1})$.

7.3. The following assertion is of great importance, since it provides a construction of a local field of characteristic zero with prime element p and given perfect residue field K.

PROPOSITION. *Let K be a perfect field of characteristic p. For a Witt vector $\alpha = (a_0, a_1, \dots) \in W(K)$ put*

$$v(\alpha) = \min\{i : a_i \neq 0\} \quad \text{if} \quad \alpha \neq 0, \qquad v(0) = +\infty.$$

Let F_0 be the field of fractions of $W(K)$ and $v \colon F_0^\times \longrightarrow \mathbb{Z}$ the extension of v from $W(K)$ ($v(\alpha\beta^{-1}) = v(\alpha) - v(\beta)$).

Then v is a discrete valuation on F_0 and F_0 is a complete discrete valuation field of characteristic 0 with ring of integers $W(K)$, prime element p, and residue field isomorphic to K. The set of multiplicative representatives in F_0 coincides with $r_0(K)$ and the map r_0 with the Teichmüller map $K \longrightarrow W(K)$.

Proof. If $\alpha = (\underbrace{0,\dots,0}_{m \text{ times}},\dots)$, $\beta = (\underbrace{0,\dots,0}_{n \text{ times}},\dots)$, then using the properties of the polynomials $\omega_i^{(*)}$, we get

$$\alpha+\beta = (\underbrace{0,\dots,0}_{l \text{ times}},\dots), \quad \alpha\beta = (\underbrace{0,\dots,0}_{n+m \text{ times}},\dots)$$

with $l \geqslant \min(m,n)$. Hence, the extension of v to F_0 is a discrete valuation.

Note that $p = (0,1,0,\dots) \in W(K)$ and $p^n \to 0$ as $n \to +\infty$ with respect to v. Since K is perfect, by Lemma 7.2 one can write an element $\alpha = (a_0, a_1, \dots) \in W(K)$ as the convergent sum

$$\alpha = (a_0,0,0,\dots)+(0,a_1,0,\dots)+\dots = \sum_{i=0}^{\infty} r_0(a_i^{p^{-i}})p^i \tag{$*$}$$

Moreover, such expressions for Witt vectors are compatible with addition and multiplication in $W(K)$.

We also obtain that $W(K)$ is complete with respect to v, and if $v(\alpha) = 0$ for $\alpha \in W(K)$, then $\alpha^{-1} \in W(K)$. Consequently, $v(\alpha) \geqslant v(\beta)$ for α, $\beta \in W(K)$ implies $\alpha\beta^{-1} \in W(K)$, i.e., the ring of integers coincides with $W(K)$ and F_0 is complete. The maximal ideal of $W(K)$ is $\mathbf{V}W(K)$ and the residue field is isomorphic to K.

Finally, $r_0(K) = \bigcap_{n \geqslant 0} F_0^{p^n}$, and hence, using Proposition 6.1, we complete the proof. □

8. The Hensel Lemma and Henselian Fields

Let F be a valuation field with the ring of integers $\mathcal{O}$, the maximal ideal $\mathcal{M}$, and the residue field $\overline{F}$. For a polynomial $f(X) = a_nX^n + \dots + a_0 \in \mathcal{O}[X]$ we will denote the polynomial $\overline{a}_nX^n + \dots + \overline{a}_0$ by $\overline{f}(X) \in \overline{F}[X]$. We will write

$$f(X) \equiv g(X) \mod \mathcal{M}^m$$

if $f(X) - g(X) \in \mathcal{M}^m[X]$.

8.1. Let A be a commutative ring. Recall the notion of resultant (introduced in many textbooks on algebra) and its basic properties. For two polynomials $f(X) = a_nX^n + \dots + a_0$, $g(X) = b_mX^m + \dots + b_0$, their resultant is the determinant of a matrix of order $(n+m) \times (n+m)$ formed by m rows of a_i and n rows of b_j, appropriately inserted.

This determinant $R(f,g)$ is zero if and only if f and g have a common root; in general $R(f,g) = ff_1 + gg_1$ for some polynomials $f_1, g_1 \in \mathcal{O}[X]$. If $f(X) = a_n \prod_{i=1}^n (X - \alpha_i)$, $g(X) = b_m \prod_{j=1}^m (X - \beta_j)$, then their resultant

$$R(f,g) = a_n^m b_m^n \prod_{i,j} (\alpha_i - \beta_j).$$

In particular, $R(X - a, g(X)) = g(a)$.

If $f, g \in \mathcal{O}[X]$ then $R(f,g) \in \mathcal{O}$. We shall use the following easy to establish properties of the resultant: if $f \equiv f_1 \mod \mathcal{M}[X]$ then $R(f,g) \equiv R(f_1,g) \mod \mathcal{M}$; if $R(f,g) \in \mathcal{M}^s \setminus \mathcal{M}^{s+1}$ then $\mathcal{M}^s[X] \subset f\mathcal{O}[X] + g\mathcal{O}[X]$.

PROPOSITION. *Let F be a complete discrete valuation field with the ring of integers $\mathcal{O}$ and the maximal ideal $\mathcal{M}$. Let $g_0(X), h_0(X), f(X)$ be polynomials over $\mathcal{O}$ such that* $\deg f(X) = \deg g_0(X) + \deg h_0(X)$ *and the leading coefficient of $f(X)$ coincides with that of $g_0(X)h_0(X)$. Let $R(g_0, h_0) \notin \mathcal{M}^{s+1}$ and $f(X) \equiv g_0(X)h_0(X)$* $\mod \mathcal{M}^{2s+1}$ *for an integer $s \geqslant 0$.*

Then there exist polynomials $g(X), h(X)$ such that

$$\begin{aligned} &f(X) = g(X)h(X), \\ &\deg g(X) = \deg g_0(X), \quad g(X) \equiv g_0(X) \mod \mathcal{M}^{s+1}, \\ &\deg h(X) = \deg h_0(X), \quad h(X) \equiv h_0(X) \mod \mathcal{M}^{s+1}. \end{aligned}$$

Proof. We first construct polynomials $g_i(X), h_i(X) \in \mathcal{O}[X]$ with the following properties: $\deg(g_i - g_0) < \deg g_0$, $\deg(h_i - h_0) < \deg h_0$

$$g_i \equiv g_{i-1} \mod \mathcal{M}^{i+s}, \quad h_i \equiv h_{i-1} \mod \mathcal{M}^{i+s}, \quad f \equiv g_i h_i \mod \mathcal{M}^{i+2s+1}.$$

Proceeding by induction, we can assume that the polynomials $g_j(X), h_j(X)$, for $j \leqslant i-1$, have been constructed. For a prime element π put

$$g_i(X) = g_{i-1}(X) + \pi^{i+s} G_i(X), \quad h_i(X) = h_{i-1}(X) + \pi^{i+s} H_i(X)$$

with $G_i(X), H_i(X) \in \mathcal{O}[X]$, $\deg G_i(X) < \deg g_0(X)$, $\deg H_i(X) < \deg h_0(X)$. Then

$$g_i h_i - g_{i-1} h_{i-1} \equiv \pi^{i+s} \left(g_{i-1} H_i + h_{i-1} G_i \right) \mod \mathcal{M}^{i+2s+1}.$$

Since by the induction assumption $f(X) - g_{i-1}(X)h_{i-1}(X) = \pi^{i+2s} f_1(X)$ for a suitable $f_1(X) \in \mathcal{O}[X]$ of degree smaller than that of f, we deduce that it suffices for $G_i(X)$, $H_i(X)$ to satisfy the congruence

$$\pi^s f_1(X) \equiv g_{i-1}(X) H_i(X) + h_{i-1}(X) G_i(X) \mod \mathcal{M}^{s+1}.$$

However, $R(g_{i-1}(X), h_{i-1}(X)) \equiv R(g_0(X), h_0(X)) \not\equiv 0 \mod \mathcal{M}^{s+1}$. Then the properties of the resultant imply the existence of polynomials H_i, G_i satisfying the congruence. Now put $g(X) = \lim g_i(X), h(X) = \lim h_i(X)$ and get $f(X) = g(X)h(X)$. □

The following statement is often called *Hensel Lemma*. It was proved by Hensel for p-adic numbers (see 4.3 of Chapter 1), and by Rychlík for complete discrete valuation fields.

8.2. COROLLARY 1. *Let F be as in the Proposition and $\overline{F}$ the residue field of F. Suppose that $f(X), g_0(X), h_0(X)$ are monic polynomials with coefficients in $\mathcal{O}$ and $\overline{f}(X) = \overline{g}_0(X)\overline{h}_0(X)$. Suppose that $\overline{g}_0(X), \overline{h}_0(X)$ are relatively prime in $\overline{F}[X]$. Then there exist monic polynomials $g(X), h(X)$ with coefficients in $\mathcal{O}$, such that*

$$f(X) = g(X)h(X),\ \overline{g}(X) = \overline{g}_0(X),\ \overline{h}(X) = \overline{h}_0(X).$$

Proof. We have $R(f_0(X), g_0(X)) \notin \mathcal{M}$ and we can apply the previous Proposition for $s = 0$. The polynomials $g(X)$ and $h(X)$ may be assumed to be monic, as it follows from the proof of the Proposition. □

Valuation fields satisfying the assertion of Corollary 1 are said to be *Henselian*. Corollary 1 demonstrates that complete discrete valuation fields are Henselian.

COROLLARY 2. *Let F be a Henselian field and $f(X)$ a monic polynomial with coefficients in $\mathcal{O}$. Let $\overline{f}(X) \in \overline{F}[X]$ have a simple root β in $\overline{F}$. Then $f(X)$ has a simple root $\alpha \in \mathcal{O}$ such that $\overline{\alpha} = \beta$.*

Proof. Let $\gamma \in \mathcal{O}$ be such that $\overline{\gamma} = \beta$. Put $g_0(X) = X - \gamma$ in Corollary 1. □

8.3. COROLLARY 3. *Let F be a complete discrete valuation field. Let $f(X)$ be a monic polynomial with coefficients in $\mathcal{O}$. Let $f(\alpha_0) \in \mathcal{M}^{2s+1}, f'(\alpha_0) \notin \mathcal{M}^{s+1}$ for some $\alpha_0 \in \mathcal{O}$ and integer $s \geqslant 0$. Then there exists $\alpha \in \mathcal{O}$ such that $\alpha - \alpha_0 \in \mathcal{M}^{s+1}$ and $f(\alpha) = 0$.*

Proof. Put $g_0(X) = X - \alpha_0$ and write $f(X) = f_1(X)(X - \alpha_0) + \delta$ with $\delta \in \mathcal{O}$. Then $\delta \in \mathcal{M}^{2s+1}$. Put $h_0(X) = f_1(X) \in \mathcal{O}[X]$. Hence $f(X) \equiv g_0(X)h_0(X) \mod \mathcal{M}^{2s+1}$ and $f'(\alpha_0) = h_0(\alpha_0) \notin \mathcal{M}^{s+1}$. This means that $R(g_0(X), h_0(X)) \notin \mathcal{M}^{s+1}$, and the Proposition implies the existence of polynomials $g(X), h(X) \in \mathcal{O}[X]$ such that $g(X) = X - \alpha, \alpha \equiv \alpha_0 \mod \mathcal{M}^{s+1}$, and $f(X) = g(X)h(X)$. □

COROLLARY 4. *Let F be a complete discrete valuation field. For every positive integer m whose image in F is not zero there is n such that $1+\mathcal{M}^n \subset F^{\times m}$.*

Proof. Put $f_a(X) = X^m - a$ with $a \in 1+\mathcal{M}^n$. Let $m \in \mathcal{M}^s \setminus \mathcal{M}^{s+1}$. Then $f_a'(1) \in \mathcal{M}^s \setminus \mathcal{M}^{s+1}$. Therefore for every $a \in 1+\mathcal{M}^{2s+1}$ due to Corollary 3 the polynomial $f_a(X)$ has a root $\alpha \equiv 1 \mod \mathcal{M}^{s+1}$. □

8.4. The following assertion is useful.

LEMMA. *Let F be a complete discrete valuation field and let*

$$f(X) = X^n + \alpha_{n-1}X^{n-1} + \cdots + \alpha_0$$

be an irreducible polynomial with coefficients in F. Then the condition $v(\alpha_0) \geqslant 0$ implies $v(\alpha_i) \geqslant 0$ for $0 \leqslant i \leqslant n-1$.

Proof. Assume that $\alpha_0 \in \mathcal{O}$ and that j is the maximal integer such that $v(\alpha_j) = \min_{0 \leqslant i \leqslant n-1} v(\alpha_i)$. If $\alpha_j \notin \mathcal{O}$, then put

$$\begin{aligned} f_1(X) &= \alpha_j^{-1} f(X), \\ g_0(X) &= X^j + \alpha_j^{-1}\alpha_{j-1}X^{j-1} + \cdots + \alpha_j^{-1}\alpha_0, \\ h_0(X) &= \alpha_j^{-1}X^{n-j} + 1 \end{aligned}$$

We have $\overline{f}_1(X) = \overline{g}_0(X)\overline{h}_0(X)$, and $\overline{g}_0(X), \overline{h}_0(X)$ are relatively prime. Therefore, by Proposition 8.1, $f_1(X)$ and $f(X)$ are not irreducible. □

9. Extensions of Valuation Fields

9.1. Let F be a field and L an extension of F with a valuation $w\colon L \longrightarrow \Gamma'$. Then w induces the valuation $w_0 = w|_F\colon F \longrightarrow \Gamma'$ on F. In this context L/F is said to be an *extension of valuation fields*. The group $w_0(F^\times)$ is a totally ordered subgroup of $w(L^\times)$ and the index of $w_0(F^\times)$ in $w(L^\times)$ is called the *ramification index* $e(L/F, w)$. The ring of integers $\mathcal{O}_{w_0}$ is a subring of the ring of integers $\mathcal{O}_w$ and the maximal ideal $\mathcal{M}_{w_0}$ coincides with $\mathcal{M}_w \cap \mathcal{O}_{w_0}$. Hence, the residue field $\overline{F}_{w_0}$ can be considered as a subfield of the residue field $\overline{L}_w$. Therefore, if α is an element of $\mathcal{O}_{w_0}$, then its residue in the field $\overline{F}_{w_0}$ can be identified with the image of α as an element of $\mathcal{O}_w$ in the field $\overline{L}_w$. We shall denote this image of α by $\overline{\alpha}$. The degree of the extension $\overline{L}_w/\overline{F}_{w_0}$ is called the *inertia degree* or *residue degree* $f(L/F, w)$. An immediate consequence is the following Lemma.

LEMMA. *Let L be an extension of F and let w be a valuation on L. Let $L \supset M \supset F$ and let w_0 be the induced valuation on M. Then*

$$e(L/F,w) = e(L/M,w)e(M/F,w_0),$$
$$f(L/F,w) = f(L/M,w)f(M/F,w_0).$$

9.2. Assume that L/F is a finite extension and w_0 is a discrete valuation. Let elements $\alpha_1,\dots,\alpha_e \in L^\times$ $e \leqslant e(L/F,w)$ be such that $w(\alpha_1)+w(F^\times),\dots,w(\alpha_e)+w(F^\times)$ are distinct in $w(L^\times)/w(F^\times)$. If $\sum_{i=1}^e c_i\alpha_i = 0$ holds with $c_i \in F$, then, as $w(c_i\alpha_i)$ are all distinct, we get

$$w\Big(\sum_{i=1}^{e} c_i\alpha_i\Big) = \min_{1\leqslant i\leqslant e} w(c_i\alpha_i) \quad \text{and} \quad c_i = 0 \quad \text{for } 1 \leqslant i \leqslant e.$$

This shows that $\alpha_1,\dots,\alpha_e$ are linearly independent over F and hence $e(L/F,w)$ is finite. Let π be a prime element with respect to w_0. Then we deduce that there are only a finite number of positive elements in $w(L^\times)$ which are $\leqslant w(\pi)$. Consider the smallest positive element in $w(L^\times)$. It generates the group $w(L^\times)$, and we conclude that w is a discrete valuation. Thus, we have proved the following result.

LEMMA. *Let L/F be a finite extension and w_0 discrete for a valuation w on L. Then w is discrete.*

9.3. Hereafter we shall consider discrete valuations. Let F and L be fields with discrete valuations v and w respectively and $F \subset L$. The valuation w is said to be an *extension of the valuation* v, if the topology defined by w_0 is equivalent to the topology defined by v. We shall write $w|v$ and use the notations $e(w|v)$, $f(w|v)$ instead of $e(L/F,w)$, $f(L/F,w)$. If $\alpha \in F$ then $w(\alpha) = e(w|v)v(\alpha)$.

LEMMA. *Let L be a finite extension of F of degree n; then*

$$e(w|v)f(w|v) \leqslant n.$$

Proof. Let $e = e(w|v)$ and let f be a positive integer such that $f \leqslant f(w|v)$. Let $\theta_1,\dots,\theta_f$ be elements of $\mathcal{O}_w$ such that their residues in $\overline{L}_w$ are linearly independent over $\overline{F}_v$. It suffices to show that $\{\theta_i\pi_w^j\}$ are linearly independent over F for $1 \leqslant i \leqslant f, 0 \leqslant j \leqslant e-1$. Assume that

$$\sum_{i,j} c_{ij}\theta_i\pi_w^j = 0$$

for $c_{ij} \in F$ and not all $c_{ij} = 0$.

Multiplying the coefficients c_{ij} by a suitable power of π_v, we may assume that $c_{ij} \in \mathcal{O}_v$ and not all $c_{ij} \in \mathcal{M}_v$. Note that if $\sum_i c_{ij}\theta_i \in \mathcal{M}_w$, then $\sum_i \overline{c}_{ij}\overline{\theta}_i = 0$ and $c_{ij} \in \mathcal{M}_v$. Therefore, there exists an index j such that $\sum_i c_{ij}\theta_i \notin \mathcal{M}_w$. Let j_0 be the minimal such index. Then $j_0 = w(\sum c_{ij}\theta_i\pi_w^j)$, which is impossible. We conclude that all $c_{ij} = 0$. Hence, $ef \leqslant n$ and $e(w|v)f(w|v) \leqslant n$. □

For instance, let $\widehat{F}$ be the completion of a discrete valuation field F with the discrete valuation $\widehat{v}$. Then $e(\widehat{v}|v) = 1, f(\widehat{v}|v) = 1$. Note that if F is not complete, then $|\widehat{F} : F| \neq e(\widehat{v}|v)f(\widehat{v}|v)$. On the contrary, in the case of complete discrete valuation fields we have

9.4. PROPOSITION. *Let L be an extension of F and let F, L be complete with respect to discrete valuations v, w. Let $w|v, f = f(w|v)$ and $e = e(w|v) < \infty$. Let $\pi_w \in L$ be a prime element with respect to w and $\theta_1, \dots, \theta_f$ elements of $\mathcal{O}_w$ such that their residues form a basis of $\overline{L}_w$ over $\overline{F}_v$. Then $\{\theta_i\pi_w^j\}$ is a basis of the F-space L and of the $\mathcal{O}_v$-module $\mathcal{O}_w$, with $1 \leqslant i \leqslant f, 0 \leqslant j \leqslant e-1$. If $f < \infty$, then L/F is a finite extension of degree*

$$n = ef.$$

Proof. Let R be a set of representatives for F. Then the set

$$R' = \{\sum_{i=1}^{f} a_i\theta_i : a_i \in R \text{ and almost all } a_i = 0\}$$

is the set of representatives for L. For a prime element π_v with respect to v put $\pi_m = \pi_v^k\pi_w^j$, where $m = ek + j, 0 \leqslant j < e$. Using Proposition 4.2 we obtain that an element $\alpha \in L$ can be expressed as a convergent series

$$\alpha = \sum_m \eta_m\pi_m \qquad \text{with} \qquad \eta_m \in R'.$$

Writing

$$\eta_m = \sum_{i=1}^{f} \eta_{m,i}\theta_i \qquad \text{with} \qquad \eta_{m,i} \in R,$$

we get

$$\alpha = \sum_{i,j}\big(\sum_k \eta_{ek+j,i}\pi_v^k\big)\,\theta_i\pi_w^j.$$

Thus, α can be expressed as $\sum \rho_{i,j}\theta_i\pi_w^j$ with

$$\rho_{i,j} = \sum_k \eta_{ek+j,i}\pi_v^k \in F, \quad 1 \leqslant i \leqslant f, 0 \leqslant j \leqslant e-1.$$

By the proof of the previous Lemma this expression for α is unique. We conclude that $\left\{\theta_i\pi_w^j\right\}$ form a basis of L over F and of $\mathcal{O}_w$ over $\mathcal{O}_v$. □

9.5. Further we shall assume that $v(F^\times) = \mathbb{Z}$ for a discrete valuation v. Then $e(w|v) = |\mathbb{Z} : w(F^\times)|$ for an extension w of v.

THEOREM. *Let F be a complete field with respect to a discrete valuation v and L a finite extension of F. Then there is precisely one extension w on L of the valuation v and $w = \frac{1}{f} v \circ N_{L/F}$ with $f = f(w|v)$. The field L is complete with respect to w.*

Proof. Let $w' = v \circ N_{L/F}$. First we verify that w' is a valuation on L. It is clear that $w'(\alpha) = +\infty$ if and only if $\alpha = 0$ and $w'(\alpha\beta) = w'(\alpha) + w'(\beta)$. Assume that $w'(\alpha) \geqslant w'(\beta)$ for $\alpha, \beta \in L^\times$, then

$$w'(\alpha+\beta) = w'(\beta) + w'(1 + \frac{\alpha}{\beta})$$

and it suffices to show that if $w'(\gamma) \geqslant 0$, then $w'(1+\gamma) \geqslant 0$. Let

$$f(X) = X^m + a_{m-1}X^{m-1} + \cdots + a_0$$

be the monic irreducible polynomial of γ over F. Then $(-1)^m a_0 = N_{F(\gamma)/F}(\gamma)$ and if $s = |L : F(\gamma)|$, then $((-1)^m a_0)^s = N_{L/F}(\gamma)$. We deduce that $v(a_0) \geqslant 0$, and making use of 8.4, we get $v(a_i) \geqslant 0$ for $0 \leqslant i \leqslant m-1$. However,

$$(-1)^m N_{F(\gamma)/F}(1+\gamma) = f(-1) = (-1)^m + a_{m-1}(-1)^{m-1} + \cdots + a_0,$$

hence

$$v\left(N_{F(\gamma)/F}(1+\gamma)\right) \geqslant 0 \quad \text{and} \quad v\left(N_{L/F}(1+\gamma)\right) \geqslant 0,$$

i.e., $w'(1+\gamma) \geqslant 0$. Thus, we have shown that w' is a valuation on L.

Let $n = |L : F|$; then $w'(\alpha) = nv(\alpha)$ for $\alpha \in F^\times$. Hence, the valuation $(1/n)w'$ is an extension of v to L (note that $(1/n)w'(L^\times) \neq \mathbb{Z}$ in general). Let $e = e(L/F, (1/n)w')$. By Lemma 9.3 e is finite. Put $w = (e/n)w' : L^\times \to \mathbb{Q}$, hence $w(L^\times) = w(\pi_w)\mathbb{Z} = \mathbb{Z}$ with a prime element π_w with respect to w. Therefore, $w = (e/n)v \circ N_{L/F}$ is at once a discrete valuation on L and an extension of v.

Let $\gamma_1, \dots, \gamma_n$ be a basis of the F-vector space L. By induction on r, $1 \leqslant r \leqslant n$, we shall show that

$$\sum_{i=1}^{r} a_i^{(m)} \gamma_i \to 0, \quad m \to \infty \Longleftrightarrow a_i^{(m)} \to 0 \quad m \to \infty \quad \text{for } i = 1, \dots, r$$

where $a_i^{(m)} \in F$.

The left arrow and the case $r = 1$ are clear. For the induction step we can assume that $a_1^{(m)} \not\to 0$. Therefore we can assume that $v(a_1^{(m)})$ is bounded. Hence

$$\gamma_1 + \sum_{i=2}^{r} b_i^{(m)} \gamma_i = \left(a_1^{(m)}\right)^{-1} \sum_{i=1}^{r} a_i^{(m)} \gamma_i \to 0,$$

where $b_i^{(m)} = (a_1^{(m)})^{-1} a_i^{(m)}$. Then $\sum_{i=2}^r (b_i^{(m)} - b_i^{(m+1)})\gamma_i \to 0$, and the induction hypothesis shows that $b_i^{(m)} - b_i^{(m+1)} \to 0$ for $i = 2, \dots, r$. Thus, each $(b_i^{(m)})_m$ converges to, say, $b_i \in F$. So the sequence $\gamma_1 + \sum_{i=2}^r b_i^{(m)} \gamma_i$ converges both to 0 and to $\gamma_1 + \sum_{i=2}^r b_i \gamma_i$, so

$$0 = \gamma_1 + \sum_{i=2}^r b_i \gamma_i$$

which contradicts the choice of γ_i.

Similarly one shows that a sequence $\sum_{i=1}^r a_i^{(m)} \gamma_i$ is fundamental if and only if $a_i^{(m)}$ is fundamental for each $i = 1, \dots, r$.

Thus, the completeness of F implies the completeness of its finite extension L with respect to any extension of v. We also have the uniqueness of the extension. □

9.6. Now we treat extensions of discrete valuations in the general case.

THEOREM. *Let F be a field with a discrete valuation v. Let $\widehat{F}$ be the completion of F, and $\widehat{v}$ the discrete valuation of $\widehat{F}$. Suppose that $L = F(\alpha)$ is a finite extension of F and $f(X)$ the monic irreducible polynomial of α over F. Let $f(X) = \prod_{i=1}^k g_i(X)^{e_i}$ be the decomposition of the polynomial $f(X)$ into irreducible monic factors in $\widehat{F}[X]$. For a root α_i of the polynomial $g_i(X)$ $(\alpha_1 = \alpha)$ put $L_i = \widehat{F}(\alpha_i)$. Let $\widehat{w}_i$ be the discrete valuation on L_i, the unique extension of $\widehat{v}$.*

Then L is embedded as a dense subfield in the complete discrete valuation field L_i under $F \hookrightarrow \widehat{F}$, $\alpha \to \alpha_i$, and the restriction w_i of $\widehat{w}_i$ on L is a discrete valuation on L which extends v. The valuations w_i are distinct and every discrete valuation which is an extension of v to L coincides with some w_i for $1 \leqslant i \leqslant k$.

Proof. First let w be a discrete valuation on L which extends v. Let $\widehat{L}_w$ be the completion of L with respect to w. By Proposition 3.2 there exists an embedding $\sigma\colon \widehat{F} \longrightarrow \widehat{L}_w$ over F. As $\alpha \in \widehat{L}_w$, we get $\sigma(\widehat{F})(\alpha) \subset \widehat{L}_w$. Since $\sigma(\widehat{F})(\alpha)$ is a finite extension of $\sigma(\widehat{F})$, Theorem 9.5 shows that $\sigma(\widehat{F})(\alpha)$ is complete. Therefore, $\widehat{L}_w \subset \sigma(\widehat{F})(\alpha)$ and, moreover, $\widehat{L}_w = \sigma(\widehat{F})(\alpha)$. Let $g(X)$ be the monic irreducible polynomial of α over $\sigma(\widehat{F})$. Then $\sigma^{-1}g(X)$ divides $f(X)$ and $\sigma^{-1}g(X) = g_i(X)$ for some $1 \leqslant i \leqslant k$, $w = w_i$.

Conversely, assume that $g(X) = g_i(X)$ and $\widehat{w}_i$ is the unique discrete valuation on $L_i = \widehat{F}(\alpha_i)$ which extends $\widehat{v}$. Since F is dense in $\widehat{F}$, we deduce that the image of L is dense in L_i and w_i extends v.

If $w_i = w_j$ for $i \neq j$ then there is an isomorphism between $\widehat{F}(\alpha_i)$ and $\widehat{F}(\alpha_j)$ over $\widehat{F}$ which sends α_i to α_j, but this is impossible. □

COROLLARY. *Let L/F be a purely inseparable finite extension. Then there is precisely one extension to L of the discrete valuation v of F.*

Proof. Assume $L = F(\alpha)$. Then $f(X)$ is decomposed as $(X-\alpha)^{p^m}$ in the fixed algebraic closure F^{alg} of F. Therefore, $k=1$ and there is precisely one extension of v to L. If there were two distinct extensions w_1, w_2 of v to L in the general case of a purely inseparable extension L/F, we would find $\alpha \in L$ such that $w_1(\alpha) \neq w_2(\alpha)$, and hence the restriction of w_1 and w_2 on $F(\alpha)$ would be distinct. This leads to contradiction. □

9.7. REMARKS.

1. More precisely, the Theorem should be formulated as follows.

The tensor product $L \otimes_F \widehat{F}$ may be treated as an L-module and $\widehat{F}$-algebra. Then the quotient of $L \otimes_F \widehat{F}$ by its radical decomposes into the direct sum of complete fields which correspond to the discrete valuations on L that are extensions of v. Under the conditions of the Theorem $L \otimes_F \widehat{F} = \widehat{F}[X]/(f(X))$, and we have the surjective homomorphism

$$L \otimes_F \widehat{F} = \widehat{F}[X]/(f(X)) \longrightarrow \bigoplus_{i=1}^{k} \widehat{F}[X]/(g_i(X)) \simeq \bigoplus_{i=1}^{k} \widehat{F}(\alpha_i) = \bigoplus_{w_i|v} \widehat{L}_{w_i}$$

with the kernel $\left(\prod_{i=1}^k g_i(X)\right)\widehat{F}[X]/f(X)\widehat{F}[X]$, where $\widehat{L}_{w_i} = \widehat{F}(\alpha_i)$. Note that this kernel coincides with the radical of $L \otimes_F \widehat{F}$. Under the conditions of the previous Theorem, if L/F is separable, then all e_i are equal to 1 and the kernel is trivial.

2. Assume that L/F is as in the Theorem and, in addition, L/F is Galois. Then $\widehat{F}(\alpha_i)/\widehat{F}$ is Galois. Let $G = \mathrm{Gal}(L/F)$. Note that if w is a valuation on L, then $w \circ \sigma$ is a valuation on L for $\sigma \in G$. Put

$$H_i = \{\sigma \in G : w_1 \circ \sigma = w_i\} \quad \text{for } 1 \leqslant i \leqslant k.$$

Then it is easy to show that G is a disjoint union of the H_i and $H_i = H_1\sigma_i$ for $\sigma_i \in H_i$. Theorem 9.6 implies that H_i coincides with $\{\sigma \in G : \sigma g_i(X) = g_1(X)\}$, whence $\{\sigma \in G : \sigma g_i(X) = g_i(X)\} = \sigma_i^{-1} H_1 \sigma_i$. Then $\deg g_i(X) = \deg g_1(X)$, $e_i = 1$. The subgroup H_1 is said to be the *decomposition group* of w_1 over F. The fixed field $M = L^{H_1}$ is said to be the decomposition field of w_1 over F. Note that the field M is obtained from F by adjoining coefficients of the polynomial $g_1(X)$. We get $L = M(\alpha_1)$, and $g_1(X) \in M[X]$ is irreducible over $\widehat{F} = \widehat{M}$. Theorem 9.6 shows that w_1 is the unique extension to L of $w_1|_M$; there are k distinct discrete valuations on M which extend v.

EXAMPLE. Let $E = F(X)$. Recall that the discrete valuations on E which are trivial on F are in one-to-one correspondence with irreducible monic polynomials

$p(X)$ over F: $p(X) \to v_{p(X)}$, $v \to p_v(X)$ and there is the valuation $-\deg$ with a prime element $\frac{1}{X}$. If a_n is the leading coefficient of $f(X)$, then

$$f(X) = a_n \prod_{v \neq -\deg} p_v(X)^{v(f(X))}.$$

Let F_1 be an extension of F. Then a discrete valuation on $E_1 = F_1(X)$, trivial on F_1, is an extension of some discrete valuation on $E = F(X)$, trivial on F. Let $p(X) = p_v(X)$ be an irreducible monic polynomial over F. Let $p(X)$ be decomposed into irreducible monic factors over $F_1 : p(X) = \prod_{i=1}^{k} p_i(X)^{e_i}$. Then one immediately deduces that the $w_i = w_{p_i(X)}$, $1 \leqslant i \leqslant k$, are all discrete valuations, trivial on F_1, which extend the valuation $v_{p(X)}$. We also have $e\left(w_{p_i(X)} | v_{p(X)}\right) = e_i$. There is precisely one extension w of $-\deg$. Thus, for every v

$$p_v(X) = \prod_{w_i | v} p_{w_i}(X)^{e(w_i|v)}$$

and we have the surjective homomorphism $F(\alpha) \otimes_F F_1 \longrightarrow \bigoplus F_1(\alpha_i)$, where α is a root of $p(X)$ and α_i is a root of $p_i(X)$. Here the kernel of this homomorphism also coincides with the radical of $F(\alpha) \otimes_F F_1$.

9.8. Finally we treat extensions of Henselian discrete valuation fields.

LEMMA. (*Gauß*) *Let F be a discrete valuation field, $\mathcal{O}$ its ring of integers. Then if a polynomial $f(X) \in \mathcal{O}[X]$ is not irreducible in $F[X]$, it is not irreducible in $\mathcal{O}[X]$.*

Proof. Assume that $f(X) = g(X)h(X)$ with $g(X), h(X) \in F[X]$. Let

$$g(X) = \sum_{i=0}^{n} b_i X^i, \quad h(X) = \sum_{i=0}^{m} c_i X^i, \quad f(X) = \sum_{i=0}^{n+m} a_i X^i.$$

Let

$$j_1 = \min\left\{i : v(b_i) = \min_{0 \leqslant k \leqslant n} v(b_k)\right\}, \quad j_2 = \min\left\{i : v(c_i) = \min_{0 \leqslant k \leqslant m} v(c_k)\right\}.$$

Then $v(b_i c_{j_1+j_2-i}) > v(b_{j_1} c_{j_2})$ for $i \neq j_1$; hence $v(a_{j_1+j_2}) = v(b_{j_1}) + v(c_{j_2})$. If $c = v(b_{j_1}) < 0$, then we obtain $v(c_{j_2}) \geqslant -v(b_{j_1})$, and one can write $f(X) = (\pi^{-c} g(X))(\pi^c h(X))$, as desired. □

THEOREM. *Let v be a discrete valuation on F. The following conditions are equivalent:*

(i) *F is a Henselian field with respect to v.*

(ii) *The discrete valuation v has a unique extension to every finite algebraic extension L of F.*

(iii) *If L is a finite separable extension of F of degree n, then*

$$n = e(w|v)f(w|v),$$

where w is an extension of v on L.

(iv) *F is separably closed in* $\widehat{F}$.

Proof.

(i)⇒(ii). Using Corollary 9.6, we can assume that L/F is separable. Moreover, it suffices to verify (ii) for the case of a Galois extension. Let $L = F(\alpha)$ be Galois, $f(X)$ be the irreducible polynomial of α over F. Let $f = g_1 \dots g_k$ be the decomposition of f over $\widehat{F}$ as in 9.6. Let H_1 and $M = L^{H_1}$ be as therein. Put $w_i' = w_i|_M$ for $1 \leqslant i \leqslant k$ and suppose that $k \geqslant 2$. Since w_1 is the discrete valuation on L, which is the unique extension of w_1', we conclude that the topology induced by w_1' is not equivalent to the topology induced by w_i' for $2 \leqslant i \leqslant k$. We get $w_i' = w_1 \circ \sigma_i|_M$ for $\sigma_1, \dots, \sigma_l \in G, \sigma_1 = 1$. Taking into account the proof of Proposition 2.8, one can find an element $\beta \in M$ such that

$$-c = w_1'(\beta) < 0, \quad w_2'(\beta) > c, \dots, \quad w_k'(\beta) > c.$$

Let $\tau_1, \dots, \tau_r$ ($\tau_1 = 1$) be the maximal set of elements of $G = \mathrm{Gal}(L/F)$ for which the elements $\beta, \tau_2(\beta), \dots, \tau_r(\beta)$ are distinct. Then $\tau_2, \dots, \tau_r \notin H_1$, and $w_1(\beta) = -c$, $w_1(\tau_i(\beta)) > c$ for $2 \leqslant i \leqslant r$.

Let $h(X) = X^r + b_{r-1}X^{r-1} + \dots + b_0$ be the irreducible monic polynomial of β over F. Then

$$w_1(b_0) = \sum_{i=1}^{r} w_1(\tau_i(\beta)) > 0.$$

Similarly one checks that $w_1(b_i) > 0$ for $i < r-1$. We also obtain that

$$w_1(b_{r-1}) = \min_{1 \leqslant i \leqslant r} w_1(\tau_i(\beta)) = -c < 0.$$

Hence, $v(b_i) > 0$ for $0 \leqslant i < r-1$ and $v(b_{r-1}) < 0$. Put $h_1(X) = b_{r-1}^{-r} h(b_{r-1}X)$. Then $h_1(X)$ is a monic polynomial with integer coefficients. Since $\overline{h}_1(X) = (X+1)X^{r-1}$, by the Hensel Lemma 8.2, we obtain that $h_1(X)$ is not irreducible, implying the same for $h(X)$, and we arrive at a contradiction. Thus, $k = 1$, and the discrete valuation v is uniquely extended on L.

(ii)⇒(iii). Let $L = F(\alpha)$ be a finite separable extension of F and let L/F be of degree n. Since v has the unique extension w to L, we deduce from Theorem 9.6 that $f(X) = g_1(X)$ is the decomposition of the irreducible monic polynomial $f(X)$ of α over F in $\widehat{F}[X]$. Therefore, the extension $\widehat{F}(\alpha)/\widehat{F}$ is of degree n. We have also $e(w|v) = e(\widehat{w}|\widehat{v})$, $f(w|v) = f(\widehat{w}|\widehat{v})$, because $e(\widehat{w}|w) = 1$, $f(\widehat{w}|w) = 1$, $e(\widehat{v}|v) = 1$,

$f(\widehat{v}|v) = 1$; see 9.3. Proposition 9.4 shows that $n = e(\widehat{w}|\widehat{v})f(\widehat{w}|\widehat{v})$. Hence $n = e(w|v)f(w|v)$.

(iii)⇒(iv). Let $\alpha \in \widehat{F}$ be separable over F. Put $L = F(\alpha)$ and $n = |L : F|$. Let w be the discrete valuation on L which induces the same topology on L as $\widehat{v}|_L$. Then $e(w|v) = f(w|v) = 1$, and hence $n = 1, \alpha \in F$.

(iv)⇒(i). Let $f(X), g_0(X), h_0(X)$ be monic polynomials with coefficients in $\mathcal{O}$. Let $\overline{f}(X) = \overline{g}_0(X)\overline{h}_0(X)$ and $\overline{g}_0(X), \overline{h}_0(X)$ be relatively prime in $\overline{F}_v[X]$. The field $\widehat{F}$ is Henselian according to 8.1. Then there exist monic polynomials $g(X)$, $h(X)$ over the ring of integers $\widehat{\mathcal{O}}$ in $\widehat{F}$, such that $f(X) = g(X)h(X)$ and $\overline{g}(X) = \overline{g}_0(X), \overline{h}(X) = \overline{h}_0(X)$. The polynomials $g_0(X), h_0(X)$ are relatively prime in $\mathcal{O}[X]$ because their residues possess this property. Consequently, they are relatively prime in $F[X]$ by the previous Lemma. The roots of the polynomial $f(X)$ are algebraic over F, hence the roots of the polynomials $g(X), h(X)$ are algebraic over F and the coefficients of $g(X), h(X)$ are algebraic over F. Since F is separably closed in $\widehat{F}$, we obtain that $g(X)^{p^m}, h(X)^{p^m} \in F[X]$ for some $m \geqslant 0$. Then $f(X)^{p^m}$ is the product of two relatively prime polynomials in $F[X]$. We conclude that $g(X)^{p^m} = g_1(X)^{p^m}$ and $h(X)^{p^m} = h_1(X)^{p^m}$ for some polynomials $g_1(X), h_1(X) \in F[X]$ and, finally, the polynomial $g(X)$ coincides with $g_1(X) \in \mathcal{O}[X]$, the polynomial $h(X)$ coincides with $h_1(X) \in \mathcal{O}[X]$. □

9.9. COROLLARY 1. *Let F be a Henselian discrete valuation field and L an algebraic extension of F. Then there is precisely one valuation $w: L^\times \longrightarrow \mathbb{Q}$ (not necessarily discrete), such that the restriction $w|_F$ coincides with the discrete valuation v on F. Moreover, L is Henselian with respect to w.*

Proof. Let M/F be a finite subextension of L/F, and let, in accordance with the previous Theorem, $w_M: M^\times \longrightarrow \mathbb{Q}$ be the unique valuation on M for which $w_M|_F = v$. For $\alpha \in L^\times$ we put $w(\alpha) = w_M(\alpha)$ with $M = F(\alpha)$. It is a straightforward exercise to verify that w is a valuation on L and that $w|_F = v$. If there were another valuation w' on L with the property $w'|_F = v$, we would find $\alpha \in L$ with $w(\alpha) \neq w'(\alpha)$, and hence $w|_{F(\alpha)}$ and $w'|_{F(\alpha)}$ would be two distinct valuations on $F(\alpha)$ with the property $w|_F = w'|_F = v$. Therefore, there exists exactly one valuation w on L for which $w|_F = v$. To show that L is Henselian we note that polynomials $f(X) \in \mathcal{O}_w[X], g_0(X) \in \mathcal{O}_w[X], h_0(X) \in \mathcal{O}_w[X]$ belong in fact to $\mathcal{O}_1[X]$, where $\mathcal{O}_1$ is the ring of integers for some finite subextension M/F in L/F. Clearly, the polynomials $\overline{g}_0(X), \overline{h}_0(X)$ are relatively prime in $\overline{M}_{w_M}[X]$, hence there exist polynomials $g(X), h(X) \in \mathcal{O}_1[X]$, such that $f(X) = g(X)h(X)$, $\overline{g}(X) = \overline{g}_0(X)$ and $\overline{h}(X) = \overline{h}_0(X)$. □

COROLLARY 2. *Let F be a Henselian discrete valuation field, and let L/F be a finite separable extension. Let v be the valuation on F and w the extension of v to L. Let $e, f, \pi_w, \theta_1, \ldots, \theta_f$ be as in Proposition 2.4. Then $\theta_i \pi_w^j$ is a basis of the F-space L and of the $\mathcal{O}_v$-module $\mathcal{O}_w$, with $1 \leqslant i \leqslant f, 0 \leqslant j \leqslant e-1$. In particular, if $e = 1$, then*

$$\mathcal{O}_w = \mathcal{O}_v[\{\theta_i\}], \quad L = F(\{\theta_i\}),$$

and if $f = 1$, then

$$\mathcal{O}_w = \mathcal{O}_v[\pi_w], \quad L = F(\pi_w).$$

Proof. One can show, similarly to the proof of Lemma 2.3, that the elements $\theta_i \pi_w^j$ for $1 \leqslant i \leqslant f, 0 \leqslant j \leqslant e-1$ are linearly independent over F. As $n = ef$, these elements form a basis of $\mathcal{O}_w$ over $\mathcal{O}_v$ and of L over F. □

COROLLARY 3. *Let F be a Henselian discrete valuation field, and L/F a finite separable extension. Let w be the discrete valuation on L and $\sigma: L \longrightarrow F^{\mathrm{alg}}$ an embedding over F. Then $w \circ \sigma^{-1}$ is the discrete valuation on σL and $\mathcal{M}_{\sigma L} = \sigma \mathcal{M}_L, \mathcal{O}_{\sigma L} = \sigma \mathcal{O}_L$.*

COROLLARY 4. *If F is a Henselian discrete valuation field, then Proposition 8.1, Corollary 3 and 4 of 8.3, and Lemma 8.4 hold for F.*

Proof. In terms of Proposition 8.1 we obtain that there exist polynomials $g, h \in \widehat{\mathcal{O}}[X]$ (where $\widehat{\mathcal{O}}$ is the ring of integers of $\widehat{F}$), such that $f = gh$, $g \equiv g_0 \mod \widehat{\mathcal{M}}^{s+1}$, $h \equiv h_0 \mod \widehat{\mathcal{M}}^{s+1}$, $\deg g = \deg g_0$, $\deg h = \deg h_0$ (where $\widehat{\mathcal{M}}$ is the maximal ideal of $\widehat{\mathcal{O}}$). Proceeding now analogously to the part (iv)⇒(i) of the proof of Theorem 9.8, we conclude that g^{p^m} and h^{p^m} belong to $\mathcal{O}[X]$ for some $m \geqslant 0$. As $g_0(X), h_0(X)$ are relatively prime in $F[X]$ because $R(g_0(X), h_0(X)) \neq 0$, we obtain that $g(X) = g_0(X), h(X) = h_0(X)$ and Proposition 8.1 holds for F. Corollary 3 of 8.3 and Lemma 8.4 for F are formally deduced from the latter. □

The separable closure of F in $\widehat{F}$ is called the *Henselisation* of F (this is a least Henselian field containing F). For example, the separable closure of $\mathbb{Q}$ in $\mathbb{Q}_p$ is a Henselian countable field with respect to the p-adic valuation.

10. Unramified and Ramified Extensions

The field F has the unique surjective discrete valuation $F^\times \longrightarrow \mathbb{Z}$ with respect to which it is Henselian; we shall denote it from now on by v_F.

Let L/F be an algebraic extension. If v_L is the unique discrete valuation on L which extends the valuation $v = v_F$ on F, then we shall write $e(L|F), f(L|F)$

instead of $e(v_L|v_F)$, $f(v_F|v_F)$. We shall write $\mathcal{O}$ or $\mathcal{O}_F, \mathcal{M}$ or $\mathcal{M}_F, U$ or U_F, π or $\pi_F, \overline{F}$ for the ring of integers $\mathcal{O}_v$, the maximal ideal $\mathcal{M}_v$, the group of units U_v, a prime element π_v with respect to v, and the residue field $\overline{F}_v$, respectively.

10.1. LEMMA. *Let L/F be a finite extension. Let $\alpha \in \mathcal{O}_L$ and let $f(X)$ be the monic irreducible polynomial of α over F. Then $f(X) \in \mathcal{O}_F[X]$. Conversely, let $f(X)$ be a monic polynomial with coefficients in $\mathcal{O}_F$. If $\alpha \in L$ is a root of $f(X)$, then $\alpha \in \mathcal{O}_L$.*

Proof. It is well known that $\beta = \alpha^{p^m}$ is separable over F for some $m \geqslant 0$. Let M be a finite Galois extension of F with $\beta \in M$. Then, in fact, $\beta \in \mathcal{O}_M$ and the monic irreducible polynomial $g(X)$ of β over F can be written as

$$g(X) = \prod_{i=1}^{r}(X - \sigma_i\beta), \quad \sigma_i \in \operatorname{Gal}(M/F),\ \sigma_1 = 1.$$

Since $\beta \in \mathcal{O}_M$ we get $\sigma_i\beta \in \mathcal{O}_M$ using Corollary 3 of 9.9. Hence we obtain $g(X) \in \mathcal{O}_F[X]$ and $f(X) = g\left(X^{p^m}\right) \in \mathcal{O}_F[X]$. If $\alpha \in L$ is a root of the polynomial $f(X) = X^n + a_{n-1}X^{n-1} + \cdots + a_0 \in \mathcal{O}_F[X]$ and $\alpha \notin \mathcal{O}_L$, then $1 = -a_{n-1}\alpha^{-1} - \cdots - a_0\alpha^{-n} \in \mathcal{M}_L$, contradiction. Thus, $\alpha \in \mathcal{O}_L$. $\square$

A finite extension L of a Henselian discrete valuation field F is called *unramified* if $\overline{L}/\overline{F}$ is a separable extension of the same degree as L/F. We deduce from 9.4 that if L/F is unramified then $e(L|F) = 1, f(L|F) = |L : F|$.

A finite extension L/F is called *totally ramified* if $f(L|F) = 1$.

A finite totally ramified extension L/F is called *wildly ramified* if $p|e(L|F)$ where $p = \operatorname{char}(\overline{F}) > 0$.

A finite extension L/F is called *tamely ramified* if $\overline{L}/\overline{F}$ is a separable extension and $p \nmid e(L|F)$ where $p = \operatorname{char}(\overline{F}) > 0$.

Unramified extensions are tamely ramified.

10.2. First we treat the case of unramified extensions.

PROPOSITION.

(1) *Let L/F be a finite unramified extension, and $\overline{L} = \overline{F}(\theta)$ for some $\theta \in \overline{L}$. Let $\alpha \in \mathcal{O}_L$ be such that $\overline{\alpha} = \theta$. Then $L = F(\alpha)$, and L is separable over F, $\mathcal{O}_L = \mathcal{O}_F[\alpha]$; θ is a simple root of the polynomial $\overline{f}(X)$ irreducible over $\overline{F}$, where $f(X)$ is the monic irreducible polynomial of α over F.*

(2) *Let $f(X)$ be a monic polynomial over $\mathcal{O}_F$, such that its residue is a monic separable polynomial over $\overline{F}$. Let α be a root of $f(X)$ in F^{alg}, and let $L = F(\alpha)$. Then the extension L/F is unramified and $\overline{L} = \overline{F}(\theta)$ for $\theta = \overline{\alpha}$.*

Proof. (1) By the preceding Lemma $f(X) \in \mathcal{O}_F[X]$. We have $f(\alpha) = 0$ and $\overline{f}(\overline{\alpha}) = 0$, $\deg f(X) = \deg \overline{f}(X)$. Furthermore,

$$|L:F| \geqslant |F(\alpha):F| = \deg f(X) = \deg \overline{f}(X) \geqslant |\overline{F}(\theta):\overline{F}| = |L:F|.$$

It follows that $L = F(\alpha)$ and θ is a simple root of the irreducible polynomial $\overline{f}(X)$. Therefore, $\overline{f}'(\theta) \neq 0$ and $f'(\alpha) \neq 0$, i.e., α is separable over F. It remains to use Corollary 2 of 9.9 to obtain $\mathcal{O}_L = \mathcal{O}_F[\alpha]$.

(2) Let $f(X) = \prod_{i=1}^n f_i(X)$ be the decomposition of $f(X)$ into irreducible monic factors in $F[X]$. Lemma 9.8 shows that $f_i(X) \in \mathcal{O}_F[X]$. Suppose that α is a root of $f_1(X)$. Then $g_1(X) = \overline{f}_1(X)$ is a monic separable polynomial over $\overline{F}$. The Henselian property of F implies that $g_1(X)$ is irreducible over $\overline{F}$. We get $\alpha \in \mathcal{O}_L$ by Lemma 10.1. Since $\theta = \overline{\alpha} \in \overline{L}$, we obtain $\overline{L} \supset \overline{F}(\theta)$ and

$$\deg f_1(X) = |L:F| \geqslant |\overline{L}:\overline{F}| \geqslant |\overline{F}(\theta):\overline{F}| = \deg g_1(X) = \deg f_1(X).$$

Thus, $\overline{L} = \overline{F}(\theta)$, and L/F is unramified. □

COROLLARY.

(1) *If $L/F, M/L$ are unramified, then M/F is unramified.*

(2) *If L/F is unramified, M is an algebraic extension of F and M is the discrete valuation field with respect to the extension of the valuation of F, then ML/M is unramified.*

(3) *If $L_1/F, L_2/F$ are unramified, then L_1L_2/F is unramified.*

Proof. (1) follows from Lemma 9.1.

To verify (2) let $L = F(\alpha)$ with $\alpha \in \mathcal{O}_L$, $f(X) \in \mathcal{O}_F[X]$ as in the first part of the Proposition. Then $\alpha \notin \mathcal{M}_L$ because $\overline{L} = \overline{F}(\overline{\alpha})$. Observing that $ML = M(\alpha)$, we denote the irreducible monic polynomial of α over M by $f_1(X)$. By the Henselian property of M we obtain that $\overline{f}_1(X)$ is a power of an irreducible polynomial over $\overline{M}$. However, $\overline{f}_1(X)$ divides $\overline{f}(X)$, hence $\overline{f}_1(X)$ is irreducible separable over $\overline{M}$. Applying the second part of the Proposition, we conclude that ML/M is unramified.

(3) follows from (1) and (2). □

An algebraic extension L of a Henselian discrete valuation field F is called *unramified* if $L/F, \overline{L}/\overline{F}$ are separable extensions and $e(w|v) = 1$, where v is the discrete valuation on F, and w is the unique extension of v on L. For finite extensions this is compatible with the previous definition.

The third assertion of the Corollary shows that the compositum of all finite unramified extensions of F in a fixed algebraic closure F^{alg} of F is unramified.

This extension is a Henselian discrete valuation field. It is called the *maximal unramified extension* F^{ur} of F. Its maximality implies $\sigma F^{ur} = F^{ur}$ for any automorphism of the separable closure F^{sep} over F. Thus, F^{ur}/F is Galois.

10.3. PROPOSITION.

(1) *Let L/F be an unramified extension and let $\overline{L}/\overline{F}$ be a Galois extension. Then L/F is Galois.*

(2) *Let L/F be an unramified Galois extension. Then $\overline{L}/\overline{F}$ is Galois. For an automorphism $\sigma \in \mathrm{Gal}(L/F)$ let $\overline{\sigma}$ be the automorphism in $\mathrm{Gal}(\overline{L}/\overline{F})$ satisfying the relation $\bar{\sigma}\bar{\alpha} = \overline{\sigma\alpha}$ for every $\alpha \in \mathcal{O}_L$. Then the map $\sigma \mapsto \overline{\sigma}$ induces an isomorphism of $\mathrm{Gal}(L/F)$ onto $\mathrm{Gal}(\overline{L}/\overline{F})$.*

Proof. (1) It suffices to verify the first assertion for a finite unramified extension L/F. Let $\overline{L} = \overline{F}(\theta)$ and let $g(X)$ be the irreducible monic polynomial of θ over $\overline{F}$. Then

$$g(X) = \prod_{i=1}^{n}(X - \theta_i),$$

with $\theta_i \in \overline{L}, \theta_1 = \theta$. Let $f(X)$ be a monic polynomial over $\mathcal{O}_F$ of the same degree as $g(X)$ and $\overline{f}(X) = g(X)$. The Henselian property (Corollary 2 in 8.2) implies

$$f(X) = \prod_{i=1}^{n}(X - \alpha_i),$$

with $\alpha_i \in \mathcal{O}_L, \overline{\alpha}_i = \theta_i$. Proposition 10.2 shows that $L = F(\alpha_1)$, and we deduce that L/F is Galois.

(2) Note that the automorphism $\overline{\sigma}$ is well defined. Indeed, if $\beta \in \mathcal{O}_L$ with $\overline{\beta} = \overline{\alpha}$, then we have $\sigma(\alpha - \beta) \in \mathcal{M}_L$ by Corollary 3 in 9.9 and $\overline{\sigma\alpha} = \overline{\sigma\beta}$. It suffices to verify the second assertion for a finite unramified Galois extension L/F. Let $\alpha, \theta, f(X)$ be as in the first part of Proposition 10.2. Since all roots of $f(X)$ belong to L, we obtain that all roots of $\overline{f}(X)$ belong to $\overline{L}$ and $\overline{L}/\overline{F}$ is Galois. The homomorphism $\mathrm{Gal}(L/F) \longrightarrow \mathrm{Gal}(\overline{L}/\overline{F})$ defined by $\sigma \mapsto \overline{\sigma}$ is surjective because the condition $\overline{\sigma}\theta = \theta_i$ implies $\sigma\alpha = \alpha_i$ for the root α_i of $f(X)$ with $\overline{\alpha}_i = \theta_i$. Since $\mathrm{Gal}(L/F)$, $\mathrm{Gal}(\overline{L}/\overline{F})$ are of the same order, we conclude that $\mathrm{Gal}(L/F)$ is isomorphic to $\mathrm{Gal}(\overline{L}/\overline{F})$. □

COROLLARY. *The residue field of F^{ur} coincides with the separable closure $\overline{F}^{sep}$ of $\overline{F}$ and $\mathrm{Gal}(F^{ur}/F) \simeq \mathrm{Gal}(\overline{F}^{sep}/\overline{F})$.*

Proof. Let $\theta \in \overline{F}^{sep}$, let $g(X)$ be the monic irreducible polynomial of θ over $\overline{F}$, and $f(X)$ as in the second part of Proposition 10.2. Let $\{\alpha_i\}$ be all the roots of $f(X)$ and $L = F(\{\alpha_i\})$. Then $L \subset F^{ur}$ and $\theta = \overline{\alpha}_i \in \overline{F^{ur}}$ for a suitable i. Hence, $\overline{F^{ur}} = \overline{F}^{sep}$. □

10.4. Let L be an algebraic extension of F, and let L be a discrete valuation field. We will assume that the algebraic closures are chosen well so that $F^{\mathrm{alg}} = L^{\mathrm{alg}}$.

PROPOSITION. *Let L be an algebraic extension of F and let L be a discrete valuation field. Then $L^{\mathrm{ur}} = LF^{\mathrm{ur}}$, and $L_0 = L \cap F^{\mathrm{ur}}$ is the maximal unramified subextension of F which is contained in L. Moreover, $\overline{L}/\overline{L_0}$ is a purely inseparable extension.*

Proof. The second part of Corollary 10.2 implies $L^{\mathrm{ur}} \supset LF^{\mathrm{ur}}$. Since the residue field of LF^{ur} contains the compositum of the fields $\overline{L}$ and $\overline{F}^{\mathrm{sep}}$, which coincides with $\overline{L}^{\mathrm{sep}}$ because $\overline{L}/\overline{F}$ is algebraic, we deduce $L^{\mathrm{ur}} = LF^{\mathrm{ur}}$. An unramified subextension of F in L is contained in L_0, and L_0/F is unramified. Let $\theta \in \overline{L}$ be separable over $\overline{F}$, and let $g(X)$ be the monic irreducible polynomial of θ over $\overline{F}$. Let $f(X)$ be a monic polynomial with coefficients in $\mathcal{O}_F$ of the same degree as $g(X)$, and $\overline{f}(X) = g(X)$. Then there exists a root $\alpha \in \mathcal{O}_L$ of the polynomial $f(X)$ with $\overline{\alpha} = \theta$ because of the Henselian property. Proposition 10.2 shows that $F(\alpha)/F$ is unramified, and hence $\theta \in \overline{L_0}$. □

COROLLARY. *Let L be a finite separable (resp. finite) extension of a Henselian (resp. complete) discrete valuation field F, and let $\overline{L}/\overline{F}$ be separable. Then L is a totally ramified extension of L_0, L^{ur} is a totally ramified extension of F^{ur}, and $|L : L_0| = |L^{\mathrm{ur}} : F^{\mathrm{ur}}|$.*

Proof. Theorem 9.8 and Proposition 9.4 show that $f(L|L_0) = 1$, and $e(L|L_0) = |L : L_0|$. At the same time, Lemma 9.1 implies

$$e(L^{\mathrm{ur}}|F^{\mathrm{ur}}) = e(L^{\mathrm{ur}}|F) = e(L|L_0).$$

Since $|L : L_0| \geqslant |L^{\mathrm{ur}} : F^{\mathrm{ur}}|$, we obtain that $|L : L_0| = |L^{\mathrm{ur}} : F^{\mathrm{ur}}|$, $e(L^{\mathrm{ur}}|F^{\mathrm{ur}}) = |L^{\mathrm{ur}} : F^{\mathrm{ur}}|$, and therefore $f(L^{\mathrm{ur}}|F^{\mathrm{ur}}) = 1$. □

10.5. We treat the case of tamely ramified extensions.

PROPOSITION.

(1) *Let L be a finite separable (resp. finite) tamely ramified extension of a Henselian (resp. complete) discrete valuation field F and let L_0/F be the maximal unramified subextension in L/F. Then $L = L_0(\pi)$ and $\mathcal{O}_L = \mathcal{O}_{L_0}[\pi]$ with a prime element π in L satisfying the equation $X^e - \pi_0 = 0$ for some prime element π_0 in L_0, where $e = e(L|F)$.*

(2) *Let L_0/F be a finite unramified extension, $L = L_0(\alpha)$ with $\alpha^e = \beta \in L_0$. Let $p \nmid e$ if $p = \mathrm{char}(\overline{F}) > 0$. Then L/F is separable tamely ramified.*

Proof. (1) The Corollary of Proposition 10.4 shows that L/L_0 is totally ramified. Let π_1 be a prime element in L_0, then $\pi_1 = \pi_L^e \varepsilon$ for a prime element π_L in L and $\varepsilon \in U_L$ according to 9.3. Since $\overline{L} = \overline{L}_0$, there exists $\eta \in \mathcal{O}_{L_0}$ such that $\overline{\eta} = \overline{\varepsilon}$. Hence $\pi_1 \eta^{-1} = \pi_L^e \rho$ for the principal unit $\rho = \varepsilon \eta^{-1} \in \mathcal{O}_L$. For the polynomial $f(X) = X^e - \rho$ we have $f(1) \in \mathcal{M}_L$, $f'(1) = e$. Now Corollary 2 of 8.2 shows the existence of an element $v \in \mathcal{O}_L$ with $v^e = \rho$, $\overline{v} = 1$. Therefore, $\pi_0 = \pi_1 \eta^{-1}$, $\pi = \pi_L v$ are the elements desired for the first part of the Proposition. It remains to use Corollary 2 of 9.9.

(2) Let $\beta = \pi_1^a \varepsilon$ for a prime element π_1 in L_0 and a unit $\varepsilon \in U_{L_0}$. The polynomial $g(X) = X^e - \overline{\varepsilon}$ is separable in $\overline{L}_0[X]$ and we can apply Proposition 10.2 to $f(X) = X^e - \varepsilon$ and a root $\eta \in F^{\text{sep}}$ of $f(X)$. We deduce that $L_0(\eta)/L_0$ is unramified and hence it suffices to verify that M/M_0 for $M = L(\eta), M_0 = L_0(\eta)$, is tamely ramified. We get $M = M_0(\alpha_1)$ with $\alpha_1 = \alpha\eta^{-1}$, $\alpha_1^e = \pi_1^a$. Put $d =$ g.c.d.(e,a). Then

$$M \subset M_0(\alpha_2, \zeta)$$

with $\alpha_2^{e/d} = \pi_1^{a/d}$ and a primitive eth root ζ of unity. Since $M_0(\zeta)/M_0$ is unramified (this can be verified by the same arguments as above), π_1 is a prime element in $M_0(\zeta)$. Let v be the discrete valuation on $M_0(\alpha_2, \zeta)$. Then $(a/d)v(\pi_1) \in (e/d)\mathbb{Z}$ and $v(\pi_1) \in (e/d)\mathbb{Z}$, because a/d and e/d are relatively prime. This shows that $e(M_0(\alpha_2,\zeta) \mid M_0(\zeta)) \geqslant e/d$. However, $|M_0(\zeta,\alpha_2) : M_0(\zeta)| \leqslant e/d$, and we conclude that $M_0(\zeta,\alpha_2)/M_0(\zeta)$ is tamely and totally ramified. Thus, $M_0(\zeta,\alpha_2)/M_0$ and M/M_0 are tamely ramified extensions. □

COROLLARY.

(1) *If $L/F, M/L$ are separable tamely ramified, then M/F is separable tamely ramified.*

(2) *If L/F is separable tamely ramified, M/F is an algebraic extension, and M is discrete, then ML/M is separable tamely ramified.*

(3) *If $L_1/F, L_2/F$ are separable tamely ramified, then L_1L_2/F is separable tamely ramified.*

If F is complete, then all the assertions hold without the assumption of separability.

Proof. It is carried out similarly to the proof of Corollary 10.2. To verify (2) one can find the maximal unramified subextension L_0/F in L/F. Then it remains to show that ML/ML_0 is tamely ramified. Put $L = L_0(\pi)$ with $\pi^e = \pi_0$. Then we get $ML = ML_0(\pi)$, and the second part of the Proposition yields the required assertion. □

10.6. Finally we treat the case of totally ramified extensions. Let F be a Henselian discrete valuation field. A polynomial

$$f(X) = X^n + a_{n-1}X^{n-1} + \cdots + a_0 \quad \text{over } \mathcal{O}$$

is called an *Eisenstein polynomial* if $a_0, \ldots, a_{n-1} \in \mathcal{M}, \quad a_0 \notin \mathcal{M}^2$.

PROPOSITION.

(1) *The Eisenstein polynomial $f(X)$ is irreducible over F. If α is a root of $f(X)$, then $F(\alpha)/F$ is a totally ramified extension of degree n, and α is a prime element in $F(\alpha)$, $\mathcal{O}_{F(\alpha)} = \mathcal{O}_F[\alpha]$.*

(2) *Let L/F be a separable totally ramified extension of degree n, and let π be a prime element in L. Then π is a root of an Eisenstein polynomial over F of degree n.*

Proof. (1) Let α be a root of $f(X)$, $L = F(\alpha)$, $e = e(L|F)$. Then

$$nv_L(\alpha) = v_L\left(\sum_{i=0}^{n-1} a_i\alpha^i\right) \geqslant \min_{0\leqslant i\leqslant n-1} (ev_F(a_i) + iv_L(\alpha)),$$

where v_F and v_L are the discrete valuations on F and L. It follows that $v_L(\alpha) > 0$. Since $ev_F(a_0) < ev_F(a_i) + iv_L(\alpha)$ for $i > 0$, one has $nv_L(\alpha) = ev_F(a_0) = e$. Lemma 9.3 implies $v_L(\alpha) = 1, n = e, f = 1$, and $\mathcal{O}_L = \mathcal{O}_F[\alpha]$ similarly to Corollary 2 of 9.9.

(2) Let π be a prime element in L. Then $L = F(\pi)$ by Corollary 2 of 9.9. Let

$$f(X) = X^n + a_{n-1}X^{n-1} + \cdots + a_0$$

be the irreducible polynomial of π over F. Then

$$n = e, \quad nv_L(\pi) = \min_{0\leqslant i\leqslant n-1} (nv_F(a_i) + i),$$

hence $v_F(a_i) > 0$, and $n = nv_F(a_0)$, $v_F(a_0) = 1$. □

11. Galois Extensions and Ramification Groups

Ramification theory was first studied by Dedekind and Hilbert. In this section F is a Henselian discrete valuation field.

11.1. LEMMA. *Let L be a finite Galois extension of F. Then $v \circ \sigma = v$ for the discrete valuation v on L and $\sigma \in \mathrm{Gal}(L/F)$. If π is a prime element in L, then $\sigma\pi$ is a prime element and $\sigma\mathcal{O}_L = \mathcal{O}_L$, $\sigma\mathcal{M}_L = \mathcal{M}_L$.*

Proof. It follows from Corollary 3 of 9.9. □

PROPOSITION. *Let L be a finite Galois extension of F and let L_0/F be the maximal unramified subextension in L/F. Then L_0/F and $\overline{L}_0/\overline{F}$ are Galois, and the map $\sigma \mapsto \overline{\sigma}$ defined in Proposition* 10.3 *induces the surjective homomorphism* $\mathrm{Gal}(L/F) \longrightarrow \mathrm{Gal}(L_0/F) \longrightarrow \mathrm{Gal}(\overline{L}_0/\overline{F})$. *If, in addition, $\overline{L}/\overline{F}$ is separable, then $\overline{L} = \overline{L}_0$ and $\overline{L}/\overline{F}$ is Galois, and L/L_0 is totally ramified.*

The extension L^{ur}/F is Galois and the group $\mathrm{Gal}(L^{\mathrm{ur}}/L_0)$ *is isomorphic with* $\mathrm{Gal}(L^{\mathrm{ur}}/L) \times \mathrm{Gal}(L^{\mathrm{ur}}/F^{\mathrm{ur}})$, *and*

$$\mathrm{Gal}(L^{\mathrm{ur}}/F^{\mathrm{ur}}) \simeq \mathrm{Gal}(L/L_0), \quad \mathrm{Gal}(L^{\mathrm{ur}}/L) \simeq \mathrm{Gal}(F^{\mathrm{ur}}/L_0).$$

Proof. Recall that in 10.4 we got an agreement $F^{\mathrm{alg}} = L^{\mathrm{alg}}$. Let $\sigma \in \mathrm{Gal}(L/F)$. Corollary 3 of 9.9 implies that σL_0 is unramified over F, hence $L_0 = \sigma L_0$ and L_0/F is Galois. The surjectivity of $\mathrm{Gal}(L/F) \longrightarrow \mathrm{Gal}(\overline{L}_0/\overline{F})$ follows from Proposition 10.3. Since L/F and F^{ur}/F are Galois extensions, we obtain that LF^{ur}/F is a Galois extension. Then $L^{\mathrm{ur}} = LF^{\mathrm{ur}}$ by Proposition 10.4. The remaining assertions are easily deduced by using Galois theory. □

Thus, a Galois extension L/F induces the Galois extension $L^{\mathrm{ur}}/F^{\mathrm{ur}}$. The converse statement can be formulated as follows.

11.2. PROPOSITION. *Let M be a finite extension of F^{ur} of degree n. Then there exist a finite unramified extension L_0 of F and an extension L/L_0 of degree n such that $L \cap F^{\mathrm{ur}} = L_0$, $LF^{\mathrm{ur}} = M$. If M/F^{ur} is separable* (*Galois*) *then one can find L_0 and L, such that L/L_0 is separable* (*Galois*).

Proof. Assume that L_0 is a finite unramified extension of F, L is a finite extension of L_0 of the same degree as M/F^{ur} and $M = LF^{\mathrm{ur}}$. Then for a finite unramified extension N_0 of L_0 and $N = N_0L$ we get $|M : F^{\mathrm{ur}}| \leqslant |N : N_0| \leqslant |L : L_0|$, hence $|N : N_0| = |L : L_0|$ and $|N : L| = |N_0 : L_0|$. This shows $L \cap F^{\mathrm{ur}} = L_0$ and L_0, L are such as desired. Moreover, N_0, N are also valid for the Proposition. Therefore, it suffices to consider a case of $M = F^{\mathrm{ur}}(\alpha)$.

Let $f(X) \in F^{\mathrm{ur}}[X]$ be the irreducible monic polynomial of α over F^{ur}. In fact, its coefficients belong to some finite subextension L_0/F in F^{ur}/F. Put $L = L_0(\alpha)$. Then $f(X)$ is irreducible over L_0, L is the finite extension of L_0 of the same degree as M/F^{ur} and $M = LF^{\mathrm{ur}}$. This proves the first assertion of the Proposition. If α is separable over F^{ur}, then it is separable over L_0. If M/F^{ur} is a Galois extension, then $M = F^{\mathrm{ur}}(\alpha)$ for a suitable α and $\sigma_i(\alpha)$ for $\sigma_i \in \mathrm{Gal}(M/F^{\mathrm{ur}})$ can be expressed as polynomials in α with coefficients in F^{ur}. All these coefficients belong to some finite extension L_0' of L_0 in F^{ur}. The pair L_0', $L' = L_0'(\alpha)$ is the desired one. □

COROLLARY. *If $\overline{M} = \overline{F^{\mathrm{ur}}}$, then L/L_0 and M/F^{ur} are totally ramified.*

Proof. It follows from Proposition 10.4. □

11.3. Let L be a finite Galois extension of F, $G = \mathrm{Gal}(L/F)$. Put

$$G_i = \{\sigma \in G : \sigma\alpha - \alpha \in \mathscr{M}_L^{i+1} \text{ for all } \alpha \in \mathscr{O}_L\}, \qquad i \geqslant -1.$$

Then $G_{-1} = G$ by Lemma 11.1 and G_{i+1} is a subset of G_i.

Let v_L be the discrete valuation of L. For a real number x define

$$G_x = \{\sigma \in G : v_L(\sigma\alpha - \alpha) \geqslant x + 1 \text{ for all } \alpha \in \mathscr{O}_L\}.$$

Certainly each of G_x is equal to G_i with the least integer $i \geqslant x$.

LEMMA. *G_i are normal subgroups of G.*

Proof. Let $\sigma \in G_i, \alpha \in \mathscr{O}_L$. Then $\sigma\alpha - \alpha \in \mathscr{M}_L^{i+1}$. Hence $\alpha - \sigma^{-1}(\alpha) \in \sigma^{-1}(\mathscr{M}_L^{i+1}) = \mathscr{M}_L^{i+1}$ by Lemma 11.1, i.e., $\sigma^{-1} \in G_i$. Let $\sigma, \tau \in G_i$. Then

$$\sigma\tau(\alpha) - \alpha = \sigma(\tau(\alpha) - \alpha) + \sigma(\alpha) - \alpha \in \mathscr{M}_L^{i+1},$$

i.e., $\sigma\tau \in G_i$. Furthermore, let $\sigma \in G_i, \tau \in G$. Then $\tau(\alpha) \in \mathscr{O}_L$ for $\alpha \in \mathscr{O}_L$ and $\sigma(\tau\alpha) - \tau\alpha \in \mathscr{M}_L^{i+1}$, $\tau^{-1}\sigma\tau(\alpha) - \alpha \in \mathscr{M}_L^{i+1}$, $\tau^{-1}\sigma\tau \in G_i$. □

The groups G_x are called *(lower) ramification groups* of $G = \mathrm{Gal}(L/F)$.

PROPOSITION. *Let L be a finite Galois extension of F, and let $\overline{L}$ be a separable extension of $\overline{F}$. Then $G_0 = \mathrm{Gal}(L/L_0)$ and the ith ramification groups of G_0 and G coincide for $i \geqslant 0$. Moreover,*

$$G_i = \{\sigma \in G_0 : \sigma\pi - \pi \in \mathscr{M}_L^{i+1}\}$$

for a prime element π in L, and $G_i = \{1\}$ for sufficiently large i.

Proof. Note that $\sigma \in G_0$ if and only if $\overline{\sigma} \in \mathrm{Gal}(\overline{L}/\overline{F})$ is trivial. Then G_0 coincides with the kernel of the homomorphism $\mathrm{Gal}(L/F) \longrightarrow \mathrm{Gal}(\overline{L}/\overline{F})$. Proposition 11.1 and Proposition 10.3 imply that this kernel is equal to $\mathrm{Gal}(L/L_0)$. Since G_i is a subgroup of G_0 for $i \geqslant 0$, we get the assertion about the ith ramification group of G_0. Finally, using Corollary 2 of 9.9 we obtain $\mathscr{O}_L = \mathscr{O}_{L_0}[\pi]$. Let

$$\alpha = \sum_{m=0}^{n} a_m \pi^m$$

be an expansion of $\alpha \in \mathscr{O}_L$ with coefficients in $\mathscr{O}_{L_0}$. As $\sigma a_m = a_m$ for $\sigma \in G_0$ it follows that

$$\sigma\alpha - \alpha = \sum_{m=0}^{n} a_m(\sigma(\pi^m) - \pi^m).$$

Now we deduce the description of G_i, since $\sigma(\pi^m)-\pi^m \in \mathcal{M}_L^{i+1}$. Now we deduce the description of G_i, since $\sigma(\pi^i)-\pi^i \in G_i$. If $i \geqslant \max\{v_L(\sigma\pi-\pi) : \sigma \in G\}$, then $G_i=\{1\}$. □

The group G_0 is called the *inertia group* of G, and the field L_0 is called the *inertia subfield* of L/F.

11.4. PROPOSITION. *Let L be a finite Galois extension of F, $\overline{L}$ a separable extension of $\overline{F}$, and π a prime element in L. Introduce the maps*

$$\psi_0\colon G_0 \longrightarrow \overline{L}^{\times}, \qquad \psi_i\colon G_i \longrightarrow \overline{L} \quad (i>0)$$

by the formulas $\psi_i(\sigma)=\lambda_i(\sigma\pi/\pi)$, where the maps

$$\lambda_0\colon U_L \longrightarrow \overline{L}^{\times}, \qquad \lambda_i\colon 1+\mathcal{M}_L^i \longrightarrow \overline{L}$$

were defined in Proposition 4.4. *Then ψ_i is a homomorphism with the kernel G_{i+1} for $i \geqslant 0$.*

Proof. The proof follows from the congruence

$$\frac{\sigma\tau(\pi)}{\pi}=\sigma\left(\frac{\tau\pi}{\pi}\right)\cdot\frac{\sigma\pi}{\pi}\equiv\frac{\tau\pi}{\pi}\cdot\frac{\sigma\pi}{\pi} \quad \bmod U_{i+1}$$

for $\sigma,\tau \in G_i$ and Proposition 4.4. The kernel of ψ_i consists of those automorphisms $\sigma \in G_i$, for which $\sigma\pi/\pi \in 1+\mathcal{M}_L^{i+1}$, i.e., $\sigma\pi-\pi \in \mathcal{M}_L^{i+2}$. □

COROLLARY 1. *Let L be a finite Galois extension of F, and $\overline{L}$ a separable extension of $\overline{F}$. If* $\operatorname{char}(\overline{F})=0$, *then $G_1=\{1\}$ and G_0 is cyclic. If* $\operatorname{char}(\overline{F})=p>0$, *then the group G_0/G_1 is cyclic of order relatively prime to p, G_i/G_{i+1} are abelian p-groups if $i>0$, and G_1 is the maximal p-subgroup of G_0.*

Proof. The previous Proposition permits us to transform the assertions of this Corollary into the following: a finite subgroup in $\overline{L}^{\times}$ is cyclic (of order relatively prime to $\operatorname{char}(\overline{L})$ when $\operatorname{char}(\overline{L}) \neq 0$); there are no non-trivial finite subgroups in the additive group of $\overline{L}$ if $\operatorname{char}(\overline{L})=0$; if $\operatorname{char}(\overline{L})=p>0$ then a finite subgroup in $\overline{L}$ is a p-group. □

COROLLARY 2. *Let L be a finite Galois extension of F and $\overline{L}$ a separable extension of $\overline{F}$. Then the group G_1 coincides with* $\operatorname{Gal}(L/L_1)$, *where L_1/F is the maximal tamely ramified subextension in L/F.*

Proof. The extension L_1/L_0 is totally ramified by Proposition 11.1 and is the maximal subextension in L/L_0 of degree relatively prime with $\operatorname{char}(\overline{F})$. Now Corollary 1 implies $G_1=\operatorname{Gal}(L/L_1)$. □

COROLLARY 3. *Let L be a finite Galois extension of F and $\overline{L}$ a separable extension of $\overline{F}$. Then G_0 is a solvable group. If, in addition, $\overline{L}/\overline{F}$ is a solvable extension, then L/F is solvable.*

Proof. It follows from Corollary 1. □

11.5. DEFINITION. Let L/F be a finite Galois extension with separable residue field extension; let $G = \mathrm{Gal}(L/F)$. Integers i such that $G_i \neq G_{i+1}$ are called *ramification numbers of L/F* or *lower ramification jumps of L/F*.

One of the first properties of ramification numbers if supplied by the following

PROPOSITION. *Let L/F be a finite Galois extension with separable residue field extension. Let $\sigma \in G_i \setminus G_{i+1}$ and $\tau \in G_j \setminus G_{j+1}$ with $i, j \geqslant 1$. Then*

$$\sigma\tau\sigma^{-1}\tau^{-1} \in G_{i+j+1}, \quad \textit{and } i \equiv j \mod p.$$

Proof. Let π_L be a prime element of L. Then

$$\frac{\sigma\pi_L}{\pi_L} = 1 + \alpha\pi_L^i, \quad \frac{\tau\pi_L}{\pi_L} = 1 + \beta\pi_L^j \qquad \text{with } \alpha, \beta \in \mathcal{O}_L^{\times}.$$

Therefore

$$\begin{aligned} \sigma\tau\pi_L &= \sigma\pi_L + (\sigma\beta)(\sigma\pi_L)^{j+1} \\ &\equiv \pi_L + \alpha\pi_L^{i+1} + \beta\pi_L^{j+1} + (j+1)\alpha\beta\pi_L^{i+j+1} \mod \mathcal{M}_L^{i+j+2}. \end{aligned}$$

Hence $(\sigma\tau - \tau\sigma)\pi_L \equiv (j-i)\alpha\beta\pi_L^{i+j+1} \mod \mathcal{M}_L^{i+j+2}$. Substituting instead of π_L the other prime element $\sigma^{-1}\tau^{-1}\pi_L$ of L we deduce that

$$\frac{\sigma\tau\sigma^{-1}\tau^{-1}\pi_L}{\pi_L} \equiv 1 + (j-i)\alpha\beta\pi_L^{i+j} \mod \mathcal{M}_L^{i+j+1}.$$

Now if j is the maximal ramification number of L/F, then $G_{j+1} = \{1\}$. Therefore the last formula in the previous paragraph shows that every positive ramification number i of L/F is congruent to j modulo p. Therefore every two positive ramification number of L/F are congruent to each other modulo p. Finally, from the same formula we deduce that $\sigma\tau\sigma^{-1}\tau^{-1} \in G_{i+j+1}$. □

12. Structure Theorems for Complete Discrete Valuation Fields

Lemma 2.2 shows that there are three cases: two equal-characteristic cases, when $\mathrm{char}(F) = \mathrm{char}(\overline{F}) = 0$ or $\mathrm{char}(F) = \mathrm{char}(\overline{F}) = p > 0$, and one unequal-characteristic case, when $\mathrm{char}(F) = 0, \mathrm{char}(\overline{F}) = p > 0$.

12.1. LEMMA. *The ring of integers $\mathcal{O}_F$ contains a non-trivial field M if and only if* $\mathrm{char}(F) = \mathrm{char}(\overline{F})$.

Proof. Since $M \cap \mathcal{M}_F = (0)$, M is mapped isomorphically onto the field $\overline{M} \subset \overline{F}$, therefore $\mathrm{char}(F) = \mathrm{char}(\overline{F})$. Conversely, let A be the subring in $\mathcal{O}_F$ generated by 1. Then A is a field if $\mathrm{char}(F) = p$, and $A \cap \mathcal{M}_F = (0)$ if $\mathrm{char}(\overline{F}) = 0$. Hence, the quotient field of A is the desired one. □

A field $M \subset \mathcal{O}_F$, that is mapped isomorphically onto the residue field $\overline{F} = \overline{M}$ is called a *coefficient field* in $\mathcal{O}_F$. Such a field, if it exists, is a set of representatives of $\overline{F}$ in $\mathcal{O}_F$, see 4.1. Proposition 4.2 implies immediately that in this case F is isomorphic (algebraically and topologically) with the field $M((X))$: a prime element π in F corresponds to X. Note that this isomorphism depends on the choice of a coefficient field (which is sometimes unique, see below) and the choice of a prime element of F.

We shall show below that a coefficient field exists in an equal-characteristic case.

12.2. The simplest case is that of $\mathrm{char}(F) = \mathrm{char}(\overline{F}) = 0$.

PROPOSITION. *Let* $\mathrm{char}(\overline{F}) = 0$*. Then there exists a coefficient field in $\mathcal{O}_F$. A coefficient field can be selected in infinitely many ways if and only if $\overline{F}$ is not algebraic over $\mathbb{Q}$.*

Proof. Let M be a maximal subfield in $\mathcal{O}_F$, in other words, M be not properly contained in any other larger subfield of $\mathcal{O}_F$. We assert that $\overline{M} = \overline{F}$, i.e., M is a coefficient field. Indeed, if $\theta \in \overline{F}$ is algebraic over $\overline{M}$, then θ is separable over $\overline{M}$ and we can apply the arguments of the proof of Proposition 10.4 to show that there exists an element $\alpha \in \mathcal{O}_F$ which is algebraic over M and such that $\overline{\alpha} = \theta$. Since $M(\alpha) = M$ by the maximality of M, we get $\alpha \in M, \theta \in \overline{M}$.

Furthermore, let $\theta \in \overline{F}$ be transcendental over $\overline{M}$. Let $\alpha \in \mathcal{O}_F$ be such that $\overline{\alpha} = \theta$. Then α is not algebraic over M, because if $\sum_{i=0}^n a_i\alpha^i = 0$ with $a_i \in M$, then $\sum_{i=0}^n \overline{a}_i\theta^i = 0$. Hence, $\overline{a}_i = 0$ and $a_i = 0$ (M is mapped isomorphically onto $\overline{M}$). By the same reason $M[\alpha] \cap \mathcal{M} = (0)$. Hence, the quotient field $M(\alpha)$ is contained in $\mathcal{O}_F$ and $M \neq M(\alpha)$, contradiction.

Thus, a coefficient field exists.

If $\overline{F}$ is not algebraic over $\mathbb{Q}$, choose an element $\alpha \in \mathcal{O}_F$ transcendental over $\mathbb{Q}$. Then the maximal subfield in $\mathcal{O}_F$, which contains $\mathbb{Q}(\alpha+\varepsilon)$ with $\varepsilon \in \mathcal{M}_F$, is a coefficient field and it will be different from the coefficient field containing $\mathbb{Q}(\alpha)$ if $\varepsilon \neq 0$.

If $\overline{F}$ is algebraic over $\mathbb{Q}$, then M is algebraic over $\mathbb{Q}$ and is uniquely determined by the previous constructions. □

12.3. To treat the case $\mathrm{char}(\overline{F}) = p$ we consider the following notion: elements θ_i of $\overline{F}$ are called a *p-basis* of $\overline{F}$ if

$$\overline{F} = \overline{F}^p[\{\theta_i\}] \quad \text{and} \quad |\overline{F}^p[\theta_1,\dots,\theta_n] : \overline{F}^p| = p^n$$

for every distinct elements $\theta_1,\dots,\theta_n$. The empty set is a p-basis if and only if $\overline{F}$ is perfect. For an imperfect $\overline{F}$, a p-basis $\Theta = \{\theta_i\}$ exists by Zorn's Lemma, because every maximal set of elements θ_i satisfying the second condition possesses the first property. The definition of a p-basis implies that $\overline{F} = \overline{F}^{p^n}[\{\theta_i\}]$ for $n \geqslant 1$.

LEMMA. *Let F be a complete discrete valuation field with the residue field $\overline{F}$ of characteristic p, and $\Theta = \{\theta_i\}$ be a p-basis of $\overline{F}$. Let $\alpha_i \in \mathcal{O}_F$ be such that $\overline{\alpha}_i = \theta_i$. Then there exists an extension L/F with $e(L|F) = 1$, such that L is a complete discrete valuation field, $\overline{L} = \bigcup_{n\geqslant 0} \overline{F}^{p^{-n}}$ and α_i are the multiplicative representatives of θ_i in L.*

Proof. Let I be an index-set for Θ. One can put $F_n = F_{n-1}(\{\alpha_{i,n}\})$ with $\alpha_{i,n}^p = \alpha_{i,n-1}$, $i \in I$, and $F_0 = F$, $\alpha_{i,0} = \alpha_i$. Then $e(F_n|F) = 1$ and the completion of $L' = \bigcup_{n\geqslant 0} F_n$ is the desired field. Since $\alpha_i \in \bigcap_{n\geqslant 0} L^{p^n}$, we obtain that α_i is the multiplicative representative of θ_i. □

12.4. Now we treat the case $\mathrm{char}(F) = \mathrm{char}(\overline{F}) = p$.

If $\overline{F}$ is perfect, then Corollaries 1 and 2 of 6.3 show that the set of the multiplicative representatives of $\overline{F}$ in $\mathcal{O}_F$ forms a coefficient field. Moreover, this is the unique coefficient field in $\mathcal{O}_F$ because if M is such a field and $\alpha \in M$, then, as M is perfect, $\alpha \in \bigcap_{n\geqslant 0} M^{p^n}$ is the multiplicative representative of $\overline{\alpha}$.

Note that in general there are infinitely many maximal fields similarly to the case of $\mathrm{char}(\overline{F}) = 0$, therefore in general when $\mathrm{char}(F) = p$ and $\overline{F}$ is perfect a maximal field is not a coefficient field.

PROPOSITION. *Let* $\mathrm{char}(F) = p$. *If $\overline{F}$ is perfect then a coefficient field exists and is unique; it coincides with the set of multiplicative representatives of $\overline{F}$ in $\mathcal{O}_F$. If $\overline{F}$ is imperfect then there are infinitely many coefficient fields.*

Proof. If $\overline{F}$ is imperfect we apply the construction of the previous Lemma. Then $\overline{L}$ is perfect and there is the unique coefficient field N of $\overline{L}$ in $\mathcal{O}_L$. Let M be the subfield of N corresponding to $\overline{F}$.

Let $\Theta = \{\theta_i\}$ be a p-basis of $\overline{F}$. Let $\alpha_i \in \mathcal{O}_F$ be such that $\overline{\alpha}_i = \theta_i$. Let $\alpha_{i,n}$ be as in the proof of Lemma 12.3.

If $\gamma \in M$ then $\overline{\gamma} \in \overline{F}^{p^n}[\Theta]$ and there exists an element $\beta_n \in \mathcal{O}_F[\{\alpha_{i,n}\}]$ such that $\overline{\beta}_n = \overline{\gamma}^{p^{-n}}$. It follows that $\beta_n \equiv \gamma^{p^{-n}} \mod \mathcal{M}_L$, and by Lemma 6.2 we deduce $\gamma \equiv \beta_n^{p^n} \mod \mathcal{M}_L^{n+1}$. Since $\beta_n^{p^n} \in \mathcal{O}_F^{p^n}[\{\alpha_i\}] \subset \mathcal{O}_F$, we obtain $\gamma = \lim \beta_n^{p^n} \in \mathcal{O}_F$. This proves the existence of a coefficient field of $\overline{F}$ in $\mathcal{O}_F$.

If we apply this construction for another set of elements $\alpha_i' \in \mathcal{O}_F$ with $\overline{\alpha}_i' = \overline{\alpha}_i$, then we get a coefficient field M' containing α_i'. Since $\mathcal{M}_F \cap M = \mathcal{M}_F \cap M' = (0)$ we deduce $M \neq M'$. □

12.5. We conclude with the case of unequal characteristic: $\operatorname{char}(F) = 0$, $\operatorname{char}(\overline{F}) = p$. For the discrete valuation v_F such that $v_F(F^\times) = \mathbb{Z}$ recall that $e(F) = v_F(p)$ is called the absolute index of ramification of F, see 4.7. The preceding assertions show that in equal-characteristic case for an arbitrary field K there exists a complete discrete valuation field F with the residue field $\overline{F}$ isomorphic to K. Here is an analog:

PROPOSITION. *Let F be a complete discrete valuation field of characteristic 0 with residue field K of characteristic p. Let K_1 be any extension of K. Then there exists a complete discrete valuation field F_1 which is an extension of F, such that $e(F_1|F) = 1$ and $\overline{F}_1 = K_1$.*

Proof. It is suffices to consider two cases: $K_1 = K(a)$ is an algebraic extension over K and $K_1 = K(y)$ is a transcendental extension over K. If, in addition, in the first case K_1/K is separable, then let $g(X)$ be the monic irreducible polynomial of a over K, and let $f(X)$ be a monic polynomial over the ring of integers of K such that $\overline{f}(X) = g(X)$. By the Hensel Lemma 8.2 there exists a root α of $f(X)$ such that $\overline{\alpha} = a$. Then $F_1 = F(\alpha)$ is the desired extension of F. Next, if $a^p = b \in K$ and β is an element in the ring of integers of F such that $\overline{\beta} = b$, then $F_1 = F(\alpha)$ is the desired extension of F for $\alpha^p = \beta$. Finally, in the second case let w be the discrete valuation on $F(y)$ defined in Example 5 in 1.3. Then completion of $F(y)$ is the desired extension F_1 of F. □

COROLLARY. *There exists a complete discrete valuation field of characteristic 0 with any given residue field of characteristic p and the absolute index of ramification is equal to 1.*

Proof. One can set $F=\mathbb{Q}_p$ and apply the Proposition. □

12.6. PROPOSITION. *Let L be a complete discrete valuation field of characteristic 0 with the residue field $\overline{L}$ of characteristic p. Let F be a complete discrete valuation field of characteristic 0 with p as a prime element. Suppose that there is an isomorphism $\overline{\omega}\colon \overline{F}\longrightarrow\overline{L}$. Then there exists a field embedding $\omega\colon F\longrightarrow L$, such that $v_L\circ\omega=e(L)v_F$ and the image of $\omega(\alpha)\in\mathcal{O}_L$ for $\alpha\in\mathcal{O}_F$ in the residue field $\overline{L}$ coincides with $\overline{\omega}(\overline{\alpha})$.*

Proof. Assume first that $\overline{F}$ is perfect. By Corollary 1 of 6.3 any element $\theta\in\overline{F}$ has the unique multiplicative representative $r_F(\theta)$ in F and $r_L(\overline{\omega}(\theta))$ in L. Put

$$\omega\left(\sum r_F(\theta_i)p^i\right)=\sum r_L(\overline{\omega}(\theta_i))p^i.$$

Proposition 4.2 shows that the map ω is defined on F, Proposition 6.6 shows that ω is a homomorphism of fields. Evidently $v_L\circ\omega=e(L)v_F$ and $\overline{\omega(\alpha)}=\overline{\omega}(\overline{\alpha})$ for $\alpha\in\mathcal{O}_F$.

Next, assume that $\overline{F}$ is imperfect. Let $\Theta=\{\theta_i\}_{i\in I}$ be a p-basis of $\overline{F}$. Let $\mathrm{A}=\{\alpha_i\}_{i\in I}$ be a set of elements $\alpha_i\in\mathcal{O}_F$ with $\overline{\alpha}_i=\theta_i$, and let $\mathrm{B}=\{\beta_i\}_{i\in I}$ be a set of elements $\beta_i\in\mathcal{O}_L$ with $\overline{\beta}_i=\theta_i$. For a map

$$\nu\colon I\longrightarrow\{0,1,\dots,p^n-1\}$$

such that $\nu(i)=0$ for almost all $i\in I$, put

$$\Theta^\nu=\prod_{i\in I}\theta_i^{\nu(i)}.$$

The same meaning will be used for $\mathrm{A}^\nu,\mathrm{B}^\nu$. By Lemma 12.3 there exist complete discrete valuation fields F',L' for F,L, such that $e(F'|F)=e(L'|L)=1$, and $\overline{F'}$ is perfect and isomorphic to $\overline{L'}$, and α_i (resp. β_i) are multiplicative representatives of θ_i in $\mathcal{O}_{F'}$ (resp. of $\overline{\omega}(\theta_i)$ in $\mathcal{O}_{L'}$). The previous arguments show the existence of a homomorphism $\omega'\colon F'\longrightarrow L'$ with $v_{L'}\circ\omega'=e(L)v_{F'}$ and $\overline{\omega'(\alpha)}=\overline{\omega}(\overline{\alpha})$ for $\alpha\in\mathcal{O}_{F'}$. Moreover, ω' maps α_i to β_i, since they are the multiplicative representatives of θ_i and $\overline{\omega}(\theta_i)$. Let $\gamma\in\mathcal{O}_F$ and $\overline{\gamma}=\sum a_\nu^{p^n}\Theta^\nu$ with $a_\nu\in\overline{F}$. Let b_ν be an element of $\mathcal{O}_F$ with the property $\overline{b}_\nu=a_\nu$, and c_ν an element of $\mathcal{O}_L$ with the property $\overline{c}_\nu=\overline{\omega'(b_\nu)}$. Then $\gamma\equiv\sum b_\nu^{p^n}\mathrm{A}^\nu \mod p\mathcal{O}_F$, i.e.,

$$\gamma=\sum b_\nu^{p^n}\mathrm{A}^\nu+p\gamma_1$$

with $\gamma_1\in\mathcal{O}_F$. We get $\omega'(\mathrm{A}^\nu)=\mathrm{B}^\nu$ and using Lemma 6.2 we have

$$\omega'(b_\nu^{p^n})\equiv c_\nu^{p^n} \mod \mathcal{M}_{L'}^{n+1}.$$

Therefore,

$$\omega'(\gamma)\equiv\sum c_\nu^{p^n}\mathrm{B}^\nu+p\omega'(\gamma_1) \mod \mathcal{M}_{L'}^{n+1}.$$

Repeating this reasoning for γ_1, we conclude that $\omega'(\gamma) \equiv \delta_n \mod \mathcal{M}_{L'}^{n+1}$ for some $\delta_n \in \mathcal{O}_L$. Then $\omega'(\gamma) = \lim \delta_n$ and since $\mathcal{O}_L$ is complete, we deduce $\omega'(\gamma) \in \mathcal{O}_L$. Thus, ω' maps $\mathcal{O}_F$ in $\mathcal{O}_L$, and we finally put $\omega = \omega'|_F$ to obtain the desired homomorphism. □

COROLLARY 1. *Let F_1, F_2 be complete discrete valuation fields of characteristic 0 with p as a prime element. Let there be an isomorphism $\overline{\omega}$ of the residue field $\overline{F_1}$ to $\overline{F_2}$. Then there exists a field embedding $\omega\colon F_1 \longrightarrow F_2$ such that $\overline{\omega(\alpha)} = \overline{\omega}(\overline{\alpha})$ for $\alpha \in \mathcal{O}_{F_1}$.*

Proof. Apply the Proposition for $F = F_1, L = F_2$ and $F = F_2, L = F_1$. □

COROLLARY 2. *The image $\omega(F)$ is uniquely determined in the field L if and only if $\overline{F}$ is perfect or $e(L) = 1$.*

Proof. If $\overline{F}$ is perfect then its multiplicative representatives are uniquely determined in F and in L, and this is compatible with ω, hence $\omega(F)$ is uniquely determined and its image is equal to the image of the fraction field of the Witt vectors over $\overline{F}$ in L. If $e(L) = 1$ then $\omega(F) = L$.

Assume that $\overline{F}$ is imperfect and $e(L) > 1$. If $\omega(F)$ were uniquely determined in L then in the proof of the Proposition we could have replaced β_i by $\beta_i + \pi_L$ to obtain $\beta_i \in \omega(\mathcal{O}_F)$, $\beta_i + \pi_L \in \omega(\mathcal{O}_F)$ and hence $\pi_L \in \omega(\mathcal{O}_F)$; the latter is impossible because $v_L \circ \omega = e(L)v_F$. □

13. Cyclic Extensions of Prime Degree

Let F be a complete discrete valuation field and L its Galois extension of prime degree n. Then there are four possible cases:

L/F is unramified;

L/F is tamely and totally ramified;

L/F is totally ramified of degree $p = \operatorname{char}(\overline{F}) > 0$;

$\overline{L}/\overline{F}$ is inseparable of degree $p = \operatorname{char}(\overline{F}) > 0$.

The fourth case is very interesting for higher local class field theory. Here we discuss the first three cases.

13.1. LEMMA. *Let L/F be a finite Galois extension of prime degree n, $\gamma \in \mathscr{M}_L$. Then*

$$N_{L/F}(1+\gamma) = 1 + N_{L/F}(\gamma) + \mathrm{Tr}_{L/F}(\gamma) + \beta$$

with some $\beta \in \mathscr{O}_L$ such that $v_L(\beta) \geqslant 2v_L(\gamma)$.

If n equals positive characteristic of the residue field of F, then

$$N_{L/F}(1+\gamma) = 1 + N_{L/F}(\gamma) + \mathrm{Tr}_{L/F}(\gamma) + \mathrm{Tr}_{L/F}(\delta)$$

with some $\delta \in \mathscr{O}_L$ such that $v_L(\delta) \geqslant 2v_L(\gamma)$.

Proof. If $n = 2$, the assertions are obvious. When $n > 2$, we get

$$\begin{aligned} N_{L/F}(1+\gamma) &= \prod_{i=1}^{n}(1+\sigma_i(\gamma)) \\ &= 1 + \sum_{i=1}^{n} \sigma_i(\gamma) + \prod_{i=1}^{n} \sigma_i(\gamma) + \sum_{1<m<n} \sum_{1 \leqslant i_1 < \cdots < i_m \leqslant n} \sigma_{i_1}(\gamma)\cdots\sigma_{i_m}(\gamma). \end{aligned}$$

Then

$$N_{L/F}(1+\gamma) = 1 + \sum_{i=1}^{n} \sigma_i(\gamma) + \prod_{i=1}^{n} \sigma_i(\gamma) + \left(\sum_{i=1}^{n} \sigma_i\right)\delta$$

where

$$\delta = \sum_{1<m<n} \frac{1}{m}\sigma_1(\gamma) \sum_{1<j_1<\cdots<j_{m-1}\leqslant n} \sigma_{j_1}(\gamma)\cdots\sigma_{j_{m-1}}(\gamma).$$

When $n > 2$ equals positive characteristic of the residue field of F, $v_L(\delta) \geqslant 2v_L(\gamma)$. □

Below $\lambda_{i,L}$, $\lambda_{i,F}$ $(i \geqslant 0)$ will be as in Proposition 4.4 for the specific choice of π_L and π_F as stated below. We denote $U_{i,L} = 1 + \pi_L^i \mathscr{O}_L$, $U_{i,F} = 1 + \pi_F^i \mathscr{O}_F$.

13.2. PROPOSITION. *Let L/F be a Galois unramified extension of degree n. Then a prime element π_F in F is a prime element in L, so we take $\pi_L = \pi_F$.*

Then the following diagrams are commutative :

$$\begin{array}{ccc} L^\times & \xrightarrow{v_L} & \mathbb{Z} \\ {\scriptstyle N_{L/F}}\downarrow & & \downarrow{\scriptstyle \times n} \\ F^\times & \xrightarrow{v_F} & \mathbb{Z} \end{array} \qquad \begin{array}{ccc} U_L & \xrightarrow{\lambda_{0,L}} & \overline{L}^\times \\ {\scriptstyle N_{L/F}}\downarrow & & \downarrow{\scriptstyle N_{\overline{L}/\overline{F}}} \\ U_F & \xrightarrow{\lambda_{0,F}} & \overline{F}^\times \end{array} \qquad \begin{array}{ccc} U_{i,L} & \xrightarrow{\lambda_{i,L}} & \overline{L} \\ {\scriptstyle N_{L/F}}\downarrow & & \downarrow{\scriptstyle \mathrm{Tr}_{\overline{L}/\overline{F}}} \\ U_{i,F} & \xrightarrow{\lambda_{i,F}} & \overline{F} \end{array}$$

$i \geqslant 1$.

Proof. Proposition 10.3 implies that $\overline{N_{L/F}(\alpha)} = N_{\overline{L}/\overline{F}}(\overline{\alpha})$ for $\alpha \in \mathscr{O}_L$, i.e., the second diagram is commutative. By the preceding Lemma we get

$$N_{L/F}(1+\theta\pi_F^i) = 1 + (\mathrm{Tr}_{L/F}\,\theta)\pi_F^i + (N_{L/F}\theta)\pi_F^{ni} + \rho$$

with $\rho \in F$, $v_L(\rho) \geqslant 2i$ and, consequently $v_F(\rho) \geqslant 2i$. Thus, we obtain

$$N_{L/F}(1+\theta\pi_F^i) \equiv 1 + (\operatorname{Tr}_{L/F}\theta)\pi_F^i \mod \pi_F^{i+1}$$

and the commutativity of the third diagram. □

COROLLARY. *In the case under consideration* $N_{L/F}U_{1,L} = U_{1,F}$.

13.3. PROPOSITION. *Let L/F be a totally and tamely ramified cyclic extension of degree n. Then for some prime element π_L in L, the element $\pi_F = \pi_L^n$ is prime in F and $\overline{F} = \overline{L}$. Then the following diagrams*

$$\begin{array}{ccc} L^\times & \xrightarrow{v_L} & \mathbb{Z} \\ {\scriptstyle N_{L/F}}\downarrow & & \downarrow{\scriptstyle \mathrm{id}} \\ F^\times & \xrightarrow{v_F} & \mathbb{Z} \end{array} \qquad \begin{array}{ccc} U_L & \xrightarrow{\lambda_{0,L}} & \overline{L}^\times \\ {\scriptstyle N_{L/F}}\downarrow & & \downarrow{\scriptstyle \uparrow n} \\ U_F & \xrightarrow{\lambda_{0,F}} & \overline{F}^\times \end{array}$$

$$\begin{array}{ccc} U_{ni,L} & \xrightarrow{\lambda_{ni,L}} & \overline{L} = \overline{F} \\ {\scriptstyle N_{L/F}}\downarrow & & \downarrow{\scriptstyle \times\overline{n}} \\ U_{i,F} & \xrightarrow{\lambda_{i,F}} & \overline{F} \end{array}$$

are commutative, where id *is the identity map,* $\uparrow n$ *takes an element to its nth power,* $\times\overline{n}$ *is the multiplication by* $\overline{n} \in \overline{F}$, $i \geqslant 1$.

Moreover, $N_{L/F}U_{i,L} = N_{L/F}U_{i+1,L}$ *if* $n \nmid i$.

Furthermore, $U_{i,F} = N_{L/F}U_{ni,L}$ *for positive* i.

Proof. Since $\pi_L^n = \pi_F$ and L/F is Galois, then $\operatorname{Gal}(L/F)$ is cyclic of order n and $\sigma(\pi_L) = \zeta\pi_L$ for a generator σ of $\operatorname{Gal}(L/F)$, where ζ is a primitive nth root of unity, $\zeta \in F$. The first diagram is commutative in view of Theorem 9.5. Proposition 11.1 shows that $\overline{\sigma(\alpha)} = \overline{\alpha}$ for $\sigma \in \operatorname{Gal}(L/F)$, $\alpha \in \mathcal{O}_L$, and we get the commutativity of the second diagram.

We have

$$\frac{\sigma(1+\theta\pi_L^i)}{1+\theta\pi_L^i} = 1 + \theta(\zeta^i - 1)\pi_L^i \mod \pi_L^{i+1}.$$

If $n \nmid i$ then the residue of $\zeta^i - 1$ is non-zero and so $U_{i,L} \subset U_{i+1,L}\ker N_{L/F}$ and $N_{L/F}U_{i,L} = N_{L/F}U_{i+1,L}$.

If $j = ni$, then $1 + \theta\pi_L^j \in F$ for $\theta \in \mathcal{O}_F$, and

$$N_{L/F}(1+\theta\pi_L^j) = (1+\theta\pi_F^i)^n \equiv 1 + n\theta\pi_F^i \mod \pi_F^{i+1}$$

by Proposition 4.4. Applying Corollary 4.5, we deduce

$$U_{i,F} = U_{i,F}^n \subset N_{L/F}U_{ni,L},$$

and $U_{i,F} = N_{L/F}U_{ni,L}$ follows from the previous paragraph. □

COROLLARY. *In the case under consideration* $N_{L/F}U_{1,L} = U_{1,F}$.
If $\overline{F}$ *is algebraically closed then* $N_{L/F}L^\times = F^\times$.

13.4. Now we treat the most complicated case when L/F is a totally ramified Galois extension of degree $p = \operatorname{char}(\overline{F}) > 0$. Then Corollary 2 of 9.9 shows that $\mathcal{O}_L = \mathcal{O}_F[\pi_L]$, $L = F(\pi_L)$ for a prime element π_L in L, and $\overline{L} = \overline{F}$.

Let σ be a generator of $\operatorname{Gal}(L/F)$, then $\sigma(\pi_L)/\pi_L \in U_L$. One can write $\sigma(\pi_L)/\pi_L = \theta\varepsilon$ with $\theta \in U_F, \varepsilon \in 1 + \mathcal{M}_L$. Then

$$\sigma^2(\pi_L)/\pi_L = \sigma(\theta\varepsilon)\cdot\theta\varepsilon = \theta^2\varepsilon\cdot\sigma(\varepsilon),$$

and

$$1 = \sigma^p(\pi_L)/\pi_L = \theta^p\cdot\varepsilon\cdot\sigma(\varepsilon)\cdot\dots\cdot\sigma^{p-1}(\varepsilon).$$

This shows that $\theta^p \in 1 + \mathcal{M}_L$ and $\theta \in 1 + \mathcal{M}_F$, because raising to the pth power is an injective homomorphism of $\overline{F}$. Thus, we obtain $\sigma(\pi_L)/\pi_L \in 1 + \mathcal{M}_L$. Put

$$\frac{\sigma(\pi_L)}{\pi_L} = 1 + \eta\pi_L^s \qquad \text{with} \qquad \eta \in U_L,\ s = s(L|F) \geqslant 1. \tag{$*$}$$

Note that s does not depend on the choice of the prime element π_L and of the generator σ of $G = \operatorname{Gal}(L/F)$. Indeed, we have

$$\frac{\sigma^i(\pi_L)}{\pi_L} \equiv 1 + i\eta\pi_L^s \mod \pi_L^{s+1} \qquad \text{and} \qquad \frac{\sigma(\rho)}{\rho} \equiv 1 \mod \pi_L^{s+1}$$

for an element $\rho \in U_L$. We also deduce that

$$\frac{\sigma(\alpha)}{\alpha} \in U_{s,L}$$

for every element $\alpha \in L^\times$. This means that $G = G_s$, $G_{s+1} = \{1\}$ (see 11.3). Thus, s is the lower ramification number/jump of L/F.

We need the following auxiliary property.

LEMMA. *Let* $f(X)$ *be the irreducible polynomial of* π_L *over* F. *Then*

$$\operatorname{Tr}_{L/F}\left(\frac{\pi_L^j}{f'(\pi_L)}\right) = \begin{cases} 0 & \text{if} \quad 0 \leqslant j \leqslant p-2, \\ 1 & \text{if} \quad j = p-1. \end{cases}$$

Proof. Since $\sigma^i(\pi_L)$ for $0 \leqslant i \leqslant p-1$ are all the roots of the polynomial $f(X)$, we get

$$\frac{1}{f(X)} = \sum_{i=0}^{p-1} \frac{1}{f'(\sigma^i(\pi_L))\,(X - \sigma^i(\pi_L))}.$$

Let $f(X) = X^p + a_{p-1}X^{p-1} + \dots + a_0$. Putting $Y = X^{-1}$ and performing the calculations in the field $L((Y))$, we consequently deduce

$$f(X) = Y^{-p}(1 + a_{p-1}Y + \dots + a_0Y^p),$$
$$\frac{1}{f(X)} = \frac{Y^p}{1 + a_{p-1}Y + \dots + a_0Y^p} \equiv Y^p \mod Y^{p+1},$$
$$\frac{1}{X - \sigma^i(\pi_L)} = \frac{Y}{1 - \sigma^i(\pi_L)Y} = \sum_{j\geqslant 0} \sigma^i(\pi_L^j)Y^{j+1}$$

(because $1/(1-Y) = \sum_{i\geqslant 0} Y^i$ in $F((Y))$). Hence

$$\sum_{j\geqslant 0}\sum_{i=0}^{p-1} \frac{\sigma^i(\pi_L^j)Y^{j+1}}{f'(\sigma^i(\pi_L))} \equiv Y^p \mod Y^{p+1},$$

or

$$\operatorname{Tr}_{L/F}\left(\frac{\pi_L^j}{f'(\pi_L)}\right) = \sum_{i=0}^{p-1} \frac{\sigma^i(\pi_L^j)}{f'(\sigma^i(\pi_L))} = \begin{cases} 0 & \text{if } \quad 0 \leqslant j \leqslant p-2, \\ 1 & \text{if } \quad j = p-1, \end{cases}$$

as desired. □

PROPOSITION. *Let $[a]$ denote the maximal integer $\leqslant a$. For an integer $i \geqslant 0$ put $j(i) = s + 1 + [(i-1-s)/p]$. Then*

$$\operatorname{Tr}_{L/F}(\pi_L^i \mathcal{O}_L) = \pi_F^{j(i)} \mathcal{O}_F.$$

Proof. One has $f'(\pi_L) = \prod_{i=1}^{p-1}\left(\pi_L - \sigma^i(\pi_L)\right)$. From the definition of s we deduce $\sigma^i(\pi_L)/\pi_L \equiv 1 + i\eta\pi_L^s \mod \pi_L^{s+1}$. Then

$$f'(\pi_L) = (p-1)!(-\eta)^{p-1}\pi_L^{(p-1)(s+1)}\varepsilon$$

with some $\varepsilon \in 1 + \mathcal{M}_L^{(p-1)(s+1)+1}$. Since $\overline{F} = \overline{L}$, for a prime element π_F in F one has the representation $\pi_F = \pi_L^p \varepsilon'$ with $\varepsilon' \in U_L$. The previous Lemma implies

$$\operatorname{Tr}_{L/F}\left(\pi_L^{j+s+1}\varepsilon_{j+s+1}\right) = \begin{cases} 0 & \text{if } \quad 0 \leqslant j < p-1, \\ \pi_F^{s+1} & \text{if } \quad j = p-1 \end{cases}$$

for $\varepsilon_{j+s+1} = (\varepsilon')^{s+1}/\left((p-1)!(-\eta)^{p-1}\varepsilon\right)$. Since $\operatorname{Tr}_{L/F}(\pi_F^i\alpha) = \pi_F^i \operatorname{Tr}_{L/F}(\alpha)$, we can choose the units ε_{j+s+1}, for every integer $j > 0$, such that

$$\operatorname{Tr}_{L/F}(\pi_L^{j+s+1}\varepsilon_{j+s+1}) = 0 \text{ if } p \nmid (j+1) \text{ and } = \pi_F^{s+(j+1)/p} \text{ if } p \mid (j+1).$$

Thus, since the $\mathcal{O}_F$-module $\pi_L^i\mathcal{O}_L$ is generated by $\pi_L^j\varepsilon_j$, $j \geqslant i$, we conclude that $\operatorname{Tr}_{L/F}(\pi_L^i\mathcal{O}_L) = \pi_F^{j(i)}\mathcal{O}_F$. □

13.5. PROPOSITION. *Let L/F be a totally ramified Galois extension of degree $p=\operatorname{char}(\overline{F})>0$. Let π_L be a prime element in L. Then $\pi_F=N_{L/F}\pi_L$ is a prime element in F.*

Then the following diagrams are commutative:

$$\begin{array}{ccc} L^\times & \xrightarrow{v_L} & \mathbb{Z} \\ {\scriptstyle N_{L/F}}\downarrow & & \downarrow{\scriptstyle \mathrm{id}} \\ F^\times & \xrightarrow{v_F} & \mathbb{Z} \end{array} \qquad \begin{array}{ccc} U_L & \xrightarrow{\lambda_{0,L}} & \overline{L}^\times \\ {\scriptstyle N_{L/F}}\downarrow & & \downarrow{\scriptstyle \uparrow p} \\ U_F & \xrightarrow{\lambda_{0,F}} & \overline{F}^\times \end{array}$$

$$\begin{array}{ccc} U_{i,L} & \xrightarrow{\lambda_{i,L}} & \overline{L}=\overline{F} \\ {\scriptstyle N_{L/F}}\downarrow & & \downarrow{\scriptstyle \uparrow p} \\ U_{i,F} & \xrightarrow{\lambda_{i,F}} & \overline{F} \end{array} \qquad \text{if} \qquad 1\leqslant i<s,$$

$$\begin{array}{ccc} U_{s,L} & \xrightarrow{\lambda_{s,L}} & \overline{L}=\overline{F} \\ {\scriptstyle N_{L/F}}\downarrow & & \downarrow{\scriptstyle \overline{\theta}\mapsto\overline{\theta}^p-\overline{\eta}^{p-1}\overline{\theta}} \\ U_{s,F} & \xrightarrow{\lambda_{s,F}} & \overline{F} \end{array}$$

$$\begin{array}{ccc} U_{s+pi,L} & \xrightarrow{\lambda_{s+pi,L}} & \overline{L}=\overline{F} \\ {\scriptstyle N_{L/F}}\downarrow & & \downarrow{\scriptstyle \times(-\overline{\eta}^{p-1})} \\ U_{s+i,F} & \xrightarrow{\lambda_{s+i,F}} & \overline{F} \end{array} \qquad \text{if} \qquad i>0.$$

Moreover, $N_{L/F}(U_{s+i,L})=N_{L/F}(U_{s+i+1,L})$ for $i>0, p\nmid i$.

Proof. The commutativity of the first and the second diagrams can be verified similarly to the proof of Proposition 13.3.

In order to prove commutativity of the remaining diagrams, put $\varepsilon=1+\theta\pi_L^i$ with $\theta\in U_L$. Then, by Lemma 13.1 we get

$$N_{L/F}\varepsilon=1+N_{L/F}(\theta)\pi_F^i+\operatorname{Tr}_{L/F}(\theta\pi_L^i)+\operatorname{Tr}_{L/F}(\theta\delta)$$

with $v_L(\delta)\geqslant 2i$. The previous Proposition implies that

$$v_F\left(\operatorname{Tr}_{L/F}(\pi_L^i)\right)\geqslant s+1+\left[\frac{i-1-s}{p}\right],\ v_F\left(\operatorname{Tr}_{L/F}(\delta)\right)\geqslant s+1+\left[\frac{2i-1-s}{p}\right]$$

and for $i<s$

$$v_F\left(\operatorname{Tr}_{L/F}(\pi_L^i)\right)\geqslant i+1,\quad v_F\left(\operatorname{Tr}_{L/F}(\delta)\right)\geqslant i+1.$$

Therefore, the third diagram is commutative. Further, using (*) of 13.4, one can write

$$1 = N_{L/F}\left(\frac{\sigma(\pi_L)}{\pi_L}\right) \equiv 1 + N_{L/F}(\eta)\pi_F^s + \mathrm{Tr}_{L/F}(\eta\pi_L^s) \mod \pi_F^{s+1}.$$

We deduce that $\mathrm{Tr}_{L/F}(\eta\pi_L^s) \equiv -N_{L/F}(\eta)\pi_F^s \mod \pi_F^{s+1}$. Since $N_{L/F}(\eta) \equiv \eta^p \mod \pi_L$ in view of $U_L \subset U_F U_{1,L}$, we conclude that

$$N_{L/F}(1+\theta\eta\pi_L^s) - 1 - \eta^p\pi_F^s(\theta^p - \theta) \in \pi_L^{ps+1}\theta\mathcal{O}_L$$

for $\theta \in \mathcal{O}_F$. This implies the commutativity of the fourth (putting $\theta \in \mathcal{O}_F$) and the fifth (when $\theta \in \pi_F^i\mathcal{O}_F$) diagrams. Finally, if $p \nmid i, \theta \in \mathcal{O}_F$, then

$$\frac{\sigma(1+\theta\pi_L^i)}{1+\theta\pi_L^i} \equiv 1 + i\theta\eta\pi_L^{i+s} \mod \pi_L^{i+s+1}.$$

This means that $N_{L/F}(1+i\theta\eta\pi_L^{i+s}) \in N_{L/F}U_{s+i+1,L}$, and the groups $N_{L/F}(U_{s+i,L})$, $N_{L/F}(U_{s+i+1,L})$ are equal. □

REMARK. Compare the behaviour of the norm map with the behaviour of raising to the pth power in Proposition 4.7.

COROLLARY. *$U_{s+1,F} = N_{L/F}U_{s+1,L}$.*
If $\overline{F}$ is algebraically closed then $N_{L/F}L^\times = F^\times$.

Proof. It follows immediately from the last diagram of the Proposition, since the multiplication by $(-\overline{\eta})^{p-1}$ is an isomorphism of the additive group $\overline{F}$. □

14. Artin–Schreier Extensions

A theorem of Artin and Schreier asserts that every cyclic extension of degree p over a field K of characteristic p is generated by a root of the polynomial

$$X^p - X - \alpha, \quad \alpha \in K.$$

This subsection deals with a remarkable extension of this result to complete discrete valuation fields of characteristic 0 with residue field of characteristic p.

14.1. First we treat the case of unramified extensions. The polynomial $X^p - X$ is denoted by $\wp(X)$.

LEMMA. *Let L/F be an unramified Galois extension of degree $p = \mathrm{char}(\overline{F})$. Then $L = F(\lambda)$, where λ is a root of the polynomial $X^p - X - \alpha$ for some $\alpha \in U_F$ with $\overline{\alpha} \notin \wp(\overline{F})$.*

Proof. Let $\overline{L} = \overline{F}(\theta)$, where θ is a root of the polynomial $X^p - X - \eta$ for some $\eta \notin \wp(\overline{F})$. Then the polynomial $X^p - X - \alpha = 0$, with $\alpha \in U_F$, such that $\overline{\alpha} = \eta$, has a root λ in L, by Hensel Lemma 9.2. Thus, $L = F(\lambda)$. □

14.2. Now we study the case of totally ramified extensions.

Let L/F be a totally ramified Galois extension of degree $p = \mathrm{char}(\overline{F})$. Let σ be a generator of $\mathrm{Gal}(L/F), \pi_L$ a prime element in L and $s = v_L(\pi_L^{-1}\sigma(\pi_L) - 1)$.

LEMMA. *For $\beta \in L$ there exists an element $b \in F$ such that*

$$v_L(\sigma\beta - \beta) = v_L(\beta - b) + s.$$

Proof. Let $\beta = a_0 + a_1\pi_L + \cdots + a_{p-1}\pi_L^{p-1}$ with $a_i \in F$ (see Proposition 10.6). Then

$$\sigma(\beta) - \beta = a_1\pi_L\gamma + \cdots + a_{p-1}\pi_L^{p-1}\left((1+\gamma)^{p-1} - 1\right),$$

where $\gamma = \pi_L^{-1}\sigma(\pi_L) - 1$. Since $v_L(\gamma) = s > 0$, we get

$$(1+\gamma)^i - 1 \equiv i\gamma \mod \pi_L^{s+1} \quad \text{for } i \geqslant 0.$$

Hence, $v_L\left(a_i\pi_L^i((1+\gamma)^i - 1)\right)$ are distinct for $1 \leqslant i \leqslant p-1$. Put $b = a_0$. Then $v_L(\sigma(\beta) - \beta) = v_L((\beta - b)\gamma) = v_L(\beta - b) + s$, as desired. □

14.3. PROPOSITION. *Let F be a complete discrete valuation field with residue field of characteristic $p > 0$. Let L be a totally ramified Galois extension of degree p of F.*

If $\mathrm{char}(F) = p$ *then* $p \nmid s$.

If $\mathrm{char}(F) = 0$, *then $s \leqslant pe/(p-1)$, where $e = e(F)$ is the absolute index of ramification of F. In this case, if $p | s$, then a primitive pth root of unity belongs to F, and $s = pe/(p-1), L = F(\sqrt[p]{\alpha})$ with some $\alpha \in F^\times, \alpha \notin U_F F^{\times p}$.*

Proof. First let $\mathrm{char}(F) = p$. Then $(1 + \theta\pi_F^i)^p = 1 + \theta^p\pi_F^{pi}$ for $\theta \in U_F$. One can take $\pi_F = N_{L/F}\pi_L$ for a prime element π_L in L. Then it follows from 13.4 that $\pi_F \equiv \pi_L^p \mod \pi_L^{p+1}$. Assume that $s = pi$. Then $N_{L/F}U_{pi+1,L} \subset U_{pi+1,F}$, and we get the congruence $1 + \theta^p\pi_F^{pi} \equiv N_{L/F}(1 + \theta\pi_L^{pi}) \mod \pi_F^{pi+1}$ that contradicts the fourth diagram of Proposition 13.5. Hence, $p \nmid s$.

Now let $\mathrm{char}(F) = 0$. Assume that $s > pe/(p-1)$. Let $\varepsilon = 1 + \theta\pi_F^s \in U_{s,F}$ with $\theta \in U_F$. Corollary 2 of 4.8 shows that $\varepsilon = \varepsilon_1^p$ for some $\varepsilon_1 = 1 + \theta_1\pi_F^{s-e} \in U_F$ with $\theta_1 \in U_F$. Then $N_{L/F}U_{p(s-e),L} \not\subset U_{s+1,F}$, but $p(s-e) \geqslant s+1$, which is impossible because of Corollary 13.5. Hence, $s \leqslant pe/(p-1)$. By the same reasons as in the case of $\mathrm{char}(F) = p$, it is easy to verify that if $s = pi < pe/(p-1)$, then $1 + \theta^p\pi_F^{pi} \equiv N_{L/F}(1 + \theta\pi_L^{pi}) \mod \pi_F^{pi+1}$, which is impossible. Therefore, in this case we get $s = pe/(p-1)$. One can write

$\sigma(\pi_L)\pi_L^{-1} \equiv 1+\theta\pi_F^{e/(p-1)} \mod \pi_L^{pe/(p-1)+1}$. Then, acting by $N_{L/F}$, we get $1 \equiv (1+\theta\pi_F^{e/(p-1)})^p \mod \pi_F^{pe/(p-1)+1}$. But $U_{pe/(p-1)+1,F} \subset U^p_{e/(p-1)+1,F}$ (see Corollary 2 of 4.8), so we can find an element $\zeta \equiv 1+\theta\pi_F^{e/(p-1)} \mod \pi_F^{e/(p-1)+1}$, such that $\zeta^p = 1$; ζ is a primitive pth root of unity in F, hence $L = F(\sqrt[p]{\alpha})$ for some $\alpha \in F^\times$, by Kummer theory. Writing $\alpha = \pi_F^a\varepsilon_1$ with $\varepsilon_1 \in U_F$ and assuming $p|a$, we can replace α with ε_1. Since $\overline{L} = \overline{F}$ we obtain $\overline{\varepsilon}_1 \in \overline{F}^p$ (otherwise L/F would not be totally ramified) and $\varepsilon_1 \equiv \varepsilon_2^p \mod \pi_L$ for some $\varepsilon_2 \in U_F$. Replacing ε_1 with $\varepsilon_3 = \varepsilon_1\varepsilon_2^{-p}$, we get $\varepsilon_3 \in U_{1,F}$, $L = F(\eta_3)$, $\eta_3^p = \varepsilon_3$. Note that

$$\frac{\sigma(1+\rho\pi_L^i)}{1+\rho\pi_L^i} \equiv 1+\rho i\eta\pi_L^{i+pe/(p-1)} \mod \pi_L^{1+i+pe/(p-1)}$$

for $\rho \in U_F$. Hence $\eta_3^{-1}\sigma(\eta_3) \equiv 1 \mod \pi_L^{1+pe/(p-1)}$, but $\eta_3^{-1}\sigma(\eta_3)$ is a primitive pth root of unity. This contradiction proves that $\alpha \notin U_F F^{\times p}$. □

14.4. PROPOSITION. *Let F be a complete discrete valuation field with residue field of characteristic $p > 0$. Let L be a Galois totally ramified extension of degree p, $s = s(L|F)$.*

Suppose that $s \neq pe/(p-1)$ if $\operatorname{char}(F) = 0$, where $e = e(F)$. Then $L = F(\lambda)$, where λ is a root of some polynomial $X^p - X - \alpha$ with $\alpha \in F$, $v_F(\alpha) = -s$.

Proof. The previous Proposition shows that $p \nmid s$. First consider the case of $\operatorname{char}(F) = p$. Then, by Artin–Schreier theory, $L = F(\lambda)$, where λ is a root of a suitable polynomial $X^p - X - \alpha$ with $\alpha \in F$. Let σ be a generator of $\operatorname{Gal}(L/F)$. Then $(\sigma(\lambda)-\lambda)^p = \sigma\lambda - \lambda$. Since $\lambda \notin F$, we get $\sigma(\lambda) - \lambda = a$ with $a \in \{1,\dots,p-1\}$. Then $\lambda^{-1}\sigma(\lambda) = 1 + a\lambda^{-1}$, and hence Proposition 13.5 implies $1+a\lambda^{-1} \in U_{s,L}$. This shows $v_L(\lambda) \leqslant -s$ and $v_F(\alpha) \leqslant -s$. Put $t = v_F(\alpha)$. Write $\lambda \equiv \pi_L^t\theta \mod \pi_L^{t+1}$ with $\theta \in U_F$ and a prime element π_L in L. If $t = pt'$, then $\alpha \equiv \pi_L^{pt}\theta^p \equiv \pi_F^{pt'}\theta^p \mod \pi_L^{pt+1}$, where $\pi_F = N_{L/F}\pi_L \equiv \pi_L^p \mod \pi_L^{p+1}$ is a prime element in F. Replacing λ by $\lambda' = \lambda - \pi_F^{t'}\theta$ and α by $\alpha' = \alpha - \pi_F^{pt'}\theta^p + \pi_F^{t'}\theta$, we get $\lambda'^p - \lambda' = \alpha'$ and $L = F(\lambda'), v_F(\alpha') > v_F(\alpha)$. Proceeding in this way we can assume $p \nmid t$ because $v_F(\alpha') \leqslant -s$. Then it follows from 13.4 that $v_L(\lambda^{-1}\sigma(\lambda) - 1) = s$ and $v_F(\alpha) = -s$.

Now we consider the case of $\operatorname{char}(F) = 0$.

First, we will show that there is an element $\lambda_1 \in L$, such that $v_L(\lambda_1) = -s$ and $v_L(\sigma(\lambda_1) - \lambda_1 - 1) > 0$. Indeed, put $\beta = -\pi_L^{-s}\rho s^{-1}$ with $\rho \in U_F$. Then

$$\sigma(\beta) - \beta = -\pi_L^{-s}\rho s^{-1}((1+\eta\pi_L^s)^{-s} - 1) \equiv \rho\eta \mod \pi_L.$$

Hence, if we choose $\overline{\rho} = \overline{\eta}^{-1}$, then $v_L(\sigma(\beta) - \beta - 1) > 0$. Put $\lambda_1 = \beta$.

Since $s < pe/(p-1) = e(L)/(p-1)$, we get $p\lambda_1^{p-1} \equiv 0 \mod \pi_L$,

$$v_L(\sigma(\lambda_1^p) - \lambda_1^p - 1) > 0 \qquad \text{and} \qquad v_L(\sigma\wp(\lambda_1) - \wp(\lambda_1)) > 0.$$

Second, we will construct a sequence $\{\lambda_n\}$, $n \geqslant 0$, of elements in L satisfying the conditions for $n > 0$:

$$v_L(\lambda_n) = -s, \quad v_L(\lambda_{n+1} - \lambda_n) \geqslant v_L(\lambda_n - \lambda_{n-1}) + 1,$$
$$v_L(\sigma\wp(\lambda_{n+1}) - \wp(\lambda_{n+1})) \geqslant v_L\big(\sigma\wp(\lambda_n) - \wp(\lambda_n)\big) + 1.$$

Then for $\lambda = \lim \lambda_n$ we obtain $\sigma\wp(\lambda) = \wp(\lambda)$, or in other words $\lambda^p - \lambda = \alpha \in F$ and $v_F(\alpha) = -s$.

Put $\lambda_0 = 0$. Denote $\delta_n = \sigma\wp(\lambda_n) - \wp(\lambda_n)$. Then $v_L(\delta_n) > 0$. If $\delta_n = 0$, then put $\lambda_m = \lambda_n$ for $m > n$. Otherwise, by Lemma 14.2, there exists an element $c_n \in F$ such that

$$v_L(\sigma\wp(\lambda_n) - \wp(\lambda_n)) = v_L(\wp(\lambda_n) - c_n) + s.$$

Put $\mu_n = \wp(\lambda_n) - c_n$, $\lambda_{n+1} = \lambda_n + \mu_n$. Then $\sigma\mu_n = \mu_n + \delta_n$, $v_L(\sigma(\lambda_{n+1}) - \lambda_{n+1} - 1) > 0$ and $v_L(\mu_n) > -s$, $v_L(\lambda_{n+1}) = -s$. So

$$\begin{aligned} v_L(\lambda_{n+1} - \lambda_n) &= v_L(\mu_n) = -s + v_L(\sigma\wp(\lambda_n) - \wp(\lambda_n)) \\ &\geqslant -s + 1 + v_L(\sigma\wp(\lambda_{n-1}) - \wp(\lambda_{n-1})) = v_L(\lambda_n - \lambda_{n-1}) + 1 \end{aligned}$$

for $n > 1$.

For $n = 1$ from the previous arguments we get

$$v_L(\lambda_2 - \lambda_1) = -s + v_L(\sigma\wp(\lambda_1) - \wp(\lambda_1)) \geqslant v_L(\lambda_1 - \lambda_0) + 1 = 1 - s.$$

Furthermore, $\sigma\mu_n - \mu_n = \delta_n$ and

$$\sigma\wp(\mu_n) - \wp(\mu_n) = \wp(\mu_n + \delta_n) - \wp(\mu_n) = -\delta_n + \sum_{i=1}^{p} \binom{p}{i} \mu_n^{p-i}\delta_n^i.$$

Since $v_L(\mu_n) = v_L(\lambda_{n+1} - \lambda_n) \geqslant v_L(\lambda_1 - \lambda_0) = -s$ and $v_L(p\mu_n^{p-1}) = pe - (p-1)s > 0$, we get

$$v_L(\sigma\wp(\mu_n) - \wp(\mu_n) + \delta_n) \geqslant v_L(\delta_n) + 1.$$

Moreover,

$$\begin{aligned} \sigma\wp(\lambda_{n+1}) - \wp(\lambda_{n+1}) &= \sigma\wp(\lambda_n) - \wp(\lambda_n) \\ &\quad + \sigma\wp(\mu_n) - \wp(\mu_n) + \sum_{i=1}^{p-1} \binom{p}{i} \left(\sigma(\lambda_n^{p-i}\mu_n^i) - \lambda_n^{p-i}\mu_n^i\right) \end{aligned}$$

and

$$\sigma(\lambda_n^{p-i}\mu_n^i) - \lambda_n^{p-i}\mu_n^i = \lambda_n^{p-i}\mu_n^i(\varepsilon_n^{p-i}(1 + \delta_n\mu_n^{-1})^i - 1),$$

where $\lambda_n^{-1}\sigma\lambda_n = \varepsilon_n \in U_{s,L}$ since $p\not|s$, and we also have $v_L(\delta_n\mu_n^{-1}) = v_L(\delta_n) + s - v_L(\delta_n) = s$. Hence, for $1 \leqslant i \leqslant p-1$ we get

$$\begin{aligned} v_L\big(\sigma(\lambda_n^{p-i}\mu_n^i) - \lambda_n^{p-i}\mu_n^i\big) &\geqslant -(p-i)s + i(v_L(\delta_n) - s) + s \\ &\geqslant -(p-1)s + v_L(\delta_n) \geqslant -pe + v_L(\delta_n) + 1. \end{aligned}$$

Thus,

$$v_L\big(\sigma\wp(\lambda_{n+1}) - \wp(\lambda_{n+1})\big) \geqslant v_L(\delta_n) + 1,$$

which completes the proof. □

14.5. The assertions converse to Propositions 14.1 and 14.4 can be formulated as follows.

PROPOSITION. *Let F be a complete discrete valuation field with a residue field of characteristic $p > 0$.*

Then every polynomial $X^p - X - \alpha$ with $\alpha \in F$, $v_F(\alpha) > -pe/(p-1)$ if $\operatorname{char}(F) = 0$ *and $e = e(F)$, either splits completely or has a root λ which generates a cyclic extension $L = F(\lambda)$ over F of degree p. In the last case $v_L(\sigma(\lambda) - \lambda - 1) > 0$ for some generator σ of* $\operatorname{Gal}(L/F)$.

If $\alpha \in U_F, \overline{\alpha} \notin \wp(\overline{F})$, then L/F is unramified; if $\alpha \in \mathcal{M}_F$, then $\lambda \in F$; if $\alpha \notin \mathcal{O}_F$ and $p \nmid v_F(\alpha)$, then L/F is totally ramified with $s = -v_F(\alpha)$.

Proof. Let $\alpha \in \mathcal{M}_F$, $f(X) = X^p - X - \alpha$. Then $f(0) \in \mathcal{M}_F$, $f'(0) \notin \mathcal{M}_F$, and, by Hensel Lemma 8.2, for every integer a there is $\lambda \in \mathcal{M}_F$ with $f(\lambda) = 0$, $\lambda - a \in \mathcal{M}_F$. This means that $f(X)$ splits completely in F. If $\alpha \in U_F, \overline{\alpha} \notin \wp(\overline{F})$, then Proposition 10.2 shows that $F(\lambda)/F$ is an unramified extension and Proposition 10.3 shows that $F(\lambda)/F$ is Galois of degree p. The generator $\sigma \in \operatorname{Gal}(L/F)$, for which $\bar{\sigma}\bar{\alpha} = \overline{\alpha} + 1$, is the required one.

If $\alpha \notin \mathcal{O}_F$, then let λ be a root of the polynomial $X^p - X - \alpha$ in F^{alg} and $L = F(\lambda)$. Put

$$g(Y) = (\lambda + Y)^p - (\lambda + Y) - \alpha = Y^p + \binom{p}{1}\lambda Y^{p-1} + \dots + \binom{p}{p-1}\lambda^{p-1}Y - Y.$$

If $\operatorname{char}(F) = p$, then L/F is evidently cyclic of degree p when $\alpha \notin \wp(F)$. If $\operatorname{char}(F) = 0$, then $v_L(\binom{p}{i}\lambda^i) > e(L|F)(e - ei/(p-1)) \geqslant 0$ for $i \leqslant p-1$ and $\overline{g}(Y) = Y^p - Y$ over $\overline{L}$. Hence by Hensel Lemma $g(Y)$ splits completely in L. Therefore, L/F is cyclic of degree p if $f(X)$ does not split over F. Let σ be a generator of $\operatorname{Gal}(L/F)$, such that $\sigma(\lambda) - \lambda$ is a root of $g(Y)$ and is congruent to 1 mod π_L. Then $v_L(\sigma(\lambda) - \lambda - 1) > 0$. If $p \nmid v_F(\alpha)$, then the equality $pv_L(\lambda) = v_L(\alpha)$ implies $e(L|F) = p$, and L/F is totally ramified. It follows from the definition of s in 13.4 that $s = v_L(\sigma(\lambda)\cdot\lambda^{-1} - 1)$, and consequently $s = v_L(\sigma(\lambda) - \lambda) - v_L(\lambda) = -v_L(\lambda) = -v_F(\alpha)$. □

COROLLARY. *Let λ be a root of the polynomial $X^p - X + \theta^p\alpha$ with $\theta \in U_F$, $v_F(\alpha) = -s > -pe/(p-1)$, $p \nmid s$. Let $L = F(\lambda)$. Then $\alpha \in N_{L/F}L^\times$ and $1 + \theta^{-p}\wp(\mathcal{O}_F)\alpha^{-1} + \pi_F^{s+1}\mathcal{O}_F \subset N_{L/F}L^\times$, where $\wp(\mathcal{O}_F) = \{\wp(\beta) : \beta \in \mathcal{O}_F\}$.*

Proof. The preceding Proposition shows that L/F is a totally ramified extension of degree p and that $v_L(\sigma(\pi_L)\pi_L^{-1} - 1) = s$ for a generator σ of $\mathrm{Gal}(L/F)$ and a prime element π_L in L. Put $f(X) = X^p - X + \theta^p\alpha$. Then we get $N_{L/F}(-\lambda) = f(0) = \theta^p\alpha$ and $\alpha = N_{L/F}(-\lambda\theta^{-1})$. For $\beta \in \mathcal{O}_F$ put

$$g(Y) = f(\beta - Y) = (\beta - Y)^p - (\beta - Y) + \theta^p\alpha.$$

Then

$$N_{L/F}(\beta - \lambda) = g(0) = \wp(\beta) + \theta^p\alpha.$$

Therefore, $1 + \wp(\beta)\,\theta^{-p}\alpha^{-1} \subset N_{L/F}L^\times$. It remains to use Corollary 13.5. □

15. Hasse–Herbrand Function

In this section we associate to a finite separable extension L/F a certain real function $h_{L/F}$ which partially describes the behaviour of the norm map from arithmetical point of view. Then we relate the function $h_{L/F}$ which was originally introduced in a different way by Hasse and Herbrand to properties of ramification subgroups.

We maintain the hypothesis of the preceding sections concerning F, and assume in addition that all residue field extensions are separable.

15.1. PROPOSITION. *Let the residue field $\overline{F}$ be infinite. Let L/F be a finite Galois extension, $N = N_{L/F}$. Then there exists a unique function*

$$h = h_{L/F} : \mathbb{N} \longrightarrow \mathbb{N}$$

such that $h(0) = 0$ and

$$NU_{h(i),L} \subset U_{i,F}, \quad NU_{h(i),L} \not\subset U_{i+1,F}, \quad NU_{h(i)+1,L} \subset U_{i+1,F}.$$

Proof. The uniqueness of h follows immediately. Indeed, for $j > h(i)$ $NU_{j,L} \subset U_{i+1,F}$, hence if $\tilde{h}$ is another function with the required properties, then $\tilde{h}(i) \leqslant h(i), h(i) \leqslant \tilde{h}(i)$, i.e., $h = \tilde{h}$.

As for the existence of h, we first consider the case of an unramified extension L/F. Then Proposition 13.2 shows that in this case $h(i) = i$ (because $N_{\overline{L}/\overline{F}}(\overline{L}^{\times}) \neq 1$ and $\mathrm{Tr}_{\overline{L}/\overline{F}}\overline{L} = \overline{F}$). The next case to consider is a totally ramified cyclic extension L/F of prime degree. In this case Proposition 13.3 and Proposition 13.5 describe the behaviour of $N_{L/F}$. By means of the homomorphisms $\lambda_{i,L}$, the map $N_{L/F}$ is determined by some nonzero polynomials over $\overline{L}$. The image of $\overline{L}$ under the action of such a polynomial is not zero since $\overline{L}$ is infinite. Hence, we obtain

$$h(i) = |L : F|i,$$

if L/F is totally tamely ramified, and

$$h(i) = \begin{cases} i, & i \leqslant s, \\ s(1-p) + pi, & i \geqslant s, \end{cases}$$

if L/F is totally ramified of degree $p = \mathrm{char}(\overline{F}) > 0$.

Now we consider the general case. Note that if we have the functions $h_{L/M}$ and $h_{M/F}$ for the Galois extensions $L/M, M/F$, then for the extension L/F one can put $h_{L/F} = h_{L/M} \circ h_{M/F}$. Indeed,

$$N_{L/F} U_{h_{L/F}(i),L} \subset N_{M/F} U_{h_{M/F}(i),M} \subset U_{i,F}.$$

Furthermore, the behaviour of $N_{L/F}$ is determined by some nonzero polynomials (the composition of the polynomials for $N_{L/M}$ and $N_{M/F}$, the existence of which can be assumed by induction). Hence

$$N_{L/F} U_{h_{L/F}(i),L} \not\subset U_{i+1,F}.$$

Since

$$N_{L/F} U_{h_{L/F}(i)+1,L} \subset N_{M/F} U_{h_{M/F}(i)+1,M} \subset U_{i+1,M},$$

we deduce that $h = h_{L/F}$ is the desired function.

In the general case we put $h_{L/F} = h_{L/L_0}$ for $L_0 = L \cap F^{\mathrm{ur}}$ and determine h_{L/L_0} by induction using Corollary 3 of 11.4, which shows that L/L_0 is solvable. □

15.2. To treat the case of finite residue fields we need

LEMMA. *Let L/F be a finite separable totally ramified extension. Then for an element $\alpha \in L$ we get*

$$N_{L/F}(\alpha) = N_{\widehat{L^{\mathrm{ur}}}/\widehat{F^{\mathrm{ur}}}}(\alpha)$$

where $\widehat{F^{\mathrm{ur}}}$ is the completion of F^{ur}, $\widehat{L^{\mathrm{ur}}} = L\widehat{F^{\mathrm{ur}}}$.

Proof. Let $L = F(\pi_L)$ with a prime element π_L in L, and let $\alpha \in L$. Let

$$\alpha\pi_L^i = \sum_{j=0}^{n-1} c_{ij}\pi_L^j \quad \text{with } c_{ij} \in F, 0 \leqslant i \leqslant n-1, n = |L:F|.$$

Then $N_{L/F}(\alpha) = \det(c_{ij})$. Since $L^{\mathrm{ur}} = F^{\mathrm{ur}}(\pi_L)$ and

$$|L^{\mathrm{ur}} : F^{\mathrm{ur}}| = e(L^{\mathrm{ur}}|F^{\mathrm{ur}}) = e(L^{\mathrm{ur}}|F) = e(L|F) = |L:F|,$$

we get

$$N_{L^{\mathrm{ur}}/F^{\mathrm{ur}}}(\alpha) = \det(c_{ij}) = N_{L/F}(\alpha).$$

Finally, let E/F^{ur} be a finite totally ramified Galois extension, $E \supset L^{\mathrm{ur}}$. Let $G = \mathrm{Gal}(E/F^{\mathrm{ur}})$ and let $H = \mathrm{Gal}(E/L^{\mathrm{ur}})$. Let G be the disjoint union of $\sigma_i H$ with $\sigma_i \in G, 1 \leqslant i \leqslant |L^{\mathrm{ur}} : F^{\mathrm{ur}}|$. Then

$$N_{L^{\mathrm{ur}}/F^{\mathrm{ur}}}(\alpha) = \prod \sigma_i(\alpha) = N_{\widehat{L^{\mathrm{ur}}}/\widehat{F^{\mathrm{ur}}}}(\alpha),$$

because G and H are isomorphic to $\mathrm{Gal}(\widehat{E}/\widehat{F^{\mathrm{ur}}})$ and $\mathrm{Gal}(\widehat{E}/\widehat{L^{\mathrm{ur}}})$ by (iv) in Theorem 9.8. □

This Lemma shows that for a finite totally ramified Galois extension L/F the functions $h_{L/F}$ and $h_{\widehat{L^{\mathrm{ur}}}/\widehat{F^{\mathrm{ur}}}}$ coincide. Now, if L/F is a finite Galois extension, we get

$$h_{L/F} = h_{L/L_0} = h_{\widehat{L^{\mathrm{ur}}}/\widehat{F^{\mathrm{ur}}}}.$$

So, if $\overline{F}$ is finite we put $h_{L/F} = h_{\widehat{L^{\mathrm{ur}}}/\widehat{F^{\mathrm{ur}}}}$ (the residue field of $\widehat{F^{\mathrm{ur}}}$ is infinite as the separable closure of a finite field).

It is useful to extend this function to real numbers. For an unramified extension, a tamely totally ramified extension of prime degree, a totally ramified extension of degree $p = \mathrm{char}(\overline{F}) > 0$ put

$$h_{L/F}(x) = x, \quad h_{L/F}(x) = |L:F|x, \quad h_{L/F}(x) = \begin{cases} x, & x \leqslant s, \\ s(1-p)+px, & x \geqslant s \end{cases}$$

for real $x \geqslant 0$ respectively. Using the solvability of L/L_0 (Corollary 3 of 11.4) and the equality $h_{L/F} = h_{L/M} \circ h_{M/F}$ define now $h_{L/F}(x)$ as the composite of the functions for a tower of cyclic subextensions in L/L_0.

PROPOSITION. *Thus defined function $h_{L/F}\colon [0,+\infty) \longrightarrow [0,+\infty)$ is independent on the choice of a tower of subfields. The function $h_{L/F}$ is called the Hasse–Herbrand function of L/F. It is piecewise linear, continuous and increasing.*

Proof. By induction on the degree of L/F it suffices to show that if M_1/M, M_2/M are linearly disjoint cyclic extensions of prime degree, then

$$h_{E/M_1} \circ h_{M_1/M} = h_{E/M_2} \circ h_{M_2/M} \qquad (*)$$

where $E = M_1 M_2$.

Note that each of $h_{M_1/M}(x)$, $h_{M_2/M}(x)$ has at most one point at which its derivate is not continuous. Therefore there are at most two points at which the function of the left (resp. right) hand side of $(*)$ has discontinuous derivative. By looking at graphs of the functions it is obvious that at such points the derivative strictly increases and there is at most one such non-integer point for at most one of the composed functions of the left hand side and the right hand side of $(*)$. At this point (if it exists) the derivative jumps from p to p^2.

From the uniqueness in the preceding Proposition we deduce that the left and right hand sides of $(*)$ are equal at all nonnegative integers. Thus, elementary calculus shows that the left and right hand sides of $(*)$ are equal at all nonnegative real numbers. □

15.3. Let the residue field of F be perfect. For a finite separable extension L/F put

$$h_{L/F} = h_{E/L}^{-1} \circ h_{E/F},$$

where E/F is a finite Galois extension with $E \supset L$. Then $h_{L/F}$ is well defined, since if E'/F is a Galois extension with $E' \supset L$ and $E'' = E'E$, then

$$h_{E''/L}^{-1} \circ h_{E''/F} = \left(h_{E''/E'} \circ h_{E'/L}\right)^{-1} \circ \left(h_{E''/E'} \circ h_{E'/F}\right) = h_{E'/L}^{-1} \circ h_{E'/F}$$

and, similarly, $h_{E''/L}^{-1} \circ h_{E''/F} = h_{E/L}^{-1} \circ h_{E/F}$. We can easily deduce from this that the equality

$$h_{L/F} = h_{L/M} \circ h_{M/F} \qquad (*)$$

holds for separable extensions.

PROPOSITION. *Let L/F be a finite separable extension, and let $\overline{F}$ be perfect. Then $h_{L/F}(\mathbb{N}) \subset \mathbb{N}$ and the left and right derivatives of $h_{L/F}$ at any point are positive integers.*

Proof. Let E/F be a finite Galois extension with $E \supset L$. Then from Lemma 15.2 we get

$$h_{L/F} = h_{E/L}^{-1} \circ h_{E/F} = h_{\widehat{E^{\mathrm{ur}}}/\widehat{L^{\mathrm{ur}}}}^{-1} \circ h_{\widehat{E^{\mathrm{ur}}}/\widehat{F^{\mathrm{ur}}}} = h_{\widehat{L^{\mathrm{ur}}}/\widehat{F^{\mathrm{ur}}}}.$$

Put $G = \mathrm{Gal}(\widehat{E^{\mathrm{ur}}}/\widehat{F^{\mathrm{ur}}}), H = \mathrm{Gal}(\widehat{E^{\mathrm{ur}}}/\widehat{L^{\mathrm{ur}}})$. Since G is a solvable group, there exists a chain of normal subgroups

$$G \triangleright G_{(1)} \triangleright \cdots \triangleright G_{(m)} = \{1\},$$

such that $G_{(i)}/G_{(i+1)}$ is a cyclic group of prime order. Then we obtain the chain of subgroups

$$G \geqslant G_{(1)}H \geqslant \ldots \geqslant G_{(m)}H = H,$$

for which $G_{(i+1)}H$ is of prime index or index 1 in $G_{(i)}H$. This shows the existence of a tower of fields

$$\widehat{F^{\mathrm{ur}}} - M_1 - \cdots - M_{n-1} - M_n = \widehat{L^{\mathrm{ur}}},$$

such that M_{i+1}/M_i is a separable extension of prime degree. Therefore, it suffices to prove the statements of the Proposition for such an extension.

If M_{i+1}/M_i is a totally tamely ramified extension of degree l, then $\pi = \pi_1^l$ is a prime element in M_i for some prime element π_1 in M_{i+1}. Since l is relatively prime with $\operatorname{char}(\overline{F})$, we obtain, using the Henselian property of M_i and the fact that the residue field of $\widehat{M_i^{\mathrm{ur}}}$ is separably closed, that a primitive lth root of unity belongs to $\widehat{M_i^{\mathrm{ur}}}$. This means that $\widehat{M_{i+1}^{\mathrm{ur}}}/\widehat{M_i^{\mathrm{ur}}}$ is a Galois extension and

$$h_{M_{i+1}/M_i}(x) = lx.$$

If M_{i+1}/M_i is an extension of degree $p = \operatorname{char}(\overline{F}) > 0$, then let K/M_i be the smallest Galois extension, for which $K \supset M_{i+1}$. Let K_1 be the maximal tamely ramified extension of M_i in K; then $l = e(K_1|M_i) = e(K|M_{i+1})$ is relatively prime to p. Choose prime elements π and π_1 in M_{i+1} and K such that $\pi = \pi_1^l$. Let $f(X) \in M_i[X]$ be the monic irreducible polynomial of π over M_i. Then

$$f'(\pi) = \prod_{i=1}^{p-1}(\pi - \sigma^i(\pi)) = \prod_{i=1}^{p-1}(\pi_1^l - \sigma^i(\pi_1^l)),$$

where σ is a generator of $\operatorname{Gal}(K/K_1)$. Let s be defined for K/K_1 as in 13.4. Then $v_K\big(\pi_1^l - \sigma^i(\pi_1^l)\big) = l + s$ for $1 \leqslant i \leqslant p-1$, and $(p-1)(l+s) = v_K\big(f'(\pi)\big)$ is divisible by l. We deduce that $l|(p-1)s$ and

$$h_{M_{i+1}/M_i}(x) = \frac{1}{l}h_{K/K_1}(lx) = \begin{cases} x, & x \leqslant sl^{-1}, \\ s(1-p)l^{-1} + px, & x \geqslant sl^{-1}. \end{cases}$$

These considerations complete the proof. □

COROLLARY. *The function $h_{L/F}$ is piecewise linear, continuous and increasing.*

15.4. The following assertion adds more information about the relation between the Hasse–Herbrand function and the norm map.

PROPOSITION. *Let L/F be a finite separable extension. Then for $\varepsilon \in \mathcal{O}_L$*

$$h_{L/F}\Big(v_F\big(N_{L/F}(\varepsilon) - 1\big)\Big) \geqslant v_L(\varepsilon - 1).$$

If, in addition, L/F is totally ramified and if $v_L(\alpha-\beta)>0$ for $\alpha,\beta\in\mathcal{O}_L$, then

$$h_{L/F}\Big(v_F\big(N_{L/F}(\alpha)-N_{L/F}(\beta)\big)\Big)\geqslant v_L(\alpha-\beta).$$

Proof. Let's show that the second inequality is a consequence of the first one.

If $v_L(\beta)\geqslant v_L(\alpha-\beta)$, then $v_L(\alpha)\geqslant v_L(\alpha-\beta)$, and applying Theorem 9.5 we get

$$\begin{aligned}v_F\big(N_{L/F}(\alpha)-N_{L/F}(\beta)\big)&\geqslant\min\{v_F\big(N_{L/F}(\alpha)\big),v_F\big(N_{L/F}(\beta)\big)\}\\&=\min\{v_L(\alpha),v_L(\beta)\}\geqslant v_L(\alpha-\beta).\end{aligned}$$

Since $h_{L/F}(x)\geqslant x$, we obtain the second inequality.

If $v_L(\beta)<v_L(\alpha-\beta)$, then put $\varepsilon=\alpha\beta^{-1}$. Using the property of the derivatives of h in Proposition 15.3 and the first inequality we obtain

$$\begin{aligned}h_{L/F}\Big(v_F\big(N_{L/F}(\alpha)-N_{L/F}(\beta)\big)\Big)&=h_{L/F}\Big(v_F\big(N_{L/F}(\varepsilon)-1\big)+v_L(\beta)\Big)\\&\geqslant v_L(\varepsilon-1)+v_L(\beta)=v_L(\alpha-\beta).\end{aligned}$$

Now we verify the first inequality of the Proposition. By the proof of the previous Proposition, we may assume that L/F is totally ramified and $\overline{F}$ is algebraically closed. It is easy to show that if the first inequality holds for L/M and M/F, then it holds for L/F. The arguments from the proof of the previous Proposition imply now that it suffices to verify the first inequality for a separable extension L/F of prime degree. If L/F is tamely ramified, then L/F is Galois, and the inequality follows from Proposition 13.3. If $|L:F|=p=\operatorname{char}(\overline{F})>0$, then we may assume that ε is a principal unit. Proposition 13.5 implies the required inequality for the Galois case. In general, assume that E/F is the minimal Galois extension such that $E\supset L$, and let E_1 is the maximal tamely ramified subextension of F in E. Let $l=|E:L|=|E_1:F|$. Then $N_{L/F}(U_{i,L})=N_{E/F}(U_{li,E})\subset N_{E_1/F}(U_{j,E_1})$ with $j\geqslant h^{-1}_{E/E_1}(li)$. Hence, $N_{L/F}(U_{i,L})\subset U_{k,F}$ with $lk\geqslant h^{-1}_{E/E_1}(li)$, i.e., $k\geqslant h^{-1}_{L/F}(i)$, as desired. □

15.5. We will relate the Hasse–Herbrand function to ramification groups which are defined in 11.3.

If H is a subgroup of the Galois group G, then $H_x=H\cap G_x$. As for the quotients, the description is provided by the following

THEOREM. *(Herbrand theorem) Let L/F be a finite Galois extension and let M/F be a Galois subextension. Let x,y be nonnegative real numbers related by $y=h_{L/M}(x)$.*

Then the image of $\operatorname{Gal}(L/F)_y$ *in* $\operatorname{Gal}(M/F)$ *coincides with* $\operatorname{Gal}(M/F)_x$.

Proof. The cases $x \leqslant 1$ or $e(L|M) = 1$ are easy. Due to solvability of Galois groups of totally ramified extensions it is sufficient to prove the assertion in the case of a ramified cyclic extension L/M of prime degree l.

If $l \neq p$, then using Proposition 10.5 choose a prime element π of L such that $\pi_M = \pi^l$ is a prime element of M. Then for every $\tau \in \mathrm{Gal}(L/F)_1$ we have $\pi_M^{-1}\tau\pi_M = (\pi^{-1}\tau\pi)^l$ and therefore

$$v_L(\pi^{-1}\tau\pi - 1) = v_L\big((\pi^{-1}\tau\pi)^l - 1\big) = l v_M(\pi_M^{-1}\tau\pi_M - 1).$$

Consider now the most interesting case $l = p$, $x \geqslant 1$. Let π_L be a prime element of L. Put $s = s(L|M)$, see 13.4.

The element $\pi_M = N_{L/M}\pi_L$ is a prime element of M. Let $\tau \in \mathrm{Gal}(L/F)_y$. We have $\pi_M^{-1}\tau\pi_M = N_{L/M}(\pi_L^{-1}\tau\pi_L)$.

From Proposition 15.4 we get

$$h_{L/M}(v_M(\pi_M^{-1}\tau\pi_M - 1)) = h_{L/M}(v_M(N_{L/M}(\pi_L^{-1}\tau\pi_L) - 1)) \geqslant y,$$

so $\tau|_M$ belongs to $\mathrm{Gal}(M/F)_x$.

Conversely, if $\tau|_M \in \mathrm{Gal}(M/F)_x$, then $i = v_M(\pi_M^{-1}\tau\pi_M - 1) \geqslant x$. If $i \leqslant s = s(L|M)$ then applying 13.5 we deduce that $\tau \in \mathrm{Gal}(L/F)_i = \mathrm{Gal}(L/F)_y$. If $i > s$ then Proposition 11.5 and 13.5 show that $j = v_L(\pi_L^{-1}\tau\pi_L - 1) = s + pr$ for some nonnegative integer r.

If $r > 0$ then Proposition 13.5 implies that $i = s + r$ and $\tau \in \mathrm{Gal}(L/F)_j = \mathrm{Gal}(L/F)_y$. If $j = s$ then since $i > s$ from the same Proposition we deduce that

$$\frac{\tau\pi_L}{\pi_L} \equiv \frac{\sigma\pi_L}{\pi_L} \quad \mathrm{mod}\ \mathscr{M}_L^{s+1}$$

for an appropriate generator σ of $\mathrm{Gal}(L/M)$. Then $\tau\sigma^{-1}$ belongs to $\mathrm{Gal}(L/F)_k$ for $k > s$. Due to the previous discussions (view k as $j > s$ above) $k = h_{L/M}(i)$ and τ belongs to $\mathrm{Gal}(L/F)_y \mathrm{Gal}(L/M)$, as required. □

DEFINITION. Define the *upper ramification subgroup* $G(x)$ of $G = \mathrm{Gal}(L/F)$ as

$$G(x) = \mathrm{Gal}(L/F)_{h_{L/F}(x)}.$$

For a normal subgroup H of G the previous Theorem shows that

$$(G/H)(x) = G(x)H/H.$$

For an infinite Galois extension L/F define *upper ramification subgroup* of $G = \mathrm{Gal}(L/F)$ as the inverse limit (see 5.1 of Chapter 1)

$$G(x) = \varprojlim \mathrm{Gal}(M/F)(x)$$

where M/F runs through all finite Galois subextensions of L/F.

Real numbers x such that $G(x) \neq G(x+\delta)$ for every $\delta > 0$ are called *upper ramification jumps* of L/F.

For example, local class field theory for local fields with finite residue field implies that the set of upper ramification jumps of the Galois group of the maximal abelian extension is the set of natural numbers.

15.6. The following Proposition is a generalisation of results of Section 13.

Suppose that L/F is a finite totally ramified Galois extension and that $|L:F|$ is a power of $p = \operatorname{char}(\overline{F})$. Put $G = \operatorname{Gal}(L/F)$. For the chain of normal ramification groups

$$G = G_1 \geqslant G_2 \geqslant \ldots \geqslant G_n > G_{n+1} = \{1\}$$

let L_m be the fixed field of G_m; then we get the tower of fields

$$F = L_1 - L_2 - \cdots - L_n - L_{n+1} = L.$$

PROPOSITION. *Let* $1 \leqslant m \leqslant n$. *Then* $\operatorname{Gal}(L_{m+1}/L_m)$ *coincides with the ramification group* $\operatorname{Gal}(L_{m+1}/L_m)_m$, $\operatorname{Gal}(L_{m+1}/L_m)_{m+1} = \{1\}$, *and* $h_{L_{m+1}/L_m}(m) = m$.

Moreover,

if $i < m$, *then* $h_{L_{m+1}/L_m}(i) = i$ *and the homomorphism*

$$U_{i,L_{m+1}}/U_{i+1,L_{m+1}} \longrightarrow U_{i,L_m}/U_{i+1,L_m}$$

induced by N_{L_{m+1}/L_m} *is injective;*

if $i > m$, *then the homomorphism*

$$U_{h(i),L_{m+1}}/U_{h(i)+1,L_{m+1}} \longrightarrow U_{i,L_m}/U_{i+1,L_m}$$

induced by N_{L_{m+1}/L_m} *for* $h = h_{L_{m+1}/L_m}$ *is bijective.*

Furthermore, the homomorphism

$$U_{h(i),L}/U_{h(i)+1,L} \longrightarrow U_{i,F}/U_{i+1,F}$$

induced by $N_{L/F}$ *for* $h = h_{L/F}$, *is bijective if* $h(i) > n$.

Proof. Induction on m. Base of induction $m = n$. Since $\operatorname{Gal}(L/L_n)_x$ is equal to the group $\operatorname{Gal}(L/F)_x \cap \operatorname{Gal}(L/L_n)$, we deduce that $\operatorname{Gal}(L/L_n)_n = \operatorname{Gal}(L/L_n)$ and $\operatorname{Gal}(L/L_n)_{n+1} = \{1\}$, and $h_{L/L_n}(x) = x$ for $x \leqslant n$. All the other assertions for $m = n$ follow from Proposition 13.5.

Induction step $m+1 \to m$. The transitivity property of the Hasse–Herbrand function implies that $h_{L/L_{m+1}}(x) = x$ for $x \leqslant m+1$. Now from the previous Theorem

$$\operatorname{Gal}(L_{m+1}/L_m)_x = \operatorname{Gal}(L/L_m)_{h_{L/L_{m+1}}(x)} \operatorname{Gal}(L_{m+1}/L_m)/\operatorname{Gal}(L_{m+1}/L_m).$$

We deduce that $\operatorname{Gal}(L_{m+1}/L_m)_m = \operatorname{Gal}(L_{m+1}/L_m)$ and $\operatorname{Gal}(L_{m+1}/L_m)_{m+1} = \{1\}$. The rest follows from Proposition 13.5.

To deduce the last assertion note that $k = h_{L/F}(i) > n$ implies $j = h_{L_m/F}(i) > m$. □

COROLLARY. *The word "injective" in the Proposition can be replaced by "bijective" if $\overline{F}$ is perfect.*

15.7. PROPOSITION. *Let L/F be a finite Galois extension, $G = \operatorname{Gal}(L/F)$, $h = h_{L/F}$. Let h'_l and h'_r be the left and right derivatives of h. Then $h'_l(x) = |G_0 : G_{h(x)}|$, and*

$$h'_r(x) = \begin{cases} |G_0 : G_{h(x)}| & \text{if } h(x) \text{ is not integer,} \\ |G_0 : G_{h(x)+1}| & \text{if } h(x) \text{ is integer.} \end{cases}$$

Therefore

$$h_{L/F}(x) = \int_0^x |G_0 : G_{h(t)}| dt.$$

Proof. Using the equality $(*)$ of 15.3, we may assume that L/F is a totally ramified extension the degree of which is a power of $p = \operatorname{char}(\overline{F}) > 0$. Then $G = G_0 = G_1$. We proceed by induction on the degree $|L : F|$. Let L_n be identical to that of 15.6; then $|L_n : F| < |L : F|$. Since $(G/G_n)_m = G_m/G_n$ for $m \leqslant n$ due to 15.6, we deduce the following series of claims.

If $h_{L_n/F}(x) \leqslant n$, then, by Proposition 15.6, $h_{L/F}(x) = h_{L_n/F}(x)$ and

$$h'_l(x) = \big|(G/G_n) : (G/G_n)_{h(x)}\big| = |G : G_{h(x)}|.$$

If $h_{L_n/F}(x) < n$ and $h_{L/F}(x) = h_{L_n/F}(x)$ is not integer, then $h'_r(x) = |G : G_{h(x)}|$. If $h_{L_n/F}(x)$ is an integer $< n$, then

$$h'_r(x) = \big|(G/G_n) : (G/G_n)_{h(x)+1}\big| = |G : G_{h(x)+1}|.$$

Since the derivative (right derivative) of $h_{L/L_n}(x)$ for $x > n$ (resp. $x \geqslant n$) is equal to $|G_n : (G_n)_{n+1}| = |G_n|$, we deduce that if $h_{L_n/F}(x) > n$, then

$$h'_l(x) = |G_n| \cdot |G : G_n| = |G| = |G : G_{h(x)}|.$$

So if $h_{L_n/F}(x) \geqslant n$, then $h'_r(x) = |G_n| \cdot |G : G_n| = |G|$. This completes the proof. □

REMARK. The function $h_{L/F}$ often appears under the notation $\psi_{L/F}$; in which case it is defined in quite a different way by using ramification groups, not the norm map. This function is inverse to the function $\varphi_{L/F} = \int_0^x \frac{dt}{|G_0 : G_t|}$.

16. Norm and Ramification Groups

16.1. The following assertion is of general interest.

PROPOSITION. *(Hilbert "Satz 90") Let F be a field. Let L/F be a cyclic Galois extension, and let $N_{L/F}(\alpha) = 1$ for some $\alpha \in L$. Then there exists an element $\beta \in L$ such that $\alpha = \beta^{\sigma-1}$, where σ is a generator of* $\mathrm{Gal}(L/F)$.

Proof. Let $\beta(\gamma)$ denote

$$\gamma + \alpha^{-1}\sigma(\gamma) + \alpha^{-1}\sigma(\alpha^{-1})\sigma^2(\gamma) + \cdots + \alpha^{-1}\sigma(\alpha^{-1}) \cdot \cdots \cdot \sigma^{n-2}(\alpha^{-1})\sigma^{n-1}(\gamma)$$

for $\gamma \in L$, $n = |L : F|$. If $\beta(\gamma)$ were equal to 0 for all γ, then we would have a nontrivial solution $1, \alpha^{-1}, \alpha^{-1}\sigma(\alpha^{-1}), \ldots$ for the $n \times n$ system of linear equations with the matrix $\left(\sigma^i(\gamma_j)\right)_{0 \leqslant i,j \leqslant n-1}$, where $(\gamma_j)_{0 \leqslant j \leqslant n-1}$ is a basis of L over F. This is impossible because L/F is separable. Hence $\beta(\gamma) \neq 0$ for some $\gamma \in L$. Then $\beta = \beta(\gamma)$ is the desired element. □

COROLLARY. *If L is a cyclic unramified extension of F and $N_{L/F}(\alpha) = 1$ for $\alpha \in L$, then $\alpha = \gamma^{\sigma-1}$ for some element $\gamma \in U_L$.*

Proof. In this case a prime element π in F is also a prime one in L. By the Proposition, $\alpha = \beta^{-1}\sigma(\beta)$ with $\beta = \pi^i\varepsilon$, $\varepsilon \in U_L$. Then $\alpha = \varepsilon^{-1}\sigma(\varepsilon)$. □

Below in this section F is a complete discrete valuation field.

Recall that in Section 11 we employed the homomorphisms

$$\psi_i \colon G_i \longrightarrow U_{i,L}/U_{i+1,L}$$

(we put $U_{0,L} = U_L$), where $G = \mathrm{Gal}(L/F)$, π_L is a prime element in L, $i \geqslant 0$. Obviously these homomorphisms do not depend on the choice of π_L if L/F is totally ramified. The induced homomorphisms $G_i/G_{i+1} \longrightarrow U_{i,L}/U_{i+1,L}$ will be also denoted by ψ_i.

16.2. THEOREM. *Let L/F be a finite totally ramified Galois extension with group G. Let $h = h_{L/F}$. Then for every integer $i \geqslant 0$ the sequence*

$$1 \longrightarrow G_{h(i)}/G_{h(i)+1} \xrightarrow{\psi_{h(i)}} U_{h(i),L}/U_{h(i)+1,L} \xrightarrow{N_i} U_{i,F}/U_{i+1,F}$$

is exact (the right homomorphism N_i is induced by the norm map).

Proof. The injectivity of $\psi_{h(i)}$ follows from the definitions. It remains to show that if $N_{L/F}\alpha \in U_{i+1,F}$ for $\alpha \in U_{h(i),L}$, then

$$\alpha \equiv \frac{\sigma(\pi_L)}{\pi_L} \mod U_{h(i)+1,L}$$

for some $\sigma \in G_{h(i)}$.

If L/F is a tamely ramified extension of degree l, then the fourth commutative diagram of Proposition 13.3 shows that N_i is injective for $i \geqslant 1$, and the kernel of N_0 coincides with the group of lth roots of unity which is contained in F. Since $\pi_L = \sqrt[l]{\pi_F}$ is a prime element in L for some prime element π_F in F, we get $\ker(N_0) \subset \operatorname{im}(\psi_0)$, and in this case the sequence of the Theorem is commutative.

If L/F is a cyclic extension of degree $p = \operatorname{char}(\overline{F}) > 0$, then the fourth commutative diagram of Proposition 13.5 shows that $\ker(N_s) \subset \operatorname{im}(\psi_s)$ for $s = v_L(\pi_L^{-1}\sigma(\pi_L))$ and a generator σ of $\operatorname{Gal}(L/F)$. Other diagrams of Proposition 13.5 show that N_i is injective for $i \neq s$.

We proceed by induction on the degree $|L:F|$. Since we have already considered the tamely ramified case, we may assume that the maximal tamely ramified extension L_1 of F in L does not coincide with L. Since $|L:L_1|$ is a power of p, the homomorphism induced by N_{L/L_1}

$$U_{0,L}/U_{1,L} \longrightarrow U_{0,L_1}/U_{1,L_1}$$

is the raising to this power of p, and $\ker(N_0)$ is equal to the preimage under this homomorphism of the kernel of $U_{0,L_1}/U_{1,L_1} \longrightarrow U_{0,F}/U_{1,F}$. In other words $\ker(N_0)$ coincides with the group of all lth roots of unity for $l = |L_1 : F|$ which is contained in F. Hence the kernel of N_0 is contained in the image of ψ_0, since ψ_0 is injective and $|G_0 : G_1| = l$.

Now suppose $i \geqslant 1$. In this case we may assume $L_1 = F$ because the homomorphism N_i induced by $N_{L_1/F}$ is injective for $i \geqslant 1$. Let L_n be as in Proposition 15.6. Then one can express N_i as the composition

$$U_{h(i),L}/U_{h(i)+1,L} \xrightarrow{N'} U_{h_1(i),L_n}/U_{h_1(i)+1,L_n} \xrightarrow{N''} U_{i,F}/U_{i+1,F},$$

where N' and N'' are induced by N_{L/L_n} and $N_{L_n/F}$ respectively. Denote $h_1 = h_{L_n/F}$. If $h_1(i) \geqslant n$, then by Proposition 15.6 $\operatorname{Gal}(L_n/F)_{h_1(i)} = \{1\}$, and we may assume that N'' is injective. Then by the induction assumption $\ker N_i = \ker N'$ coincides with the set of elements $\pi_L^{-1}\sigma(\pi_L) \mod U_{h(i)+1,L}$, where σ runs over $\operatorname{Gal}(L/L_n)_n = G_n$. If $h_1(i) < n$ and $N_{L/F}(\alpha) \in U_{i+1,F}$ for some $\alpha \in U_{h(i),L}$, then $h(i) = h_1(i)$, and by the induction assumption,

$$N'(\alpha) \equiv \frac{\sigma(\pi_{L_n})}{\pi_{L_n}} \mod U_{h_1(i)+1,L_n}$$

for a prime element π_{L_n} in L_n and some $\sigma \in \operatorname{Gal}(L/F)$. We can take $\pi_{L_n} = N_{L/L_n}\pi_L$. Hence

$$N'(\alpha) \equiv N'\left(\frac{\sigma(\pi_L)}{\pi_L}\right) \mod U_{h_1(i)+1,L_n}.$$

The homomorphisms

$$U_{j,L}/U_{j+1,L} \longrightarrow U_{j,L_n}/U_{j+1,L_n}$$

induced by N_{L/L_n}, are injective for $j < n$ by Proposition 15.6. Therefore, the element $\pi_L^{-1}\sigma(\pi_L)$ belongs to $U_{h(i),L}$ and so $\sigma \in G_{h(i)}$,

$$\alpha \equiv \frac{\sigma(\pi_L)}{\pi_L} \mod U_{h(i)+1,L}.$$

□

16.3. Now we study ramification numbers of abelian extensions. We shall see that these satisfy much stronger congruences than those of Proposition 11.5.

THEOREM. (*Hasse–Arf theorem*) *Let L/F be a finite abelian extension, and let the residue extension $\overline{L}/\overline{F}$ be separable. Let $G = \mathrm{Gal}(L/F)$. Then $G_j \neq G_{j+1}$ for an integer $j \geqslant 0$ implies $j = h_{L/F}(j')$ for an integer $j' \geqslant 0$. In other words, upper ramification jumps of abelian extensions are integers.*

Proof. We may assume that $j > 0$ and that L/F is totally ramified. Let E/F be the maximal p-subextension in L/F, and $m = |L:E|$. Let π_L be a suitable prime element in L such that $\pi_L^m \in E$. For $\sigma \in G_j$, $\sigma \notin G_{j+1}$ we get $\pi_L^{-m}\sigma\pi_L^m = 1 + m\theta\pi_L^j$ for some $\theta \in U_L$; therefore $j = mj_1$, and $\sigma|_E \in \mathrm{Gal}(E/F)_{j_1}$, $\sigma \notin \mathrm{Gal}(E/F)_{j_1+1}$. If we verify that $j_1 = h_{E/F}(j')$ for some integer j', then $j = h_{L/F}(j')$. Thus, we may also assume $G = G_1$.

If L/F is cyclic of degree $p = \mathrm{char}(\overline{F})$, then the required assertion follows from Proposition 13.5. In the general case we proceed by induction on the degree of L/F. In terms of Proposition 15.6 it suffices to show that $n \in h_{L_n/F}(\mathbb{N})$ where $G_n \neq \{1\} = G_{n+1}$. Let $\sigma \in G_n, \sigma \neq 1$. Assume that there is a cyclic subgroup H of order p such that $\sigma \notin H$. Then denote the fixed field of H by M. For a prime element π_L in L the element $\pi_M = N_{L/M}(\pi_L)$ is prime in M, and $M = F(\pi_M)$ by Corollary 2 of 9.9. Then $\varepsilon = N_{L/M}(\pi_L^{-1}\sigma(\pi_L)) = N_{L/M}(\pi_L^{-1})\sigma(N_{L/M}(\pi_L)) \neq 1$, since $\sigma(\pi_M) \neq \pi_M$. Put $n' = v_M(\varepsilon - 1)$; then $\sigma|_M \in (G/H)_{n'}$, $\sigma|_M \notin (G/H)_{n'+1}$. By the induction hypothesis, $n' = h_{M/F}(n'')$ for some $n'' \in \mathbb{N}$. Proposition 13.5 implies $n \leqslant h_{L/M}(n')$, and we obtain $n \leqslant h_{L/F}(n'')$. If $n < h_{L/F}(n'')$, then, by Proposition 15.7 the left derivative of $h_{L/F}$ at n'' is equal to $|L:F|$, and the left derivative of $h_{L/M}$ at n' is equal to $|L:M|$. Therefore, the left derivative of $h_{M/F}$ at n'', which is equal to $|(G/H):(G/H)_{n'}|$ by Proposition 15.7, coincides with $|M:F|$. This contradiction shows that $n = h_{L/F}(n'')$.

It remains to consider the case when there are no cyclic subgroups H of order p, such that $\sigma \notin H$. This means that G is itself cyclic. Let τ be a generator of G.

The choice of n and Theorem 16.2 imply that $\sigma = \tau^{ip^{m-1}}$, where $p \nmid i, p^m = |G|$. We can assume $m \geqslant 2$ because the case of $m = 1$ has been considered above. Let $n_1 = v_L(\pi_L^{-1}\tau^{p^{m-2}}(\pi_L) - 1)$. Since $|G : G_n| = p^{m-1}$, Proposition 15.7 shows now that it suffices to prove that $p^{m-1} | (n - n_1)$. This is, in fact, a part of the third statement of the following Proposition. □

PROPOSITION. *Let L/F be a totally ramified cyclic extension of degree p^m. Let π_L be a prime element in L. For $\sigma \in \mathrm{Gal}(L/F)$ and integer k put*

$$c_k = c_k(\sigma) = v_L\left(\frac{\sigma^k(\pi_L)}{\pi_L} - 1\right).$$

Then

- *c_k depends only on $v_p(k)$, where v_p is the p-adic valuation (see section 1);*
- *there exists an element $\alpha_k \in L^\times$ such that*

$$v_L(\alpha_k) = k, \qquad v_L\left(\frac{\sigma(\alpha_k)}{\alpha_k} - 1\right) = c_k;$$

- *if $v_p(k_1 - k_2) \geqslant l$, then $v_p(c_{k_1} - c_{k_2}) \geqslant l + 1$.*

Proof. We start with several remarks.

(1) Note that c_k does not depend on the choice of a prime element in L by the same reasons as s in 13.4. Let $k = ip^j$ with $p \nmid i$, $j \geqslant 0$. Then $\sigma^k - 1 = (\rho - 1)\mu$ for $\rho = \sigma^{p^j}, \mu = \rho^{i-1} + \rho^{i-2} + \cdots + 1$. Since c_k does not depend on the choice of a prime element in L and i is prime to p, we deduce $c_k = c_{p^j}$. We also have $c_k(\sigma^p) = c_{kp}(\sigma)$.

(2) Put $\alpha_k = \prod_{i=0}^{k-1} \sigma^i(\pi_L)$ for $k > 0$, $\alpha_k = \alpha_{-k}^{-1}$ for $k < 0$ and $\alpha_0 = 1$. The elements α_k satisfy condition (2) of the Proposition.

(3) Assume, by induction, that if $v_p(k_1 - k_2) \geqslant l$ for $l \leqslant n - 2$. Then we obtain $v_p(c_{k_1}(\sigma) - c_{k_2}(\sigma)) \geqslant l + 1$ for $\sigma \in \mathrm{Gal}(L/F)$.

First we show that all the integers $c_{p^{n-1}}, k + c_k$ for $v_p(k) \leqslant n - 1$ are distinct. If $v_p(k_1) = v_p(k_2)$, $k_1 \neq k_2$, then $c_{k_1} = c_{k_2}$ and $k_1 + c_{k_1} \neq k_2 + c_{k_2}$. Let $v_p(k_1), v_p(k_2)$ be distinct and $\leqslant n - 1$, then $v_p(k_1 - k_2) \leqslant n - 2$. So if $k_1 + c_{k_1} = k_2 + c_{k_2}$ then $v_p(k_1 - k_2) = v_p(c_{k_2} - c_{k_1}) \geqslant v_p(k_1 - k_2) + 1$, and thus $k_1 = k_2$. If $v_p(k) = n - 1$ then $c_{p^{n-1}} \neq c_k + k$. If $v_p(k) < n - 1$ then $v_p(c_{p^{n-1}} - c_k) \geqslant v_p(p^{n-1} - k) + 1 > v_p(k)$ and so $c_{p^{n-1}} \neq c_k + k$.

Assume that $v_p(c_{p^{n-1}}(\tau) - c_{p^n}(\tau)) < n$ for a generator τ of $\mathrm{Gal}(L/F)$. Our purpose is to show that this leads to a contradiction. Then, obviously, $v_p(c_{k_1}(\sigma) - c_{k_2}(\sigma)) \geqslant l + 1$ for $v_p(k_1 - k_2) \geqslant l, l \leqslant n - 1$.

Put $d = c_{p^{n-1}}(\tau) - c_{p^n}(\tau)$. Since

$$v_p(d) = v_p(c_{p^{n-2}}(\tau^p) - c_{p^{n-1}}(\tau^p)) \geqslant n-1,$$

we get $v_p(d) = n-1$. By (2), there exists an element $\alpha \in L$ such that $v_L(\alpha) = d$,

$$v_L(\tau^p(\alpha) - \alpha) = d + c_d(\tau^p) = d + c_{p^n}(\tau) = c_{p^{n-1}}(\tau).$$

Put

$$\beta = (\tau^{p-1} + \tau^{p-2} + \cdots + 1)\alpha.$$

Since $v_L(\tau^p(\alpha) - \alpha) = c_{p^{n-1}}(\tau) > 0$, we get $v_L(\tau(\alpha) - \alpha) > v_L(\alpha)$ and $v_L(\beta) > d$. We also obtain $v_L(\tau(\beta) - \beta) = v_L(\tau^p(\alpha) - \alpha) = c_{p^{n-1}}(\tau)$.

Note that any element α_k as in (2) can be changed to $\theta\alpha_k$ satisfying the same property (2), with a unit $\theta \in U_F$ that has a given residue. Hence we deduce that β can be expanded as

$$\beta = \sum_{k \geqslant v_L(\beta)} \beta_k,$$

with $\beta_k \in L$ possessing the same properties with respect to τ as α_k of (2). Then

$$\tau(\beta) - \beta = \sum_{\substack{k \geqslant v_L(\beta) \\ v_p(k) < n}} (\tau(\beta_k) - \beta_k) + \sum_{\substack{k \geqslant v_L(\beta) \\ v_p(k) \geqslant n}} (\tau(\beta_k) - \beta_k).$$

The valuations of the elements of the first sum on the right-hand side are all distinct because $v_L(\tau(\beta_k) - \beta_k) = k + c_k(\tau)$ are all distinct and none of them coincides with $c_{p^{n-1}}(\tau) = v_L(\tau(\beta) - \beta)$. Therefore,

$$c_{p^{n-1}}(\tau) = v_L\Big(\sum_{\substack{k \geqslant v_L(\beta) \\ v_p(k) \geqslant n}} (\tau(\beta_k) - \beta_k)\Big).$$

In this sum

$$v_L(\tau(\beta_k) - \beta_k) = k + c_k(\tau) \geqslant v_L(\beta) + c_{p^n}(\tau) > d + c_{p^n}(\tau) = c_{p^{n-1}}(\tau),$$

a contradiction. □

REMARK. This Theorem can be naturally proved using local class field theory. In addition, there is a converse theorem: a finite Galois totally ramified extension L/F is abelian if and only if for every finite abelian totally ramified extension M/F the extension LM/F has integer upper ramification jumps, [**6**]. It is not true that if a finite Galois totally ramified extension has integer upper ramification jumps then it is abelian.

17. Field of Norms

The theory of a field of norms was started by Fontaine and Wintenberger, [**13**], [**38**]. This section may be more difficult to study than the other sections of this chapter and its material is not essential for standard class field theory; on the other hand, its material is closely related to several developments in Tate–Fontaine theory, sometimes called p-adic Hodge theory. This section can be skipped depending on the reader's interests.

In this section F is a local field with perfect residue field of characteristic $p > 0$.

17.1. DEFINITION. Let L be a separable extension of F with finite residue field extension $\overline{L}/\overline{F}$. We can view L as the union of an increasing directed family of subfields L_i, which are finite extensions of F, $i \geqslant 0$. The extension L/F is said to be *arithmetically profinite* if the composite $\cdots \circ h_{L_i/L_{i-1}} \circ \cdots \circ h_{L_0/F}(a)$ is a real number for every real $a > 0$.

In other words, taking into consideration Proposition 15.3, L/F is arithmetically profinite if and only if its residue field extension is finite and for every real $a > 0$ there exists an integer j, such that the derivative (left or right) of $h_{L_i/L_j}(x)$ for $x < h_{L_j/F}(a)$, $i > j$, is equal to 1. Equivalently, for every real $a > 0$ the derivative (left or right) of $h_{L_i/F}(x)$ is bounded for $x < a$ and all i.

Define the *Hasse–Herbrand function* of L/F as

$$h_{L/F} = \cdots \circ h_{L_i/L_{i-1}} \circ \cdots \circ h_{L_0/F}.$$

PROPOSITION. *The function $h_{L/F}$ is well defined. It is a piecewise linear, continuous and increasing function. If E/L is a finite separable extension, then E/F is arithmetically profinite. If M/F is a subextension of L/F, then M/F is arithmetically profinite. If, in addition, M/F is finite, then L/M is arithmetically profinite and*

$$h_{L/F} = h_{L/M} \circ h_{M/F}.$$

Proof. Let L_i' be another increasing directed family of subfields in L such that $L = \cup L_i'$. Let a be a real number > 0. There exist integers j and k such that

$$h_{L_i/L_j}(x) = x \qquad \text{for } x < h_{L_j/F}(a), i > j$$

and

$$h_{L_i'/L_k'}(x) = x \qquad \text{for } x < h_{L_k'/F}(a), i > k.$$

Since there exists an integer $m \geqslant j$ such that $L_j L_k' \subset L_m$, we obtain by 15.3 that

$$h_{L_jL_k'/L_j}(x) = x \qquad \text{for } x < h_{L_j/F}(a).$$

Then

$$h_{L_j/F}(x) = h_{L_jL'_k/F}(x) \qquad \text{for } x < a$$

and similarly,

$$h_{L'_k/F}(x) = h_{L_jL'_k/F}(x) \qquad \text{for } x < a.$$

Therefore,

$$h_{L_i/F}(x) = h_{L'_i/F}(x) \qquad \text{for } x < a \text{ and sufficiently large } i,$$

and the function $h_{L/F}$ is well defined.

Let $E = L(\beta)$, and let $P = L(\alpha)$ be a finite Galois extension of L with $P \supset E$. Using the same arguments as in the proof of Proposition 11.2, one can show that $L_i(\alpha) \cap L = L_i$ and $L_i(\alpha)/L_i$ is a Galois extension of the same degree as P/L for a sufficiently large i. Then $\mathrm{Gal}(L_i(\alpha)/L_i)$ and $\mathrm{Gal}(L_i(\alpha)/L_i(\beta))$ are isomorphic with $\mathrm{Gal}(P/L)$ and $\mathrm{Gal}(P/E)$ for $i > m$, respectively.

Put $E_i = L_i$ for $i \leqslant m$ and $E_i = L_i(\beta)$ for $i > m$. Then $E = \cup E_i$. If the left derivative of $h_{L_i/F}(x)$ is bounded by d for $x < a$ and $c = |E : L|$, then the left derivative of $h_{E_i/F}(x)$ is bounded by cd for $x < a$, $i > m$. This means that E/F is arithmetically profinite.

If M/F is a finite subextension of L/F, then we can take $L_0 = M$. Therefore L/M is arithmetically profinite and

$$h_{L/F} = h_{L/M} \circ h_{M/F}.$$

If M/F is a separable subextension of L/F, then there exists an increasing directed family of subfields $M_i, i \geqslant 0$, which are finite extensions of F and such that $M = \cup M_i$. If $L = \cup L_i$, then also $L = \cup L_i M_i$, and the left derivative of $h_{L_iM_i/F}(x)$ for $x < a$ is bounded. Hence, the left derivative of $h_{M_i/F}(x)$ for $x < a$ is bounded, i.e., M/F is arithmetically profinite. □

REMARKS.

1. Translating to the language of ramification groups by using the two previous sections, we deduce that a Galois extension L/F with finite residue field extension is arithmetically profinite extension if and only if its upper ramification jumps form a discrete unbounded set and for every upper ramification jump x the index of $\mathrm{Gal}(L/F)(x+\delta)$ in $\mathrm{Gal}(L/F)(x)$ is finite. Alternatively, a Galois extension L/F is arithmetically profinite if and only if for every x the upper ramification group $\mathrm{Gal}(L/F)(x)$ is open (i.e. of finite index) in $\mathrm{Gal}(L/F)$. More generally, a separable extension L/F is arithmetically profinite if and only if for every x the group $\mathrm{Gal}(F^{\mathrm{sep}}/F)(x)\,\mathrm{Gal}(F^{\mathrm{sep}}/L)$ is open in $\mathrm{Gal}(F^{\mathrm{sep}}/F)$.

Since the Hasse–Herbrand function relates upper and lower ramification filtrations, we can define lower ramification groups of an infinite Galois arithmetically profinite extension L/F as $\mathrm{Gal}(L/F)_x = \mathrm{Gal}(L/F)(h_{L/F}^{-1}(x))$.

2. Since upper ramification jumps of abelian extensions are subsets of natural numbers by Theorem 16.3, every abelian extension of a local field with finite residue field and finite residue field extension is arithmetically profinite, see Corollary of 3.4.

3. An important property of a totally ramified $\mathbb{Z}_p$-extension L/F in characteristic zero is that its upper ramification jumps form an arithmetic progression with difference $e = e(F)$ for sufficiently large jumps.

Maus–Sen's Theorem on ramification filtration of p-adic Lie extensions L/F in characteristic zero with finite residue field extension states that the p-adic Lie filtration is equivalent to the upper ramification filtration of the Galois group of such extensions. This theorem implies that every such extension is an arithmetically profinite extension. In positive characteristic the analogous result was proved by Wintenberger.

4. An important example of an arithmetically profinite extension is given by $L = \cup L_i$, $L_0 = F$, $L_i = L_{i-1}(\pi_i)$ such that $\pi_i^p = \pi_{i-1}$ is a prime element of L_{i-1}. The extension L/F is not Galois.

17.2. Let L/F be arithmetically profinite. Put

$$q(L|F) = \sup\{x \geqslant 0 : h_{L/F}(x) = x\}.$$

LEMMA.

(1) *if M/F is a subextension in L/F, then $q(L|F) \leqslant q(M|F)$.*
(2) *if M/F is a finite subextension in L/F, then $q(L|M) \geqslant q(L|F)$.*
(3) *if $L = \cup L_i$ as in* (17.1), *then $q(L_j|L_i) \to +\infty$ as $j > i$, $i, j \to +\infty$.*
(4) *$q(L|F) = +\infty$ if and only if L/F is unramified; $q(L|F) = 0$ if and only if L/F is totally and tamely ramified, and $q(L|F) \leqslant p v_F(p)/(p-1)$ if L/F is totally ramified.*

Proof. (1) Let $L = \cup L_i, M = \cup M_i$ and $L_i' = L_i M_i$. As $h_{L_i'/F}(x) \leqslant h_{L/F}(x)$ by 15.3, we get $h_{L_i'/F}(x) = x$ for $x \leqslant q(L|F)$, hence $h_{M_i/F}(x) = x$ for $x \leqslant q(L|F)$. Therefore, $q(L|F) \leqslant q(M|F)$. (2) The previous Proposition shows that

$$h_{L/M}(x) = x \qquad \text{for } x \leqslant h_{M/F}(q(L|F)).$$

This means that $q(L|M) \geqslant h_{M/F}(q(L|F))$. But by Proposition 15.3, $h_{M/F}(x) \geqslant x$, hence $q(L|M) \geqslant q(L|F)$. (3) It follows from the definition. (4) The first two assertions follow from Proposition 15.3. Proceeding as in the proof of Proposition 15.3 and using (1), it suffices to verify the last assertion for a separable totally ramified extension of degree p. Now the computations in the proof of Proposition 15.3 and Proposition 14.3 lead to the required inequality. □

17.3. Let L be an infinite arithmetically profinite extension of F, and let L_i, $i \geqslant 0$, be an increasing directed family of subfields, which are finite extensions of $F, L = \cup L_i$. Let

$$N(L|F)^\times = \varprojlim L_i^\times$$

be the inverse limit (see 5.1 of Chapter 1) of the multiplicative groups with respect to the norm homomorphisms $N_{L_i/L_j}, i \geqslant j$. Denote $N(L|F) = N(L|F)^\times \cup \{0\}$.

LEMMA. *The group $N(L|F)^\times$ does not depend on the choice of L_i.*

Proof. Let L_i' be another increasing directed family of finite extensions of F and $L = \cup L_i'$. For every i there exists an index j, such that $L_i' \subset L_j$ and $N_{L_j/F} = N_{L_i'/F} \circ N_{L_j/L_i'}$. This immediately implies the desired assertion. □

Therefore

$$N(L|F)^\times = \varprojlim{}_{M \in S_{L/F}} M^\times,$$

where $S_{L/F}$ is the partially ordered family of all finite subextensions in L/F and the inverse limit is taken with respect to the norm maps. If $\mathrm{A} = (\alpha_M) \in N(L|F)$ with $\alpha_M \in M$, then $N_{M_1/M_2}\alpha_{M_1} = \alpha_{M_2}$ for $M_2 \subset M_1$.

We will show that $N(L|F)$ is in fact a field (the *field of norms*). Moreover, one can define a natural discrete valuation on $N(L|F)$, which makes $N(L|F)$ a complete discrete valuation field of characteristic p with residue field $\overline{L}$.

17.4. The following statement plays a central role.

PROPOSITION. *Let M'/M be totally ramified of degree a power of p. Then*

$$v_M\left(N_{M'/M}(\alpha+\beta) - N_{M'/M}(\alpha) - N_{M'/M}(\beta)\right) \geqslant \frac{(p-1)q(M'|M)}{p}$$

for $\alpha, \beta \in \mathcal{O}_{M'}$.

For $\alpha \in \mathcal{O}_M$ there exists an element $\beta \in \mathcal{O}_{M'}$ such that

$$v_M\left(N_{M'/M}(\beta) - \alpha\right) \geqslant \frac{(p-1)q(M'|M)}{p}.$$

Proof. To prove the first inequality, assume first that M'/M is a cyclic extension of degree p. Then we get $q(M'|M) = s(M'|M)$ (see 13.4 and 15.1) and, by Proposition 13.4,

$$\mathrm{Tr}_{M'/M}(\mathscr{O}_{M'}) = \pi_M^r \mathscr{O}_M$$

with $r = s+1+[(-1-s)/p] \geqslant (p-1)s(M'|M)/p$. Then Lemma 13.1 shows that

$$v_M\left(N_{M'/M}(1+\gamma) - 1 - N_{M'/M}(\gamma)\right) \geqslant \frac{(p-1)q(M'|M)}{p}$$

for $\gamma \in \mathscr{O}_{M'}$. Substituting $\gamma = \alpha\beta^{-1}$ if $v_{M'}(\alpha) \geqslant v_{M'}(\beta)$ and $\beta \neq 0$, we obtain the desired inequality.

In the general case we proceed by induction on the degree of M'/M. Let E/M be a finite Galois extension with $E \supset M'$, and let E_1 be the maximal tamely ramified extension of M in E. Then E_1 and M' are linearly disjoint over M, and

$$\begin{aligned} N_{M'/M}(\alpha+\beta) &- N_{M'/M}(\alpha) - N_{M'/M}(\beta) \\ &= N_{E_1M'/E_1}(\alpha+\beta) - N_{E_1M'/E_1}(\alpha) - N_{E_1M'/E_1}(\beta). \end{aligned}$$

The group $G = \mathrm{Gal}(E/E_1)$ is a p-group, and hence for $H = \mathrm{Gal}(E/E_1M')$ there exists a chain of subgroups

$$G = G_{(0)} \geqslant G_{(1)} \geqslant \ldots \geqslant G_{(m)} = H,$$

such that $G_{(i+1)}$ is a normal subgroup of index p in $G_{(i)}$. For the fields we obtain the tower $E_1 = E_{(0)} - E_{(1)} - \cdots - E_{(m)} = E_1M'$, in which $E_{(i+1)}$ is a cyclic extension of degree p over $E_{(i)}$. Let E_2 be some $E_{(i)}$ for $1 \leqslant i < m$. By the induction assumption,

$$N_{E_1M'/E_2}(\alpha+\beta) = N_{E_1M'/E_2}(\alpha) + N_{E_1M'/E_2}(\beta) + \delta$$

with $v_{E_2}(\delta) \geqslant (p-1)q(E_1M'|E_2)/p$. We deduce also that

$$N_{E_1M'/E_1}(\alpha+\beta) = N_{E_1M'/E_1}(\alpha) + N_{E_1M'/E_1}(\beta) + N_{E_2/E_1}(\delta) + \delta'$$

with $v_{E_1}(\delta') \geqslant (p-1)q(E_2|E_1)/p$. Then

$$v_{E_1}\left(N_{E_2/E_1}(\delta)\right) \geqslant \frac{(p-1)q(E_1M'|E_2)}{p} \geqslant \frac{(p-1)q(E_1M'|E_1)}{p}$$

and

$$v_{E_1}(\delta') \geqslant \frac{(p-1)q(E_1M'|E_1)}{p}$$

by Lemma 17.2. These two inequalities imply that

$$v_M\left(N_{M'/M}(\alpha+\beta) - N_{M'/M}(\alpha) - N_{M'/M}(\beta)\right) \geqslant \frac{(p-1)q(M'|M)}{p},$$

as required.

To prove the second inequality of the Proposition, we choose a prime element π' in M' and put $\pi = N_{M'/M}\pi'$. Then π is a prime element in M. Let $n = |M' : M|$ (a power of p). Writing the element α of M as

$$\alpha = \sum_{i \geqslant a} \theta_i \pi^i$$

with multiplicative representatives θ_i, put

$$\beta = \sum_{i \geqslant a} \theta_i^{1/n} \pi'^i \in M'.$$

Then $N_{M'/M}\left(\theta_i^{1/n}\pi'\right) = \theta_i \pi$. By the first inequality of the Proposition and passing to the limit, we obtain

$$v_M(N_{M'/M}(\beta) - \alpha) \geqslant \frac{(p-1)q(M'|M)}{p},$$

as required. □

17.5. Let L/F be an arithmetically profinite extension. Let L_0 be the maximal unramified extension of F in L, and let L_1 be the maximal tamely ramified extension of F in L. Then L_0/F is finite by the definition, and L_1/F is finite because of the equality $h_{L_1/L_0}(x) = |L_1 : L_0|x$. So one can choose L_i for $i \geqslant 2$ as finite extensions of L_1 in L with $L_i \subset L_{i+1}$ and $L = \cup L_i$.

For an element $\mathrm{A} \in N(L|F)$ put

$$v(\mathrm{A}) = v_{L_0}(\alpha_{L_0}).$$

Then $v(\mathrm{A}) = v_{L_i}(\alpha_{L_i})$ for $i \geqslant 0$.

Let a be an element of the residue field $\overline{L} = \overline{L}_0$, and $\theta = r(a)$ the multiplicative representative of a in L_0 (see Section 6). Put $\theta_{L_i} = \theta^{1/n_i}$, where $n_i = |L_i : L_1|$ for $i \geqslant 1$ and $\theta_{L_0} = N_{L_1/L_0}\theta$. Then $\Theta = (\theta_{L_i})$ is an element of $N(L|F)$. Denote the map $a \mapsto \Theta$ by R.

THEOREM. *Let L/F be an infinite arithmetically profinite extension. Let $\mathrm{A} = (\alpha_M)$ and $\mathrm{B} = (\beta_M)$ be elements of $N(L|F)$, $M \in S_{L/F}$. Then the sequence $N_{M'/M}(\alpha_{M'} + \beta_{M'})$ is convergent in M when $M \subset M' \subset L, |M' : M| \to +\infty$. Let γ_M be the limit of this sequence. Then $\Gamma = (\gamma_M)$ is an element of $N(L|F)$. Put $\Gamma = \mathrm{A} + \mathrm{B}$.*

Then $N(L|F)$ is a field with respect to the multiplication and addition defined above. The map v is a discrete valuation of $N(L|F)$ and $N(L|F)$ is a complete field of characteristic p. The map R is an isomorphism of $\overline{L}$ onto a subfield in $N(L|F)$ which maps isomorphically onto the residue field of $N(L|F)$.

Proof. Let L_i be as above in 17.5, in the context of Lemma 17.3.

Let a be a positive integer and let k be an integer such that $(p-1)q(L_j|L_i)/p \geqslant a$ for $j > i \geqslant k$, see Lemma 17.2. Let $\mathrm{A} = (\alpha_{L_i}), \mathrm{B} = (\beta_{L_i})$ be elements of $N(L|F)$ and $\alpha_{L_0}, \beta_{L_0} \in \mathscr{O}_{L_0}$. Then Proposition 17.4 shows that

$$N_{L_i/L_k}(\alpha_{L_i} + \beta_{L_i}) \equiv \alpha_{L_k} + \beta_{L_k} \mod \mathscr{M}_{L_k}^a. \qquad (*)$$

Let $a_k \geqslant 0$ be a sequence of integers such that

$$a_k \leqslant a_{k+1}, \quad a_k \leqslant (p-1)q(L|L_k)/p, \quad \lim a_k = +\infty$$

(the existence of the sequence follows from Lemma 17.2). Let an index $k \geqslant 1$ be in addition such that $a_k > 1$. Suppose that β_{L_k} is a prime element in L_k. Proposition 17.4 and Lemma 17.2 show that one can construct a sequence $\beta_{L_i} \in L_i$, $i \geqslant k$, such that

$$v_{L_i}(N_{L_{i+1}/L_i}\beta_{L_{i+1}} - \beta_{L_i}) \geqslant a_i.$$

Then β_{L_i} is prime in L_i, and applying $(*)$, we get

$$v_{L_i}(N_{L_j/L_i}\beta_{L_j} - \beta_{L_i}) \geqslant a_i \qquad \text{for } j \geqslant i \geqslant k.$$

Now Proposition 15.4 and Proposition 17.1 imply that

$$v_{L_s}(N_{L_j/L_s}\beta_{L_j} - N_{L_i/L_s}\beta_{L_i}) \geqslant h^{-1}_{L_i/L_s}(a_i) \geqslant h^{-1}_{L/L_s}(a_i)$$

for $j \geqslant i \geqslant s \geqslant k$. Since $h^{-1}_{L/L_s}(a_i) \to +\infty$ as $i \to +\infty$, there exists limit $\gamma_{L_s} = \lim_{i\to+\infty} N_{L_i/L_s}\beta_{L_i}$ and γ_{L_s} is a prime element in L_s. Putting $\gamma_{L_j} = N_{L_k/L_j}\gamma_{L_k}$ for $j < k$, we get the element $\Gamma = (\gamma_{L_i}) \in N(L|F)$ with $v(\Gamma) = 1$.

Furthermore, by Proposition 15.4 and $(*)$ we obtain:

$$v_{L_j}\Big(N_{L_i/L_j}(\alpha_{L_i} + \beta_{L_i}) - N_{L_k/L_j}(\alpha_{L_k} + \beta_{L_k})\Big) \geqslant h^{-1}_{L_k/L_j}(a) \geqslant h^{-1}_{L/L_j}(a).$$

This means that the sequence $N_{L_i/L_j}(\alpha_{L_i} + \beta_{L_i})$ is convergent. In the general case let $c = v_{L_0}(\alpha_{L_0}), d = v_{L_0}(\beta_{L_0})$. Taking prime elements π_{L_i} in L_i such that $\Pi = (\pi_{L_i}) \in N(L|F)$ with $v(\Pi) = 1$ and replacing $\mathrm{A} = (\alpha_{L_i})$ by $\mathrm{A}' = (\alpha_{L_i}\pi_{L_i}^{-g})$ and $\mathrm{B} = (\beta_{L_i})$ by $\mathrm{B}' = (\beta_{L_i}\pi_{L_i}^{-g})$, where $g = \min(c,d)$, we deduce that $N_{L_i/L_j}(\alpha_{L_i} + \beta_{L_i})$ is convergent. Put $\gamma_{L_j} = \lim_{i\to+\infty} N_{L_i/L_j}(\alpha_{L_i} + \beta_{L_i})$. Obviously, $(\gamma_{L_i}) = \Gamma \in N(L|F)$ and $N(L|F)$ is a field. As

$$v(\Gamma) = v_{L_k}(\gamma_{L_k}) = \lim_{i\to+\infty} v_{L_k}(N_{L_i/L_k}(\alpha_{L_i} + \beta_{L_i})),$$

we get $v(\Gamma) \geqslant \min(v(\mathrm{A}), v(\mathrm{B}), a)$. Choosing $a \geqslant \max(v(\mathrm{A}), v(\mathrm{B}))$, we immediately obtain that $v(\Gamma) \geqslant \min(v(\mathrm{A}), v(\mathrm{B}))$. Since $1 = (1_{L_i})$, for $p = (\alpha_{L_i})$ we get that

$$\alpha_{L_j} = \lim_{i\to+\infty} N_{L_i/L_j}(p) = \lim_{i\to+\infty} p^{|L_i:L_j|} = 0.$$

Therefore, $N(L|F)$ is a discrete valuation field of characteristic p.

To verify the completeness of $N(L|F)$ with respect to v, take a Cauchy sequence $\mathrm{A}^{(n)} = (\alpha_{L_i}^{(n)}) \in N(L|F)$. We may assume $v(\mathrm{A}^{(n)}) \geqslant 0$. For any i there exists an integer n_i such that $v(\mathrm{A}^{(n)} - \mathrm{A}^{(m)}) \geqslant a_i$ for $n, m \geqslant n_i$ (a_i as above). One may assume that $(n_i)_i$ is an increasing sequence. Applying $(*)$, we get

$$v_{L_i}(\alpha_{L_i}^{(n)} - \alpha_{L_i}^{(m)}) \geqslant a_i \qquad \text{for } n, m \geqslant n_i.$$

Let α_{L_i} be an element in L_i such that

$$v_{L_i}(\alpha_{L_i} - \alpha_{L_i}^{(n_i)}) \geqslant a_i.$$

Then, by $(*)$,

$$v_{L_i}(N_{L_j/L_i}\alpha_{L_j} - \alpha_{L_i}) \geqslant a_i.$$

Propositions 15.4 and 17.1 imply now that

$$v_{L_s}(N_{L_i/L_s}\alpha_{L_i} - N_{L_j/L_s}\alpha_{L_j}) \geqslant h_{L/L_s}^{-1}(a_j) \to +\infty$$

when $i \geqslant j \to +\infty$. Putting $\alpha'_{L_s} = \lim_{i \to +\infty} N_{L_i/L_s}\alpha_{L_i}$, we obtain an element $\mathrm{A}' = (\alpha'_{L_i}) \in N(L|F)$ with $\mathrm{A}' = \lim \mathrm{A}^{(n)}$. Therefore, $N(L|F)$ is complete with respect to the discrete valuation v.

Finally, R is multiplicative. If $R(a) = \Theta$, $R(b) = \Lambda$, $R(a+b) = \Omega$, then it follows immediately from 6.3, that $\omega_{L_i} \equiv \theta_{L_i} + \lambda_{L_i} \mod p$. By the definition of a_i we get $v_{L_i}(p) \geqslant a_i$. Then by $(*)$ and Proposition 15.4 we obtain

$$v_{L_i}(\omega_{L_i} - N_{L_j/L_i}(\theta_{L_j} + \lambda_{L_j})) \to +\infty$$

as $j \to +\infty$. This means that $\Omega = \Theta + \Lambda$ and R is an isomorphism of $\overline{L}$ onto a subfield in $N(L|F)$. The latter subfield is mapped onto the residue field of $N(L|F)$, hence it is isomorphic to the residue field $\overline{N(L|F)}$. □

COROLLARY. *Let* $\mathrm{A} = (\alpha_{L_i}), \mathrm{B} = (\beta_{L_i})$ *belong to the ring of integers of* $N(L|F)$. *Let* $\Gamma = \mathrm{A} + \mathrm{B}$. *Then* $\gamma_{L_i} \equiv \alpha_{L_i} + \beta_{L_i} \mod \mathscr{M}_{L_i}^{a_i}$, *where* a_i *are those defined in the proof of the Theorem. Moreover, for any* $\alpha \in \mathscr{O}_{L_j}$ *there exists an element* $\mathrm{A} = (\alpha_{L_i})$ *in the ring of integers of* $N(L|F)$ *such that* $\alpha \equiv \alpha_{L_j} \mod \mathscr{M}_{L_j}^{a_j}$.

Proof. The first assertion follows from $(*)$ and the second from Proposition 17.4. □

17.6. An immediate consequence of the definitions is that if M/F is a finite subextension of an arithmetically profinite extension L/F, then $N(L|F) = N(L|M)$. On the other hand, if E/L is a finite separable extension, then, as shown in Proposition 17.1, E/F is an arithmetically profinite extension. Let M be a finite extension of F such that $ML = E$. Since $N_{L_jM/L_iM}(\alpha) = N_{L_j/L_i}(\alpha)$ for $\alpha \in L_j$, $j \geqslant i \geqslant m$, and sufficiently large m, we deduce that $N(L|F)$ can be identified with a subfield of $N(E|F)$: $\mathrm{A} = (\alpha_{L_i}) \mapsto \mathrm{A}' \in N(E|F)$ with $\mathrm{A}' = (\alpha'_{L_iM})$, $\alpha'_{L_iM} = \alpha_{L_i}$ for $i \geqslant m$, $\alpha'_{L_iM} = N_{L_mM/L_iM}(\alpha_{L_m})$ for $i < m$. In fact the discrete valuation topology of $N(L|F)$ coincides with the induced topology from $N(E|F)$, and $N(E|F)/N(L|F)$ is an extension of complete discrete valuation fields. For an arbitrary separable extension E/L denote by $N(E,L|F)$ the direct limit of $N(E'|F)$ for finite separable subextensions E'/L in E/L. Obviously, $N(E,L|F) = N(E|F)$ if E/L is finite.

Let L/F be infinite arithmetically profinite, and let L'/L be a finite separable extension. Let τ be an automorphism in the absolute Galois group $G_F = \mathrm{Gal}(F^{\mathrm{sep}}/F)$ (see 5.2 of Chapter 1) with $\tau(L) \subset L'$. There exists a tower of increasing subfields L'_i in L' such that L'_i/F is finite, $\tau(L)L'_i = L'$, $L' = \cup L'_i$, and $N_{L'_j/L'_i}(\tau\alpha) = \tau N_{\tau^{-1}L'_j/\tau^{-1}L'_i}(\alpha)$ for $j \geqslant i, \alpha \in \tau^{-1}L'_j$; see the proof of Proposition 17.1. Let $\mathrm{T}\colon N(L|F) \longrightarrow N(L'|F)$ denote the homomorphism of fields, which is defined for $\mathrm{A} = (\alpha_{L_i}) \in N(L|F)$ as $\mathrm{T}(\mathrm{A}) = \mathrm{A}' = (\alpha'_{L'_i})$ with $\alpha'_{L'_i} = \tau(\alpha_{\tau^{-1}L'_i})$. Then $\mathrm{A}' \in N(L'|F)$.

This notion is naturally generalised for $N(E,L|F)$ and $N(E',L|F)$ with $\tau(E) \subset E'$.

PROPOSITION. *Let E_1 and E_2 be separable extensions of L. Then the set of all automorphisms $\tau \in G_L$ with $\tau(E_1) \subset E_2$ is identified (by $\tau \mapsto \mathrm{T}$) with the set of all automorphisms $\mathrm{T} \in G_{N(L|F)}$ with $\mathrm{T}(N(E_1,L|F)) \subset N(E_2,L|F)$. In particular, if E/L is a Galois extension, then* $\mathrm{Gal}(E/L) \simeq \mathrm{Gal}(N(E,L|F)/N(L|F))$.

Proof. First we verify the second assertion for a finite Galois extension E/L. Let T act trivially on $N(E|F)$. Then $\overline{\mathrm{T}}$ acts trivially on the residue field of $N(E|F)$, which coincides with $\overline{E}$, and hence τ belongs to the inertia subgroup $\mathrm{Gal}(E/F)_0$. Let $E = L(\beta)$ and L_i form a standard tower of fields for L over F, as in (17.5). Since the coefficients of the irreducible polynomial of β over L belong to some L_m, we deduce that $L_i(\beta)/L_i$ is Galois and $\mathrm{Gal}(L_i(\beta)/L_i)$ is isomorphic to $\mathrm{Gal}(E/L)$ for $i \geqslant m$. Let $\Pi = (\pi_{L_i(\beta)})_{i>m}$ be a prime element of $N(E|F)$. Then $\mathrm{T}(\Pi) = \Pi$ and $\tau\pi_{L_i(\beta)} = \pi_{L_i(\beta)}$ for $i > m$. We obtain now that $\tau = 1$ because τ acts trivially on the residue field $\overline{L_i(\beta)} = \overline{E}$.

We conclude that $\mathrm{Gal}(E/L)$ can be identified with a subgroup of

$$\mathrm{Gal}(N(E|F)/N(L|F)).$$

Since the field of the fixed elements under the action of the image of $\mathrm{Gal}(E/L)$ is contained in $N(L|F)$, these two groups are isomorphic.

From this we easily deduce the second assertion of the Proposition for an arbitrary Galois extension E/L.

Finally, if E/L is a Galois extension such that $E_1, E_2 \subset E$, denote the Galois groups of E/E_1 and E/E_2 by H_1 and H_2. The groups H_1 and H_2 can be identified with $\mathrm{Gal}(N(E,L|F)/N(E_1,L|F))$, and $\mathrm{Gal}(N(E,L|F)/N(E_2,L|F))$ respectively. Since the set of $\tau \in G_L$ with $\tau(E_1) \subset E_2$ coincides with $\{\tau \in G_L : \tau H_1 \tau^{-1} \supset H_2\}$, the proof is completed. □

17.7. The preceding Proposition shows that the group $\mathrm{Gal}(F^{\mathrm{sep}}/L)$ can be considered as a quotient group of $\mathrm{Gal}(N(L|F)^{\mathrm{sep}}/N(L|F))$. We will show in what follows that the former group coincides with the latter.

THEOREM. *Let Q be a separable extension of $N(L|F)$. Then there exists a separable extension E/L and an $N(L|F)$-isomorphism of $N(E,L|F)$ onto Q.*

Thus, the absolute Galois group of L is naturally isomorphic to the absolute Galois group of $N(L|F)$:

$$G_L \xrightarrow{\sim} G_{N(L|F)}.$$

Proof. One can assume that $Q/N(L|F)$ is a finite Galois extension. Using the description of Galois extensions of 11.4 we must consider the following three cases: $Q/N(L|F)$ is unramified, cyclic tamely totally ramified, and cyclic totally ramified of degree $p = \mathrm{char}(\overline{F})$.

Let $\mathcal{O}_Q = \mathcal{O}_{N(L|F)}[\Gamma]$. Let $f(X)$ be the monic irreducible polynomial of Γ over $N(L|F)$. It suffices to find a separable extension E'/L such that $f(X)$ has a root in $N(E',L|F)$. Let L_i and a_i be identical to those in the proof of Theorem 17.5. By Lemma 10.1, we can write

$$f(X) = X^n + \mathrm{A}^{(n-1)}X^{n-1} + \cdots + \mathrm{A}^{(0)}$$

with $\mathrm{A}^{(m)} = (\alpha_{L_i}^{(m)}) \in \mathcal{O}_{N(L|F)}$, $n = |Q : N(L|F)|$. Denote by $f_i(X) \in \mathcal{O}_{L_i}[X]$ the polynomial $X^n + \alpha_{L_i}^{(n-1)}X^{n-1} + \cdots + \alpha_{L_i}^{(0)}$. Let α_i be a root of $f_i(X)$ and $M_i = L_i(\alpha_i), E_i = L(\alpha_i)$.

The following assertion will be useful in our considerations.

LEMMA. *Let Γ_m for $1 \leqslant m \leqslant n$ be all roots of $f(X)$ and $\Delta = \prod_{m<l}(\Gamma_m - \Gamma_l)^2$ be the discriminant of $f(X)$. Then $\Delta = (-1)^{\frac{n(n-1)}{2}} \prod_{m=1}^n \sigma_m f'(\Gamma)$ where $\sigma_1, \dots, \sigma_n$ are elements of* $\mathrm{Gal}(Q/N(L|F))$.

Let $d_i \in L_i$ be the discriminants of $f_i(X)$. Then there exists an index i_1 such that $v_{L_i}(d_i) = v(\Delta)$ for $i \geqslant i_1$.

Proof. Let $\Delta = (\delta_{L_i})$, and let i_1 be such that $a_i > v(\Delta)$ for $i \geqslant i_1$. Then $v(\Delta) = v_{L_i}(\delta_{L_i})$, and Corollary 17.5 shows that $v_{L_i}(\delta_{L_i} - d_i) \geqslant a_i$. Therefore, $v_{L_i}(d_i) = v_{L_i}(\delta_{L_i}) = v(\Delta)$ for $i \geqslant i_1$. □

This Lemma implies that M_i/L_i is separable for $i \geqslant i_1$. Now we shall verify that in the three cases under consideration, there exists an index i_2, such that M_i/L_i and L/L_i are linearly disjoint and $q(E_i|M_i) \geqslant q(L|L_i)$ for $i \geqslant i_2$.

If $Q/N(L|F)$ is unramified, then the residue polynomial $\overline{f}_i \in \overline{L}[X]$ is irreducible of degree n and M_i/L_i is an unramified extension of the same degree. Hence, M_i/L_i and L/L_i are linearly disjoint and $h_{E_i/M_i}(x) = h_{L/L_i}(x)$, and so $q(E_i|M_i) = q(L|L_i)$.

If $Q/N(L|F)$ is totally and tamely ramified, then one can take $f(X) = X^n - \Pi$, where Π is a prime element in $N(L|F)$ (see 10.5). Hence, M_i/L_i is tamely and totally ramified of degree n for $i \geqslant 1$. We deduce that $L \cap M_i = L_i$ and $h_{E_i/M_i}(nx) = nh_{L/L_i}(x)$, and hence $q(E_i|M_i) \geqslant nq(L|L_i)$ for $i \geqslant 1$.

If $Q/N(L|F)$ is totally ramified of degree $n = p = \operatorname{char}(\overline{F})$, then one may assume that $f(X)$ is an Eisenstein polynomial (see 10.6). Then $f_i(X)$ is a separable Eisenstein polynomial in $L_i[X]$, and α_i is prime in M_i. Let N_i be the minimal finite extension of M_i such that N_i/L_i is Galois, and M_i' the maximal tamely unramified extension of L_i in N_i. Then $|N_i : L_i| \leqslant p!$. One has $N_i = M_i'(\alpha_i)$ and $s_i = s(N_i|M_i') = v_{N_i}(\sigma\alpha_i - \alpha_i) - v_{N_i}(\alpha_i)$ for a generator σ of $\operatorname{Gal}(N_i/M_i')$ (see 13.4 and the proof of Proposition 15.3). Note that

$$v_{N_i}(\sigma\alpha_i - \alpha_i) = \frac{1}{p(p-1)} v_{N_i}(d_i) \leqslant \frac{p!}{p(p-1)} v_{L_i}(d_i) = (p-2)! v(\Delta)$$

for $i \geqslant i_1$. Furthermore, in the same way as in the proof of Proposition 15.3, we get $h_{M_i/L_i}(x) = l^{-1} h_{N_i/M_i'}(lx)$, where $l = e(M_i'|L_i)$. Consequently,

$$q(M_i|L_i) = s_i l^{-1} < (p-2)! v(\Delta).$$

Since $h_{L_j(\alpha_i)/M_i} \circ h_{M_i/L_i} = h_{L_j(\alpha_i)/L_j} \circ h_{L_j/L_i}$ for $j \geqslant i$, we deduce $q(E_i|M_i) = h_{M_i/L_i}(q(L|L_i)) \geqslant q(L|L_i)$.

Now we construct the desired field E'. Let $v\colon N(L|F)^{\mathrm{sep}\times} \longrightarrow \mathbb{Q}$ be the extension of the discrete valuation $v\colon N(L|F)^\times \longrightarrow \mathbb{Z}$ (see Corollary 1 of 9.9). According to Corollary 17.5 there is an element $\mathrm{B}^{(j)} = (\beta^{(j)}_{L_i(\alpha_j)})_{i \geqslant j} \in N(E_j|F)$ such that $v_{M_j}(\alpha_j - \beta^{(j)}_{M_j}) \geqslant b_j$, where b_j is the maximal integer $\leqslant (p-1)q(E_j|M_j)/p$.

Since $q(E_j|M_j) \geqslant q(L|L_j)$, we obtain $b_j \geqslant a_j$. We claim that $v(f(\mathrm{B}^{(j)})) \to +\infty$ as $j \to +\infty$.

Indeed, E_j/M_j is totally ramified. Therefore, if $f(\mathrm{B}^{(j)}) = (\rho_{L_i(\alpha_j)})_{i \geqslant j}$ then we have $v(f(\mathrm{B}^{(j)})) \geqslant v_{M_j}(\rho_{M_j})/n$.

By using Corollary 17.5 we deduce

$$v_{M_j}(\rho_{M_j} - f_j(\beta^{(j)}_{M_j})) \geqslant (p-1)q(E_j|M_j)/p \geqslant a_j.$$

This means that

$$v(f(\mathrm{B}^{(j)})) \geqslant \frac{a_j}{n} \qquad \text{for } j \geqslant i_2.$$

Since $a_j \to +\infty$ when $j \to +\infty$, we conclude that $v(f(\mathrm{B}^{(j)})) \to +\infty$.

By the same arguments we obtain that for $f'(\mathrm{B}^{(j)}) = (\mu_{L_i(\alpha_j)})_{i \geqslant j}$

$$v(f'(\mathrm{B}^{(j)})) \leqslant v_{M_j}(\mu_{M_j}), \quad v_{M_j}(\mu_{M_j} - f'_j(\alpha_j)) \geqslant a_j, \quad v_{M_j}(f'_j(\alpha_j)) \leqslant nv(\Delta)$$

for $j \geqslant i_2$. This implies that for a sufficiently large j

$$v(f'(\mathrm{B}^{(j)})) \leqslant nv(\Delta) < \frac{1}{2} v(f(\mathrm{B}^{(j)})).$$

Corollary 3 of 8.3 shows the existence of a root of $f(X)$ in $N(E_j|F)$. Putting $E' = E_j$ we complete the proof of the Theorem. □

DEFINITION. *The functor of fields of norms* associates to every arithmetically profinite extension L over F its field of norms $N(L|F)$, to every separable extension E of L the field $N(E,L|F)$ and to every element of G_F the corresponding element of the group of automorphisms of the field $N(L|F)^{\mathrm{sep}}$ (so that elements of $G_L \leqslant G_F$ are mapped isomorphically to elements of $G_{N(L|F)}$).

REMARKS.

1. If L/F is an arithmetically profinite extension, it is easy to show that for a separable extension E/L (not necessarily finite), E/F is an arithmetically profinite extension if and only if the extension $N(E,L|F)/N(L|F)$ is arithmetically profinite. In this case the field $N(E|F)$ can be identified with $N(N(E,L|F)|N(L|F))$ and

$$h_{E/F} = h_{N(E,L|F)/N(L|F)} \circ h_{L/F}.$$

If, in addition, E/F and E/L are Galois extensions, then

$$\mathrm{Gal}(N(E,L|F)/N(L|F))(h_{L/F}(x)) = \mathrm{Gal}(E/F)(x) \cap \mathrm{Gal}(N(E,L|F)/N(L|F))$$

where we identified $\mathrm{Gal}(N(E,L|F)/N(L|F))$ with $\mathrm{Gal}(E/L)$.

2. Fields of norms are related to various rings introduced by Fontaine in his study of Galois representations over local fields.

3. A local field F with finite residue field $\mathbb{F}_q$ has infinitely many *wild automorphisms*, i.e., continuous homomorphisms $\sigma\colon F \longrightarrow F$ such that $\pi_F^{-1}\sigma(\pi_F) \in U_1$, if and only if F is of positive characteristic. The group R of wild automorphisms of F has a natural filtration $R_i = \{\sigma \in R : \pi_F^{-1}\sigma\pi_F \in U_i\}$ and R is isomorphic to $\varprojlim R/R_i$. Therefore the group R is a pro-p-group. It is called the

Nottingham group by group theorists. It has finitely many generators. One can check that every non-trivial closed normal subgroup of an open subgroup of R is open; so R is a so-called hereditarily just infinite pro-p-group, [**7**].

Every Galois totally ramified and arithmetically profinite p-extension of a local field with residue field $\mathbb{F}_q$ is mapped via the functor of fields of norms to a closed subgroup of R. Using this functor and realisability of pro-p-groups as Galois groups of arithmetically profinite extensions in positive characteristic one can easily show that every finitely generated pro-p-group is isomorphic to a closed subgroup of R, [**7**].

For integer $r \geqslant 1$ define a closed subgroup $T = T[r]$ of R

$$T[r] = \{\sigma \in R\colon \pi_F^{-1}\sigma\pi_F = f(\pi_F) \quad \text{with } f(X) \in \mathbb{F}_q[[X^{p^r}]]\ \}.$$

For $p > 2, r \geqslant 1$ the group T is hereditarily just infinite (i.e. every non-trivial normal closed subgroup of every open subgroup is open), T does not have infinite subquotients isomorphic to p-adic Lie groups, and the group $T[r]$ for $r > 1$ can be realised as the Galois group of an arithmetically profinite extension of a finite extension of $\mathbb{Q}_p$, [**7**].

4. General ramification theory of infinite extensions of local fields is far from being complete, despite many deep investigations.

A satisfactory ramification theory of complete discrete valuation fields with imperfect residue field is also still missing. See review [**39**] for several developments. Such a theory is expected to have analogues of the Hasse–Herbrand function and satisfy analogues of three key properties of ramification theory of local fields, i.e. Herbrand Theorem, Hasse–Arf Theorem and compatibility of upper ramification filtration of finite abelian extensions with appropriate filtration on objects describing abelian extensions via an appropriate higher local reciprocity map.

18. Local Fields with Finite Residue Fields

18.1. Let F be a local field with finite residue field $\overline{F} = \mathbb{F}_q$, $q = p^f$ elements. The number f is called the *absolute residue degree* of F. Since $\text{char}(\mathbb{F}_q) = p$, Lemma 2.2 shows that F is of characteristic 0 or of characteristic p.

In the first case $v(p) > 0$ for the discrete valuation v in F, hence the restriction of v on $\mathbb{Q}$ is equivalent to the p-adic valuation. Then we can view the field $\mathbb{Q}_p$ of p-adic numbers as a subfield of F; another way to show this is to use the quotient field of the Witt ring of a finite field and Proposition 12.6.

Let $e = v(p) = e(F)$ be the absolute ramification index of F. Then by Proposition 9.4 we obtain that F is a finite extension of $\mathbb{Q}_p$ of degree $n = ef$. We call F a *local number field.*

In the second case Propositions 12.4 and 12.1 show that F is isomorphic (with respect to the field structure and the discrete valuation topology) to the field of formal power series $\mathbb{F}_q((X))$ with prime element X. We call F a *local functional field.*

The topology of the multiplicative group $F^\times$ of a local field is the induced topology from the topology of F. It is the product of the discrete topology on the infinite cyclic group generated by a prime element and the induced from F topology on the group of units U. Equivalently, the topology of $F^\times$ is induced from the topology of $F \times F$ via the embedding $\alpha \mapsto (\alpha, \alpha^{-1})$.

LEMMA. *The additive group of F is a complete locally compact topological space with respect to the discrete valuation topology. Open subgroups $\mathcal{M}^n$ of F form a base of neighbourhoods of 0. The additive group of the ring of integers $\mathcal{O}$ and of the maximal ideal $\mathcal{M}$ are compact.*

The multiplicative group $F^\times$ is locally compact, and the group of units U is compact. Open subgroups $1 + \mathcal{M}^n$ of $F^\times$ form a base of neighbourhoods of 1.

Proof. Assume that the additive group of $\mathcal{O}$ is not compact. Let $(V_i)_{i\in I}$ be a covering by open subsets in $\mathcal{O}$, i.e., $\mathcal{O} = \cup V_i$, such that $\mathcal{O}$ is not covered by a finite union of V_i. Let π be a prime element of $\mathcal{O}$. Since $\mathcal{O}/\pi\mathcal{O}$ is finite, there exists an element $\theta_0 \in \mathcal{O}$ such that the set $\theta_0 + \pi\mathcal{O}$ is not contained in the union of a finite number of V_i. Similarly, there exist elements $\theta_1, \dots, \theta_n \in \mathcal{O}$ such that $\theta_0 + \theta_1\pi + \dots + \theta_n\pi^n + \pi^{n+1}\mathcal{O}$ is not contained in the union of a finite number of V_i. However, the element $\alpha = \lim_{n\to+\infty} \sum_{m=0}^{n} \theta_m\pi^m$ belongs to some V_i, a contradiction. Hence, $\mathcal{O}$ is compact and U, as the union of $\theta + \pi\mathcal{O}$ with $\overline{\theta} \neq 0$, is compact. □

18.2. LEMMA. *The Galois group of every finite extension of F is solvable.*

Proof. Follows from Corollary 3 of 11.4. □

PROPOSITION. *For every $n \geqslant 1$ there exists a unique unramified extension L of F of degree n: $L = F(\mu_{q^n-1})$.*

The extension L/F is cyclic and the maximal unramified extension F^{ur} of F is a Galois extension.

The group $\mathrm{Gal}(F^{\mathrm{ur}}/F)$ *is isomorphic to $\widehat{\mathbb{Z}}$ and topologically generated by an automorphism φ_F, such that*

$$\varphi_F(\alpha) \equiv \alpha^q \mod \mathcal{M}_{F^{\mathrm{ur}}} \qquad \textit{for } \alpha \in \mathcal{O}_{F^{\mathrm{ur}}}.$$

The automorphism φ_F is called the Frobenius automorphism of F.

Proof. First we note that, by Corollary 1 of 6.3, F contains the group μ_{q-1} of $(q-1)$th roots of unity which coincides with the set of nonzero multiplicative representatives of $\overline{F}$ in $\mathscr{O}$. Moreover, Proposition 4.4 and Section 6 imply that the unit group U_F is isomorphic to $\mu_{q-1} \times U_{1,F}$.

The field $\mathbb{F}_q$ has the unique extension $\mathbb{F}_{q^n}$ of degree n, which is cyclic over $\mathbb{F}_q$. Propositions 10.2 and 10.3 show that there is a unique unramified extension L of degree n over F and hence $L = F(\mu_{q^n-1})$.

Now let E be an unramified extension of F and $\alpha \in E$. Then $F(\alpha)/F$ is of finite degree. Therefore, F^{ur} is contained in the union of all finite unramified extensions of F. We have

$$\mathrm{Gal}(F^{\mathrm{ur}}/F) \simeq \varprojlim \mathrm{Gal}(\mathbb{F}_{q^n}/\mathbb{F}_q) \simeq \widehat{\mathbb{Z}}.$$

It is well known that $\mathrm{Gal}(\mathbb{F}_q^{\mathrm{sep}}/\mathbb{F}_q)$ is topologically generated by the automorphism σ such that $\sigma(a) = a^q$ for $a \in \mathbb{F}_q^{\mathrm{sep}}$. Hence, $\mathrm{Gal}(F^{\mathrm{ur}}/F)$ is topologically generated by the Frobenius automorphism φ_F. □

REMARK. If $\theta \in \mu_{q^n-1}$, then

$$\varphi_F(\theta) \equiv \theta^q \mod \mathscr{M}_L$$

and $\varphi_F(\theta) \in \mu_{q^n-1}$. The uniqueness of the multiplicative representative for $\overline{\theta}^q \in \overline{F}$ implies now that $\varphi_F(\theta) = \theta^q$.

18.3. In order to describe the group $U_1 = U_{1,F}$ of principal units we can apply assertions of Sections 4 and 5.

If $\mathrm{char}(F) = p$, then Proposition 5.2 shows that every element $\alpha \in U_1$ can be uniquely written as the convergent product

$$\alpha = \prod_{\substack{p\nmid i\\ i>0}} \prod_{j\in J} (1+\theta_j \pi_i)^{a_{ij}},$$

where the index-set J numerates f elements in $\mathscr{O}_F$, such that their residues form a basis of $\mathbb{F}_q$ over $\mathbb{F}_p$, and the elements θ_j belong to this set of f elements; π_i are elements of $\mathscr{O}_F$ with $v(\pi_i) = i$, and $a_{ij} \in \mathbb{Z}_p$. Thus, U_1 has the infinite topological basis $\{1+\theta_j\pi_i\}$.

If $\mathrm{char}(F) = 0$, 5.4 and 5.5 show that every element $\alpha \in U_1$ can be written as a convergent product

$$\alpha = \prod_{i\in I} \prod_{j\in J} (1+\theta_j \pi_i)^{a_{ij}} \omega_*^a$$

where $I = \{1 \leqslant i < pe/(p-1),\, p\nmid i\}$, $e = e(F)$; the index-set J numerates f elements in $\mathscr{O}_F$, such that their residues form a basis of $\mathbb{F}_q$ over $\mathbb{F}_p$, and the elements

θ_j belong to this set of f elements; π_i are elements of $\mathcal{O}_F$ with $v(\pi_i) = i$, and $a_{ij} \in \mathbb{Z}_p$.

If a primitive pth root of unity does not belong to F, then $\omega_* = 1, a = 0$ and the above expression for α is unique; U_1 is a free $\mathbb{Z}_p$-module of rank $n = ef = |F : \mathbb{Q}_p|$.

If a primitive pth root of unity belongs to F, then $\omega_* = 1 + \theta_* \pi_{pe/(p-1)}$ is a principal unit such that $\omega_* \notin F^{\times p}$, and $a \in \mathbb{Z}_p$. In this case the above expression for α is not unique. Subsections 4.7 and 4.8 imply that U_1 is isomorphic to the product of n copies of $\mathbb{Z}_p$ and the p-torsion group μ_{p^r}, where $r \geqslant 1$ is the maximal integer such that $\mu_{p^r} \subset F$.

LEMMA. *If* $\mathrm{char}(F) = 0$, *then* $F^{\times n}$ *is an open subgroup of finite index in* $F^\times$ *for* $n \geqslant 1$. *If* $\mathrm{char}(F) = p$, *then* $F^{\times n}$ *is an open subgroup of finite index in* $F^\times$ *for* $p \nmid n$. *If a primitive nth root is in* F *then* $|F^\times : F^{\times n}| = n^2 q^{v(n)}$.

If $\mathrm{char}(F) = p$ *and* $p|n$, *then* $F^{\times n}$ *is not open and is not of finite index in* $F^\times$.

Proof. Everything except the formula follows from Proposition 4.9 and the previous considerations. To obtain the formula for $|F^\times : F^{\times n}|$, first it is $= n|U : U^n|$. Write $n = p^r m$ with integer m prime to p. The integer r can be positive only when F is of characteristic zero. We have $|\mathbb{F}_q^\times : \mathbb{F}_q^{\times n}| = |\mathbb{F}_q^\times : \mathbb{F}_q^{\times m}| = m$; $|U_1 : U_1^n| = 1$ in characteristic p and $= |U_1 : U_1^{p^r}| = p^{rd+1}$ in characteristic 0, where $d = |F : \mathbb{Q}_p|$. Hence $|F^\times : F^{\times n}| = n^2 p^{rd}$ and $p^{rd} = q^{v(n)}$. □

18.4. Now we have a look at the norm group $N_{L/F} L^\times$ for a finite extension L of F. Recall that the norm map

$$N_{\mathbb{F}_{q'}/\mathbb{F}_q} : \mathbb{F}_{q'}^\times \longrightarrow \mathbb{F}_q^\times$$

is surjective when $\mathbb{F}_{q'} \supset \mathbb{F}_q$. Then the second and third diagrams of Proposition 3.2 show that $N_{L/F} U_L = U_F$ in the case of an unramified extension L/F. Further, the first diagram there implies that

$$N_{L/F} L^\times = \langle \pi^n \rangle \times U_F,$$

where π is a prime element in F, $n = |L : F|$. This means, in particular, that $F^\times / N_{L/F} L^\times$ is a cyclic group of order n in the case under consideration. Conversely, every subgroup of finite index in $F^\times$ that contains U_F coincides with the norm group $N_{L/F} L^\times$ for a suitable unramified extension L/F.

The next case is a totally and tamely ramified Galois extension L/F of degree n. Since L/F is Galois, we get $\mu_n \subset F^\times$ and $n|(q-1)$. Proposition 13.3 and its Corollary show that

$$N_{L/F} U_{1,L} = U_{1,F}, \quad \pi \in N_{L/F} L^\times,$$

for a suitable prime element π in F such that $L = F(\sqrt[n]{-\pi})$, and $\theta \in N_{L/F}L^\times$ for $\theta \in U_F$ if and only if $\overline{\theta} \in \mathbb{F}_q^{\times n}$. Since $n|(q-1)$, the quotient group $\mathbb{F}_q^\times/\mathbb{F}_q^{\times n}$ is cyclic of order n. We conclude that

$$N_{L/F}L^\times = \langle \pi \rangle \times \langle \theta \rangle \times U_{1,F}$$

with an element $\theta \in U_F$, such that its residue $\overline{\theta}$ generates $\mathbb{F}_q^\times/\mathbb{F}_q^{\times n}$. In particular, $F^\times/N_{L/F}L^\times$ is cyclic of order n. Conversely, every subgroup of index n relatively prime to $\operatorname{char}(\overline{F})$ coincides with the norm group $N_{L/F}L^\times$ for a suitable cyclic extension L/F.

The last case to be considered is the case of a totally ramified Galois extension L/F of degree p. Preserving the notations of 13.4 we apply Proposition 13.5. The right vertical homomorphism of the fourth diagram

$$\overline{\theta} \to \overline{\theta}^p - \overline{\eta}^{p-1}\overline{\theta}$$

has a kernel of order p; therefore its cokernel is also of order p. Let $\theta_* \in U_F$ be such that $\overline{\theta}_*$ does not belong to the image of this homomorphism. Since $\overline{F}$ is perfect, we deduce, using the third and fourth diagrams, that $1 + \theta_*\pi_F^s \notin N_{L/F}U_{1,L}$. The other diagrams imply that $F^\times/N_{L/F}L^\times$ is a cyclic group of order p and generated by

$$1 + \theta_*\pi_F^s \mod N_{L/F}L^\times.$$

If $\operatorname{char}(F) = 0$, then, by Proposition 14.3, $s \leqslant pe/(p-1)$, where $e = e(F)$. That Proposition also shows that if $p|s$, then $s = pe/(p-1)$ and a primitive pth root of unity ζ_p belongs to F, and $L = F(\sqrt[p]{\pi})$ for a suitable prime element π in F. In this case $F^\times/N_{L/F}L^\times$ is generated by $\omega_* \mod N_{L/F}L^\times$.

Conversely, note that every subgroup of index p in the additive group $\mathbb{F}_q$ can be written as $\overline{\eta}\wp(\mathbb{F}_q)$ for some $\overline{\eta} \in \mathbb{F}_q$. Let N be an open subgroup of index p in $F^\times$ such that some prime element $\pi_F \in N$ and $\omega_* \in N$ (if $\operatorname{char}(F) = 0$). Then, in terms of the cited Corollary 14.5, if s is the maximal integer relatively prime to p such that $U_{s,F} \not\subset N$ and $U_{s+1,F} \subset N$, then $1 + \eta\wp(\mathcal{O}_F)\pi^s + \pi^{s+1}\mathcal{O}_F \subset N$ for some element $\eta \in \mathcal{O}_F$. By that Corollary we obtain that $1 + \eta\wp(\mathcal{O}_F)\pi^s + \pi^{s+1}\mathcal{O}_F \subset N_{L/F}L^\times$, where $L = F(\lambda)$ and λ is a root of the polynomial $X^p - X + \theta^p\alpha$, with $\alpha = \theta^{-p}\eta^{-1}\pi^{-s}$ for a suitable $\theta \in U_F$. Since $s = s(L|F)$ (the same notation as in 13.4), we get $U_{i,F} \subset U_{i+1,F}N_{L/F}U_L$ for $i < s$, by Proposition 13.5. In terms of the homomorphism λ_i of Section 4 we obtain that

$$\lambda_i((N \cap U_{i,F})U_{i+1,F}/U_{i+1,F}) = \lambda_i\left((N_{L/F}L^\times \cap U_{i,F})U_{i+1,F}/U_{i+1,F}\right)$$

for $i \geqslant 0$. If $\omega_* \notin N$ and $\operatorname{char}(F) = 0$, then one can put $L = F(\sqrt[p]{\pi})$ to obtain the same relations for N and $N_{L/F}L^\times$ as just above.

When F is of positive characteristic p, the Artin–Schreier extension L/F generated by a root of the polynomial $X^p - X + \theta^p\alpha$ with $v_F(\alpha) = -s < 0$ and not divisible by p has its ramification jump is s (see Section 14). Proposition 13.5 implies that $U_{i,F} \subset U_{i+1,F}N_{L/F}U_L$ for $i < s$ and $U_{s+1,F} \subset N_{L/F}U_L$. Since $|U_F : N_{L/F}U_L| = p$, and by Corollary 14.5 the units $1 + \theta^{-p}\wp(\mathcal{O}_F)\alpha^{-1}$ are in the norm group of L/F, we deduce that $1 + \theta^{-p}\rho\alpha^{-1} \notin N_{L/F}U_L$ for any unit $\rho \in U_F$ such that $\overline{\rho} \notin \wp(\overline{F})$. Hence every open subgroup of index p in $F^\times$ is the norm group of the appropriate Artin–Schreier extension.

Later we will show that every open subgroup N of finite index in $F^\times$, $N = N_{L/F}L^\times$ for a suitable abelian extension L/F.

18.5. The following property will be useful in motivating the Neukirch's approach to class field theory.

PROPOSITION. *Let L/F be a finite Galois extension and $\sigma \in \mathrm{Gal}(L/F)$. There is a finite separable extension K/F such that $M = KL$ is a finite unramified Galois extension of K and of L, $K^{\mathrm{ur}} = L^{\mathrm{ur}} = M^{\mathrm{ur}}$, and the image of the Frobenius automorphism $\varphi_K \in \mathrm{Gal}(L^{\mathrm{ur}}/K)$ with respect to the restriction on L is σ.*

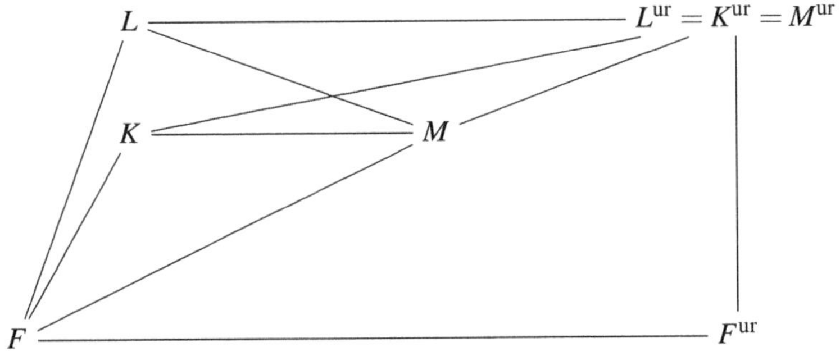

Proof. The restriction of σ on $L_0 = L \cap F^{\mathrm{ur}}$ is $\varphi_F^n|_{L_0}$ for some $n > 0$. Let $\varphi \in \mathrm{Gal}(L^{\mathrm{ur}}/F)$ be an extension of φ_F. The product of σ and the restriction of φ^{-n} on L is an element of $\mathrm{Gal}(L/L_0)$, let $\tau \in \mathrm{Gal}(L^{\mathrm{ur}}/F^{\mathrm{ur}})$ correspond to it via the canonical isomorphism with $\mathrm{Gal}(L/L_0)$. Then $\tilde{\sigma} = \tau\varphi^n$ has the property: $\tilde{\sigma}|_L = \sigma(\varphi^{-n}\varphi^n)|_L = \sigma$, $\tilde{\sigma}|_{F^{\mathrm{ur}}} = \varphi_F^n$.

Let K by the fixed field of $\tilde{\sigma}$. Since $F \subset K \subset L^{\mathrm{ur}}$ we deduce that $F^{\mathrm{ur}} \subset K^{\mathrm{ur}} \subset L^{\mathrm{ur}}$. The Galois group of L^{ur}/K is topologically generated by $\tilde{\sigma}$ and is isomorphic to $\widehat{\mathbb{Z}}$, therefore it does not have non-trivial closed subgroups of finite order. The group $\mathrm{Gal}(L^{\mathrm{ur}}/K^{\mathrm{ur}})$ being a subgroup of the finite group $\mathrm{Gal}(L^{\mathrm{ur}}/F^{\mathrm{ur}})$ is trivial, so $L^{\mathrm{ur}} = K^{\mathrm{ur}}$. Due to the latter, M/K is a subextension of K^{ur}/K and M/L is a subextension of L^{ur}/L, hence those are unramified extensions.

The degree of the extension K/F is the product of the degree of the extension K/K_0, $K_0 = K \cap F^{\mathrm{ur}}$, whose Galois group is isomorphic to $\mathrm{Gal}(L^{\mathrm{ur}}/F^{\mathrm{ur}})$ and the degree of K_0/F equal to n, so it is finite.

In the unramified extension L^{ur}/K the automophism $\tilde{\sigma}$ is a power of φ_K and their restrictions to F^{ur} are equal: $\varphi_K|_{F^{\mathrm{ur}}} = \varphi_F^{|K_0:F|}|_{F^{\mathrm{ur}}} = \varphi_F^n|_{F^{\mathrm{ur}}} = \tilde{\sigma}|_{F^{\mathrm{ur}}}$, so $\tilde{\sigma} = \varphi_K$. □

Chapter 3

Class Field Theory

This chapter is a short presentation of class field theory of local fields with finite residue field and of global fields, both of characteristic zero and of positive characteristic. Algebraic topics such as central division algebras and Galois cohomology groups, as well as Lubin–Tate formal groups, all of which are not necessary to derive the main theorems of class field theory, are not included. The presentation of global class field theory is based on the use of abstract class field theory mechanism discovered by Neukirch and described in Section 2 in a slightly simple form than the original presentation. This mechanism is natural from the point of view of the theory of local fields and local class field theory, as explained in Sections 1 and 2. The miracle is that this mechanism, so much flavoured by local considerations, also works for global fields. The underlying topological group theoretical algorithm is of anabelian geometry character. Section 3 applies abstract class field theory to local fields with finite residue field and derives the main results of local class field theory; its various local generalisations are mentioned. The study of adeles and ideles of global fields starts in Section 4, with the aim of checking two axioms of class field theory for global fields. Zeta integrals that lift usual zeta functions to an adelic level are studied in Section 5, following the Iwasawa and Tate approaches. This theory is applied to deduce fundamental properties of zeta functions of global fields and to complete the proof of one of axioms of class field theory (computation of the index of the norm map image of the idele class group of a global field). Main theorems of global class field theory are established in Section 6.

1. Main Results of Local Class Field Theory

1.1. Let k be a local field with finite residue field. The main results of local class field theory in this case are

1. For every finite separable extension F/k and finite Galois extension L/F there is a surjective homomorphism

$$\Upsilon_{L/F}\colon \mathrm{Gal}(L/F) \longrightarrow F^{\times}/N_{L/F}L^{\times}$$

whose kernel is $\mathrm{Gal}(L/E)$, where E is the maximal abelian subextension of F in L (hence $N_{L/F}L^\times = N_{E/F}E^\times$), such that:

(a) if L/F is unramified then

$$\Upsilon_{L/F}(\varphi_{L/F}) \equiv \pi_F \mod N_{L/F}L^\times$$

where $\varphi_{L/F}$ is the restriction of the Frobenius automorphism φ_F on L, π_F is a prime element of F;

(b) if $M/F, E/L$, F/k, L/k are finite separable extensions, and L/F and E/M are finite Galois extensions, then the diagram

$$\begin{array}{ccc} \mathrm{Gal}(E/M) & \xrightarrow{\Upsilon_{E/M}} & M^\times/N_{E/M}E^\times \\ \big\downarrow & & \big\downarrow N^*_{M/F} \\ \mathrm{Gal}(L/F) & \xrightarrow{\Upsilon_{L/F}} & F^\times/N_{L/F}L^\times \end{array}$$

is commutative, where the left vertical map is the restriction of Galois automorphisms and the right vertical map is induced by the norm map $N_{M/F}$;

(c) if M/k is a finite separable extension and L/M is a finite Galois extension, then for $\sigma \in \mathrm{Gal}(k^{\mathrm{sep}}/k)$ the diagram

$$\begin{array}{ccc} \mathrm{Gal}(L/M) & \xrightarrow{\Upsilon_{L/M}} & M^\times/N_{L/M}L^\times \\ \sigma^*\big\downarrow & & \big\downarrow \sigma \\ \mathrm{Gal}(\sigma L/\sigma M) & \xrightarrow{\Upsilon_{\sigma L/\sigma M}} & (\sigma M)^\times/N_{\sigma L/\sigma M}(\sigma L)^\times \end{array}$$

is commutative, where $\sigma^*(\tau) = \sigma\tau\sigma^{-1}$.

2. Denote the maximal abelian extension of F by F^{ab}.

For every finite separable extension F/k, passing to the inverse limit (see 5.1 of Chapter 1), we get

$$\Psi_F \colon F^\times \longrightarrow \varprojlim F^\times/N_{L/F}L^\times \longrightarrow \varprojlim \mathrm{Gal}(L/F)^{\mathrm{ab}} = \mathrm{Gal}(F^{\mathrm{ab}}/F)$$

where L runs through all finite Galois (or all finite abelian) extensions of F, and the second arrow is the inverse isomorphism to $\Upsilon_{L/F}$. The homomorphism Ψ_F is called the *local reciprocity map*.

(a) Ψ_F is injective and continuous, its image is dense in $\mathrm{Gal}(F^{\mathrm{ab}}/F)$.

(b) Compatibility with 0-dimensional class field theory (for finite fields): the restriction of the image of Ψ_F on F^{ur} coincides with $\alpha \mapsto \varphi_F^{v_F(\alpha)}$, i.e. the diagram is commutative

$$\begin{array}{ccc} F^\times & \xrightarrow{\Psi_F} & \mathrm{Gal}(F^{\mathrm{ab}}/F) \\ \downarrow{\scriptstyle v_F} & & \downarrow \\ \mathbb{Z} & \xrightarrow{1\mapsto\varphi_F} & \mathrm{Gal}(F^{\mathrm{ur}}/F) \\ \downarrow{\scriptstyle =} & & \downarrow{\scriptstyle \simeq} \\ \mathbb{Z} & \xrightarrow{1\mapsto\varphi_F} & \mathrm{Gal}(\mathbb{F}_q^{\mathrm{sep}}/\mathbb{F}_q) \simeq \widehat{\mathbb{Z}} \end{array}$$

(c) Compatibility with ramification theory for abelian extensions for $n \geqslant 0$:

$$\Psi_F : U_{n,F} \twoheadrightarrow \mathrm{Gal}(F^{\mathrm{ab}}/F)(n).$$

Note that there is no analog of this property in class field theory of global fields or higher local fields.

(d) For every finite separable extension F/k, if L is a finite separable extension of F, and σ is an automorphism of $\mathrm{Gal}(k^{\mathrm{sep}}/k)$, then the diagrams

$$\begin{array}{ccc} L^\times & \xrightarrow{\Psi_L} & \mathrm{Gal}(L^{\mathrm{ab}}/L) \\ \downarrow{\scriptstyle N_{L/F}} & & \downarrow \\ F^\times & \xrightarrow{\Psi_F} & \mathrm{Gal}\left(F^{\mathrm{ab}}/F\right) \end{array}$$

$$\begin{array}{ccc} F^\times & \xrightarrow{\Psi_F} & \mathrm{Gal}(F^{\mathrm{ab}}/F) \\ \downarrow & & \downarrow{\scriptstyle \mathrm{Ver}} \\ L^\times & \xrightarrow{\Psi_L} & \mathrm{Gal}\left(L^{\mathrm{ab}}/L\right) \end{array}$$

$$\begin{array}{ccc} L^\times & \xrightarrow{\Psi_L} & \mathrm{Gal}(L^{\mathrm{ab}}/L) \\ \downarrow{\scriptstyle \sigma} & & \downarrow{\scriptstyle \sigma^*} \\ (\sigma L)^\times & \xrightarrow{\Psi_{\sigma L}} & \mathrm{Gal}\left((\sigma L)^{\mathrm{ab}}/\sigma L\right) \end{array}$$

are commutative, where $\sigma^*(\tau) = \sigma\tau\sigma^{-1}$, the right vertical homomorphism of the second diagram is the restriction and $\mathrm{Ver}\colon \mathrm{Gal}(F^{\mathrm{ab}}/F) = \mathrm{Gal}(F^{\mathrm{sep}}/F)^{\mathrm{ab}} \longrightarrow \mathrm{Gal}(F^{\mathrm{sep}}/L)^{\mathrm{ab}} = \mathrm{Gal}(L^{\mathrm{ab}}/L)$ is induced by the transfer map $\mathrm{Ver}\colon G^{\mathrm{ab}} \longrightarrow H^{\mathrm{ab}}$ for a subgroup H of finite index in a group G.

3. *Existence Theorem*, the correspondence between open subgroups of finite index in $F^\times$ and the norm subgroups of finite abelian extensions L/F: $N \longleftrightarrow N_{L/F}L^\times, N = \Psi_F^{-1}(\mathrm{Gal}(F^{\mathrm{ab}}/L))$, is an order reversing bijection between the lattice of open subgroups of finite index in $F^\times$ (with respect to the intersection $N_1 \cap N_2$ and the product N_1N_2) and the lattice of finite abelian extensions of F (with respect to the compositum L_1L_2 and intersection $L_1 \cap L_2$).

1.2. *Neukirch's method* in class field theory constructs $\Upsilon_{L/F}$ by using desired properties 1a, 1b and Proposition 18.5 of Chapter 2. In other words, one uses desired functoriality with respect to the base change to reduce to the case of finite unramified extensions, in order to get an explicit formula for the map $\Upsilon_{L/F}$, [**29**], [**30**].

For a finite Galois extension L/F one can try to define

$$\Upsilon_{L/F}\colon\ \mathrm{Gal}(L/F) \longrightarrow F^\times/N_{L/F}L^\times$$

by finding for a $\sigma \in \mathrm{Gal}(L/F)$ any $\tilde{\sigma} \in \mathrm{Gal}(L^{\mathrm{ur}}/F) = \varphi_K$ as in the proof of Proposition 18.5 of Chapter 2, and then applying 1b, 1a to deduce that

$$\Upsilon_{L/F}(\sigma) = N_{K/F}(\Upsilon_{KL/K}(\varphi_K)) = N_{K/F}\pi_K \mod N_{L/F}L^\times.$$

So it is natural to define $\Upsilon_{L/F}(\sigma)$ as $N_{K/F}\pi_K \mod N_{L/F}L^\times$ where π_K is any prime element of the field K which is the fixed field of any lift $\tilde{\sigma} \in \mathrm{Gal}(L^{\mathrm{ur}}/F)$ such that $\tilde{\sigma}_L = \sigma$ and $\tilde{\sigma}|_{F^{\mathrm{ur}}}$ is a positive integer power of φ_K. There are two indeterminacies in relation to the choice of $\tilde{\sigma}$ and the choice of π_K.

In order for everything to work fine, two axioms of class field theory have to be satisfied. Typically, for one dimensional fields they are: for cyclic extensions of prime degree with a generator σ the kernel of the norm map $N_{L/F}$ is the image of $1-\sigma$ and the index of the norm group equals the degree of L/F.

Neukirch's mechanism derives $\Upsilon_{L/F}$ and its properties in the situations when these two class field theory axioms are satisfied for a chosen class of fields and abelian groups associated to them, such as the multiplicative group of a local field with finite residue field or the idele class group of a global field. This mechanism is universal in the sense that it also works, with some small modifications, for higher local, global-local and global fields as well.

This explicit and clear mechanism is purely topological group theoretical, while to verify the two axioms for a specific class of fields and associated abelian groups one has to use their ring structure.

Thus, these class field theory axioms separate a topological group theoretical part of class field theory from its part that uses ring structures and cannot be established by working with groups only. Such separation is important in anabelian geometry, one of generalisations of class field theory.

2. Neukirch's Abstract Class Field Theory

This section does not depend on any of the other sections in this book. We will refer to the previous theory of local fields for illustration and motivation, but the reader can skip such references. A local field with finite residue field will be abbreviated in this section as LFF.

We would like to have a purely group-theoretical deduction of the main theorems of class field theory from as few as possible as mild as possible conditions. Two conditions C1 and C2 are about continuous homomorphisms deg and v. They and two class field theory axioms A1 and A2, first unramified and then for all cyclic extensions of prime degree, will imply, via a purely topological group theoretical mechanism, main theorems of class field theory.

2.1. Let k be a field. It does not need to be related to algebraic number fields or to local fields.

C1. *Assume that the absolute Galois group $G_k = \mathrm{Gal}(k^{\mathrm{sep}}/\tilde{k})$ of k is sufficiently large, namely that there is a surjective continuous homomorphism of topological profinite groups*

$$\deg\colon G_k \longrightarrow \widehat{\mathbb{Z}}.$$

Here $\widehat{\mathbb{Z}}$ is the additive group endowed with its profinite topology.

REMARK. Instead of the field k with its absolute Galois group G_k (see 5.2 of Chapter 1), one can start with a profinite group G which has a $\widehat{\mathbb{Z}}$-quotient. All the following use of subfields of k^{sep} in abstract class field theory can obviously be rewritten in the language of closed subgroups of G.

Denote the kernel of deg by $G_{\tilde{k}} = \mathrm{Gal}(k^{\mathrm{sep}}/\tilde{k})$, with $\tilde{k}/k$ the corresponding $\widehat{\mathbb{Z}}$-extension.

For any finite separable extension F of k, denote $\tilde{F} = F\tilde{k}$.

Extensions of F in $\tilde{F}$ will be called *unramified* in this section.

Denote $F_0 = F \cap \tilde{k}$ and $f_F = |F_0 : k|$.

For LFF $\tilde{k} = k^{\mathrm{ur}}$ and f_F is the inertia degree of F/k.

The morphism deg induces a surjective morphism

$$\deg_F = f_F^{-1} \deg \colon G_F \longrightarrow \widehat{\mathbb{Z}}.$$

Then for a finite separable extension L/F the following diagram is commutative

$$\begin{array}{ccc} G_L & \xrightarrow{\deg_L} & \widehat{\mathbb{Z}} \\ \downarrow & & \downarrow{\scriptstyle f_L f_K^{-1}} \\ G_F & \xrightarrow{\deg_F} & \widehat{\mathbb{Z}} \end{array}$$

Call any element of G_F which is sent by $\deg_F$ to $1 \in \widehat{\mathbb{Z}}$ a *Frobenian of* F.

The restriction of any Frobenian of F on $\tilde{F}$ is called the *Frobenius* φ_F of F, it is uniquely determined by F. So $\deg_F(\tau) = n \in \widehat{\mathbb{Z}}$ where $\tau|_{\tilde{F}} = \varphi_F^n$.

For LFF the Frobenius of F is the Frobenius automorphisms of F.

For a finite Galois extension L/F denote by φ an extension in $\mathrm{Gal}(\tilde{L}/F)$ of φ_F. Then every element τ of $\mathrm{Gal}(\tilde{L}/F)$ is uniquely written as $\rho\varphi^n$ with $\rho \in \mathrm{Gal}(\tilde{L}/\tilde{F})$ and $n = \deg_F(\tau) \in \widehat{\mathbb{Z}}$.

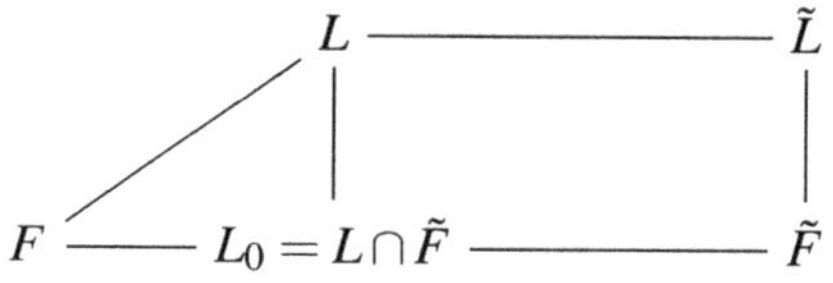

If $L_0 = F$, then $\mathrm{Gal}(\tilde{L}/F)$ is the product of $\mathrm{Gal}(\tilde{L}/\tilde{F})$ and $\mathrm{Gal}(\tilde{L}/L)$.

2.2. DEFINITION. Introduce a subset

$$\mathrm{Frob}(L/F) = \{\tau \in \mathrm{Gal}(\tilde{L}/F) : \deg_F(\tau) \text{ is a positive integer}\}$$

of $\mathrm{Gal}(\tilde{L}/F)$.

Compare the following Proposition with Proposition 18.5 of Chapter 2.

PROPOSITION. *The set* $\mathrm{Frob}(L/F)$ *is closed with respect to multiplication; it is not closed with respect to inversion, and* $1 \notin \mathrm{Frob}(L/F)$.

The fixed field K *of* $\tau \in \mathrm{Frob}(L/F)$ *is a finite extension of* F, $\tau = \varphi_K$, $\tilde{K} = \tilde{L}$.

The field $M = KL$ *is a finite unramified extension of* K *and of* L.

The set $\mathrm{Frob}(L/F)$ *consists of the Frobeniuses* φ_K *of finite extensions* K *of* F *in* $\tilde{L}$.

The map $\mathrm{Frob}(L/F) \longrightarrow \mathrm{Gal}(L/F), \tau \mapsto \tau|_L$ *is surjective.*

Proof. The first assertion is obvious.

Since $F \subset K \subset \tilde{L}$ we deduce that $\tilde{F} \subset \tilde{K} \subset \tilde{L}$. The Galois group of $\tilde{L}/K$ is topologically generated by τ and isomorphic to $\widehat{\mathbb{Z}}$, therefore it does not have non-trivial closed subgroups of finite order. So the group $\mathrm{Gal}(\tilde{L}/\tilde{K})$ being a subgroup of the finite group $\mathrm{Gal}(\tilde{L}/\tilde{F})$ is trivial. So $\tilde{L} = \tilde{K}$. Due to the latter, M/K is a subextension of $\tilde{K}/K$ and M/L is a subextension of $\tilde{L}/L$, hence those are unramified extensions.

Put $K_0 = K \cap \tilde{F}$. This field is the fixed field of $\tau|_{\tilde{F}} = \varphi_F^n$, $n > 0$, therefore $|K_0 : F| = n$ is finite. We deduce that

$$|K : K_0| = |\tilde{K} : \tilde{F}| = |\tilde{L} : \tilde{F}| = |L : L_0|$$

is finite. Thus, K/F is a finite extension.

Now $\varphi_K|_{\tilde{F}} = \varphi_F^{|K_0:F|} = \varphi_F^n|_{\tilde{F}} = \tau|_{\tilde{F}}$. Therefore, $\tau = \varphi_K$.

Denote by φ an extension in $\mathrm{Gal}(\tilde{L}/F)$ of φ_F. Let $\sigma \in \mathrm{Gal}(L/F)$, then $\sigma|_{L_0}$ is equal to φ_F^m for some positive integer m. Let $\rho \in \mathrm{Gal}(\tilde{L}/\tilde{F})$ be such that $\rho|_L$ is $\sigma\varphi^{-m}|_L$ (it belongs to $\mathrm{Gal}(L/L_0)$ since $\sigma\varphi^{-m}|_L$ acts trivially on L_0). Then for $\tau = \rho\varphi^m$ we deduce that $\tau|_{\tilde{F}} = \varphi_F^m$ and $\tau|_L = \sigma$. Then the element $\tau \in \mathrm{Frob}(L/F)$ is mapped to $\sigma \in \mathrm{Gal}(L/F)$. □

REMARK. If instead of the $\widehat{\mathbb{Z}}$-extension $\tilde{k}/k$ one starts with a $\mathbb{Z}_l$-extension $\breve{k}/k$ for a prime l and the corresponding surjective homomorphism $\deg^{\smile} : G_k \longrightarrow \mathbb{Z}_l$, then the assertions of the Proposition for a finite Galois extension L/F of degree a power of l remain true, with exactly the same proof.

2.3.

C2. *Assume that there is an abelian topological group A endowed with an action by the profinite group G_k such that A is the union of fixed elements A^H where H runs through G_F for finite separable extensions F/k and A^H is the subgroup of A of elements each of which is fixed under the action of H. In other words, $A = \varinjlim A^{G_F}$ with respect to inclusions.* We will write the operation of A multiplicatively.

For a closed subgroup G_F of G_k (i.e. F/k is a separable extension) denote by A_F the G_F-fixed elements of A.

For an open subgroup G_L of a closed subgroup G_F of G_k denote by

$$N_{L/F} : A_L \longrightarrow A_F$$

the product of the action of right representatives of G_L in G_F:

$$N_{L/F}(\alpha) = \prod \sigma(\alpha), \quad G_F = \dot{\cup} G_L \sigma.$$

It is easy to check that $N_{L/F}$ is a well defined homomorphism.

It immediately follows that

$$N_{L/F} = N_{M/F} \circ N_{L/M}$$

for a subextension M/F of L/F, and

$$N_{\sigma L/\sigma F} \circ \sigma = \sigma \circ N_{L/F}$$

for $\sigma \in G_k$.

Assume that there is a continuous homomorphism

$$v\colon A_k \longrightarrow \widehat{\mathbb{Z}}$$

whose image is $\mathbb{Z}$ or $\widehat{\mathbb{Z}}$ and such that the equality

$$v(N_{F/k}A_F) = f_F v(A_k)$$

holds for every finite separable extension F/k. The group $\mathbb{Z}$ is endowed, as usual, with the discrete topology.

Partially similarly to how $\deg_F$ was defined in relation to deg, define v_F in relation to v as

$$v_F = f_F^{-1} v \circ N_{F/k}\colon A_F \longrightarrow \widehat{\mathbb{Z}},$$

then $v_F(A_F) = v(A_k)$.

For LFF $A = k^{\text{sep}\times}$, its topology is the inductive limit topology of the multiplicative groups of finite separable extensions F of k, i.e. a translation invariant topology such that a subgroup is open if and only if it is an open subgroup of some $F^\times$. Since every element of A belongs to some finite separable extension F/k, $A = \varinjlim A_{G_F}$. Then $A_F = F^\times$ and v_F is the discrete valuation of F. The required compatibility with the norm map for finite separable extensions and their inertia degree follows from Theorem 9.5 of Chapter 2.

The definition of v_F immediately implies that

$$f_L v_L = f_F v_F \circ N_{L/F}, \quad v_{\sigma F} = v_F \circ \sigma \text{ for } \sigma \in G_k.$$

Similarly to the definition of Frobenian we have

DEFINITION. An element π_F of A_F such that $v_F(\pi_F) = 1$ is called a *prime element* of F. Also define

$$U_F = \{u \in A_F : v_F(u) = 0\}.$$

So A_F is isomorphic to the direct product of U_F and the subgroup generated by π_F.

We note that if $\sigma F = F$ then $\pi_F^{\sigma - 1} \in U_F$.

2.4. Now we need two unramified axioms for the G-module A (unramified axioms of CFT):

A1$^{\sim}$. *For any finite separable extension F of k and any unramified extension L/F of prime degree*

$$\ker N_{L/F} = A_L^{\sigma-1},$$

where σ is any generator of $\mathrm{Gal}(L/F)$.

A2$^{\sim}$. *For any finite separable extension F of k and any unramified extension L/F of prime degree*

$$|A_F : N_{L/F}A_L| = |L:F|.$$

Equivalently, $A_F/N_{L/F}A_L \simeq$ $\mathrm{Gal}(L/F)$.

COROLLARY. *For any finite unramified extension L/F with a generator σ we have*

$$\ker N_{L/F} = A_L^{\sigma-1}, \qquad |A_F : N_{L/F}A_L| = |L:F|,$$

and

$$\ker N_{L/F} = U_L^{\sigma-1}, \qquad N_{L/F}U_L = U_F.$$

Proof. For any finite unramified extension L/F, π_F is a prime element of A_L and $N_{L/F}\pi_F = \pi_F^{|L:F|}$. Then $A_F/N_{L/F}A_L$ is the product of $\mathbb{Z}/|L:F|\mathbb{Z}$ and $U_F/N_{L/F}U_L$. Since π_F is a prime element of A_L, for $\alpha = \pi_F^r u \in A_L$ we have $\alpha^{\sigma-1} = (\pi_F^r u)^{\sigma-1} = u^{\sigma-1}$, $u \in U_L$. Thus, the properties in the second displayed formula hold for unramified extensions of prime degree.

We check the assertions by induction on the degree. Let M/F be a subextension of cyclic unramified extension L/F such that $|L:M|$ is a prime number. By the induction hypothesis, $N_{L/M}U_L = U_M$, $N_{M/F}U_M = U_F$, so $N_{L/F}U_L = U_F$ and then $A_F/N_{L/F}A_L \simeq \mathbb{Z}/|L:F|\mathbb{Z}$. If $\alpha \in \ker N_{L/F}$ then by the induction hypothesis $N_{L/M}\alpha = \beta^{\sigma-1}$ for some $\beta \in U_M$, so $\beta = N_{L/M}\gamma$ for some $\gamma \in U_L$ and $\alpha\gamma^{1-\sigma} \in \ker N_{L/M}$, hence $\alpha = \gamma^{\sigma-1}\delta^{\sigma'-1}$ where $\sigma' = \sigma^{|M:F|}$. Hence $\alpha \in U_L^{\sigma-1}$. □

DEFINITION. Let L/F be a finite Galois extension. Define

$$\widetilde{\Upsilon}_{L/F}\colon \mathrm{Frob}(L/F) \longrightarrow A_F/N_{L/F}A_L, \quad \tau \mapsto N_{K/F}\pi_K \mod N_{L/F}A_L,$$

where K is the fixed field of $\tau \in \mathrm{Frob}(L/F)$ and π_K is any prime element of K.

LEMMA. *The map $\widetilde{\Upsilon}_{L/F}$ is well defined. If $\tau|_L = \mathrm{id}_L$ then $\widetilde{\Upsilon}_{L/F}(\tau) = 1$.*

Proof. Let π_1, π_2 be prime elements in K. Then $\pi_1 = \pi_2\varepsilon$ with a $\varepsilon \in U_K$. Let M be the compositum of K and L. Since the extension M/K is unramified, by the previous Corollary there is $\eta \in U_M$ such that $\varepsilon = N_{M/K}\eta$. Hence

$$N_{K/F}\pi_1 = N_{K/F}(\pi_2\varepsilon) = N_{K/F}\pi_2 \cdot N_{K/F}(N_{M/K}\eta) = N_{K/F}\pi_2 \cdot N_{L/F}(N_{M/L}\eta).$$

We obtain that $N_{K/F}\pi_1 \equiv N_{K/F}\pi_2 \mod N_{L/F}A_L$.

If $\tau|_L = \mathrm{id}_L$ then $L \subset K$ and therefore $N_{K/F}\pi_K \in N_{L/F}A_L$. □

PROPOSITION. *The map $\tilde{\Upsilon}_{L/F}$ sends the product of two of its elements to the product of their images.*

Proof. Denote by ψ an extension in $\mathrm{Gal}(\tilde{L}/F)$ of φ_F. Take three elements of $\mathrm{Frob}(L/F)$ such that the third is the product of the first two. Let K_i for $i \in \{1,2,3\}$ be their fixed fields, so these elements are φ_{K_i} by the previous results. Let $\varphi_{K_i}|_{\tilde{F}} = \varphi_F^{m_i}$ for positive integer m_i, then $\tau_i = \psi^{m_i}\varphi_{K_i}^{-1} \in \mathrm{Gal}(\tilde{L}/\tilde{F})$.

Also introduce $K_4 = \psi^{m_2}K_1$ then $\varphi_{K_4}|_{\tilde{F}} = \psi^{m_2}\varphi_{K_1}\psi^{-m_2}|_{\tilde{F}} = \varphi_F^{m_1}$. Denote $\tau_4 = \psi^{m_1}\varphi_{K_4}^{-1} = \psi^{m_3}\varphi_{K_1}^{-1}\psi^{-m_2}$, since $m_3 = m_1 + m_2$. From $\varphi_{K_3} = \varphi_{K_2}\varphi_{K_1}$ we obtain $\tau_3 = \tau_4\tau_2$.

Enlarge L by replacing it with a finite Galois extension of F in $\tilde{L}$ which contains L and all K_i. Proving the Proposition for this enlarged field implies the Proposition in the general case.

Denote by N the norm map $N_{\tilde{L}/\tilde{F}}$.

For a finite extension R of F in $\tilde{L}$ such that $R\tilde{F} = \tilde{L}$ let $R_0 = R \cap \tilde{F}$, $|R_0 : F| = m$.

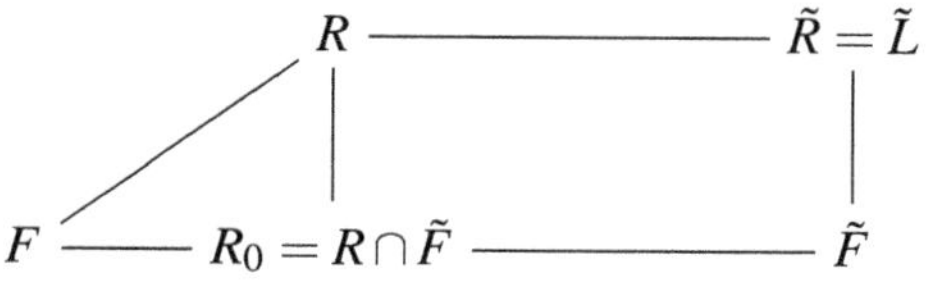

Denote $\mathscr{N}_R\colon A_R \longrightarrow A_{\tilde{L}}$, $\alpha \mapsto \alpha^{1+\psi+\dots+\psi^{m-1}}$.

For $\alpha \in A_R$, $\beta \in A_{R_0}$ we have

$$N_{R/R_0}(\alpha) = N(\alpha), \quad N_{R_0/F}(\beta) = \beta^{1+\varphi_F+\dots+\varphi_F^{m-1}} = \mathscr{N}_R(\beta),$$

and

$$N_{R/F}(\alpha) = N_{R_0/F} \circ N_{R/R_0}(\alpha) = N \circ \mathscr{N}_R(\alpha).$$

Let π_i be a prime element of K_i for $i = 1,2,3$. Put $\pi_4 = \psi^{m_2}(\pi_1)$. Then $\varphi_{K_i}\pi_i = \pi_i$ and

$$\mathscr{N}_{K_i}(\pi_i)^{\psi-1} = \pi_i^{\psi^{m_i}-1} = \pi_i^{\psi^{m_i}\varphi_{K_i}^{-1}-1} = \pi_i^{\tau_i-1}.$$

Now,

$$N_{K_3/F}\pi_3 N_{K_2/F}\pi_2^{-1} N_{K_1/F}\pi_1^{-1} = N_{K_3/F}\pi_3 N_{K_2/F}\pi_2^{-1} N_{K_4/F}\pi_4^{-1} = N\rho,$$

where $\rho = \mathcal{N}_{K_3}(\pi_3)\mathcal{N}_{K_2}(\pi_2)^{-1}\mathcal{N}_{K_4}(\pi_4)^{-1}$. Then we have $v_L(\rho) = m_3 - m_2 - m_1 = 0$, i.e. $\rho \in U_L$. Hence we deduce $\rho^{\psi-1} = \pi_3^{\tau_3-1}\pi_2^{1-\tau_2}\pi_4^{1-\tau_4}$.

Introduce three elements $\rho_2 = \pi_4\pi_2^{-1}$, $\rho_3 = \pi_3\pi_4^{-1}$, $\rho_4 = \pi_4^{\tau_2-1}$ of U_L. Then we obtain

$$\rho^{\psi-1} = \rho_2^{\tau_2-1}\rho_3^{\tau_3-1}\rho_4^{\tau_4-1}.$$

To complete the proof of the Proposition we will show that $N\rho \in N_{L/F}A_L$. It is convenient to work with yet another field M which is the fixed field of φ_L^n where $n = |L:F|$.

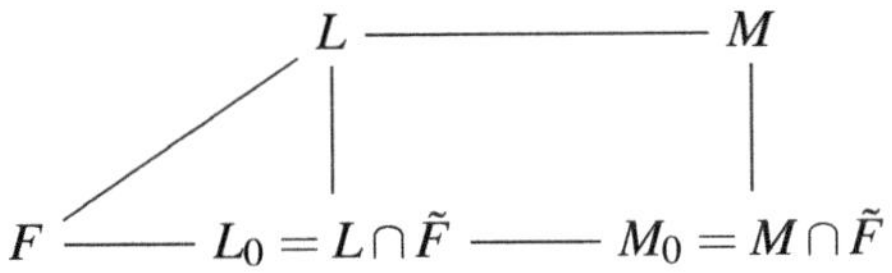

Then M/L is an unramified extension of degree n. Hence by the previous Corollary there are units $\nu, \nu_i \in U_M$ such that their images with respect to $N_{M/L}$ are equal to ρ, ρ_i. Then by the same Corollary

$$\nu^{\psi-1} = \nu_2^{\tau_2-1}\nu_3^{\tau_3-1}\nu_4^{\tau_4-1}\xi$$

where $\xi = \varepsilon^{\varphi_L-1}$ for some $\varepsilon \in U_M$.

Applying N, we obtain

$$(N\nu)^{\varphi_F-1} = (N\nu)^{\psi-1} = (N\varepsilon)^{\varphi_L-1} = (N\varepsilon)^{\varphi_F^r-1} = (\mathcal{N}_L N\varepsilon)^{\varphi_F-1}$$

where $r = |L_0 : F|$ and we use $\varphi_F^r - 1 = (\varphi_F - 1)\mathcal{N}_L$ on M_0. Thus, $N\nu = \kappa \cdot \mathcal{N}_L N\varepsilon$ with some $\kappa \in A_F$.

Applying $N_{M/L}$ and using $N_{M/L}\kappa = N_{L/F}\kappa$, we conclude $N\rho = N_{M/L}N\nu = N_{L/F}(\kappa)N_{M/F}(\varepsilon) \in N_{L/F}A_L$. □

COROLLARY. *For a finite Galois extension L/F the map $\tilde{\Upsilon}_{L/F}$ induces a well defined homomorphism*

$$\Upsilon_{L/F}\colon \operatorname{Gal}(L/F) \longrightarrow A_F/N_{L/F}A_L.$$

Proof. Let two Frobeniuses $\varphi_{K_1}, \varphi_{K_2} \in \operatorname{Gal}(\tilde{L}/F)$ have the same restriction on L. If their $\deg_F$ are the same then their restriction on $\tilde{F}$ are also the same, so they are equal. If $\deg(\varphi_{K_1}) - \deg(\varphi_{K_2})$ is positive then $\varphi_{K_1}\varphi_{K_2}^{-1}$ is a Frobenius whose restriction on L is the identity automorphism, with fixed field K_3. For prime elements

π_i of K_i by the previous Proposition we obtain $N_{K_1/F}\pi_1 \equiv N_{K_2/F}\pi_2 N_{K_3/F}\pi_3 \equiv N_{K_2/F}\pi_2 \mod N_{L/F}A_L$ since $K_3 \supset L$. □

We will denote $\Upsilon^{\mathrm{ab}}_{L/F}\colon \mathrm{Gal}(L/F)^{\mathrm{ab}} \longrightarrow A_F/N_{L/F}A_L$ the induced map from the maximal abelian quotient $\mathrm{Gal}(L/F)^{\mathrm{ab}}$ of $\mathrm{Gal}(L/F)$.

REMARK. Let L/F be a finite Galois extension such that $L\cap\tilde F = F$. Let $\sigma \in \mathrm{Gal}(\tilde L/\tilde F)$, denote by the same notation its restriction to L. Let $\varphi = \varphi_L$. Then $\Upsilon_{L/F}(\sigma) \equiv N_{K/F}\pi_K \mod N_{L/F}A_L$ where π_K is a prime element of the fixed field K of $\varphi_K = \sigma\varphi$, $K\cap\tilde F = F$. Let M be a finite Galois extension of L inside $\tilde L$ and containing K. Then for a prime element π_L of L there is $\varepsilon \in U_M$ such that $\pi_K = \pi_L\varepsilon$. Hence $\varepsilon^{1-\varphi} = \varepsilon^{1-\sigma\varphi}\varepsilon^{\sigma\phi-\varphi} = \pi_L^{\sigma\varphi-1}(\varepsilon^{\varphi})^{\sigma-1} = (\pi_L\varepsilon^{\varphi})^{\sigma-1}$, so for the prime element $\pi_M = \pi_L\varepsilon^{\varphi}$ we have

$$\varepsilon^{1-\varphi} = \pi_M^{\sigma-1}, \quad \Upsilon_{L/F}(\sigma) \equiv N_{M/M\cap\tilde F}\,\varepsilon \mod N_{L/F}A_L.$$

The equation

$$\varepsilon^{1-\varphi} = \pi_M^{\sigma-1}$$

in the very special case of cyclotomic extensions of local fields with finite residue field plays the key role in the theory of ϕ–γ modules, but, as we see, its role is much more significant in abstract class field theory, and hence, in particular, in local class field theory and in global class field theory. This equation also plays the key role in non-commutative class field theory of arithmetically profinite extensions of local fields with finite residue field, see Remark 6 in 3.7.

2.5. Now we deduce some of the properties 1(a), 1(b), 1(c) of 1.1, and more.

LEMMA. *Let L/F be a finite unramified extension of prime degree. Then*

$$\Upsilon_{L/F}(\varphi_F|_L) \equiv \pi_F \mod N_{L/F}A_L,$$

where π_F is any prime element of F, and $\Upsilon_{L/F}$ is an isomorphism of cyclic groups of order $|L:F|$.

Proof. The fixed field of $\varphi_F \in \mathrm{Frob}(L/F)$ is F. □

PROPOSITION. *If $M/F, E/L, F/k, L/k$ are finite separable extensions, and L/F and E/M are finite Galois extensions, then the diagram*

$$\begin{array}{ccc} \mathrm{Gal}(E/M) & \xrightarrow{\Upsilon_{E/M}} & A_M/N_{E/M}A_E \\ \downarrow & & \downarrow N^*_{M/F} \\ \mathrm{Gal}(L/F) & \xrightarrow{\Upsilon_{L/F}} & A_F/N_{L/F}A_L \end{array}$$

is commutative, where the left vertical map is the restriction of Galois automorphisms and the right vertical map is induced by the norm map $N_{M/F}$.

Proof. For a $\tau \in \mathrm{Frob}(E/M)$ its restriction on $\tilde{L}$ is $\sigma \in \mathrm{Frob}(L/F)$, as $\deg_F(\sigma) = \deg_M(\tau) f_M f_F^{-1}$ is a positive natural number. The intersection of the fixed field K of τ with $\tilde{L}$ is the fixed field R of σ and for a prime element π_K of K its norm $N_{K/R}\pi_K$ is a prime element of R. It remains to use $N_{M/F} \circ N_{K/M} = N_{R/F} \circ N_{K/R}$. □

COROLLARY. *Let M/F be a Galois subextension in a finite Galois extension L/F. Then the diagram of homorphisms*

$$\begin{array}{ccccccc} \mathrm{Gal}(L/M) & \longrightarrow & \mathrm{Gal}(L/F) & \longrightarrow & \mathrm{Gal}(M/F) & \longrightarrow & 1 \\ \downarrow{\scriptstyle \Upsilon_{L/M}} & & \downarrow{\scriptstyle \Upsilon_{L/F}} & & \downarrow{\scriptstyle \Upsilon_{M/F}} & & \\ A_M/N_{L/M}A_L & \xrightarrow{N^*_{M/F}} & A_F/N_{L/F}A_L & \longrightarrow & A_F/N_{M/F}A_M & \longrightarrow & 1 \end{array}$$

is commutative. Here the central homomorphism of the lower exact sequence is induced by the identity map of A_F.

Proof. An easy consequence of the preceding Proposition. □

PROPOSITION. *If M/k is a finite separable extension and L/M is a finite Galois extension, and $\sigma \in \mathrm{Gal}(k^{\mathrm{sep}}/k)$, then the diagram*

$$\begin{array}{ccc} \mathrm{Gal}(L/M) & \xrightarrow{\Upsilon_{L/M}} & A_M/N_{L/M}A_L \\ {\scriptstyle \sigma^*}\downarrow & & \downarrow{\scriptstyle \sigma} \\ \mathrm{Gal}(\sigma L/\sigma M) & \xrightarrow{\Upsilon_{\sigma L/\sigma M}} & A_{\sigma M}/N_{\sigma L/\sigma M}A_{\sigma L} \end{array}$$

is commutative, where $\sigma^*(\tau) = \sigma\tau\sigma^{-1}$.

Proof. Let $\tau' \in G_k$ be an extension of $\tau \in \mathrm{Frob}(L/M)$, then $\deg_{\sigma M}(\sigma\tau'\sigma^{-1}|_{\sigma\tilde{M}}) = \deg_M \tau$ is a positive integer. If K is the fixed field of τ' with a prime element π then σK is the fixed field of $\sigma\tau'\sigma^{-1}|_{\sigma\tilde{L}}$ with a prime element $\sigma\pi$. □

2.6. Another functorial property involves the transfer map from group theory. Recall the notion of *transfer* (*Verlagerung*). Let G be a group and let G' be its commutator subgroup (derived group). Denote the quotient group G/G' by G^{ab}; it is abelian. Let H be a subgroup of finite index in G. Let

$$G = \cup_i H\rho_i, \qquad \rho_i \in G,\ 1 \leqslant i \leqslant |G:H|$$

be the decomposition of G into the disjoint union of sets $H\rho_i$.

Define the transfer

$$\mathrm{Ver}\colon G^{\mathrm{ab}} \longrightarrow H^{\mathrm{ab}}, \quad \sigma \mod G' \mapsto \prod_i \rho_i \sigma \rho_{\sigma(i)}^{-1} \mod H',$$

where $\sigma(i)$ is determined by the condition $\rho_i \sigma \in H\rho_{\sigma(i)}$. So $\sigma(1), \ldots, \sigma(|G:H|)$ is a permutation of $1, \ldots, |G:H|$.

We shall verify that Ver is well defined. Let $\rho_i' = \kappa_i \rho_i$ with $\kappa_i \in H$. Then

$$\prod \rho_i' \sigma \rho'^{-1}_{\sigma(i)} = \prod \kappa_i \left(\rho_i \sigma \rho_{\sigma(i)}^{-1}\right) \kappa_{\sigma(i)}^{-1} \equiv \prod \rho_i \sigma \rho_{\sigma(i)}^{-1} \cdot \prod \kappa_i \cdot \prod \kappa_{\sigma(i)}^{-1} \mod H',$$

because H/H' is abelian. Hence

$$\prod \rho_i' \sigma \rho'^{-1}_{\sigma(i)} \equiv \prod \rho_i \sigma \rho_{\sigma(i)}^{-1} \mod H'.$$

Now we shall verify that Ver is a homomorphism. Let $\sigma, \tau \in G$; then

$$\rho_i \sigma \tau \rho_{\sigma\tau(i)}^{-1} \equiv \rho_i \sigma \rho_{\sigma(i)}^{-1} \rho_{\sigma(i)} \tau \rho_{\sigma\tau(i)}^{-1} \mod H'$$

and, as $\rho_i \sigma \rho_{\sigma(i)}^{-1} \in H$, $\rho_i \sigma \tau \rho_{\sigma\tau(i)}^{-1} \in H$, we get $\rho_{\sigma(i)} \tau \rho_{\sigma\tau(i)}^{-1} \in H$, i.e., $\sigma\tau(i) = \tau(\sigma(i))$. Hence

$$\prod \rho_i \sigma \tau \rho_{\sigma\tau(i)}^{-1} \equiv \prod \rho_i \sigma \rho_{\sigma(i)}^{-1} \cdot \prod \rho_i \tau \rho_{\tau(i)}^{-1} \mod H'.$$

If G is abelian then $\mathrm{Ver}(\sigma) = \sigma^{|G:H|}$.

We need another description of Ver. Let σ be an element of G. For an element $\tau_1 \in G$ let $g_1 = g(\sigma, \tau_1)$ be the maximal integer such that all the sets $H\tau_1\sigma, H\tau_1\sigma^2, \ldots, H\tau_1\sigma^{g_1}$ are distinct. Then, take an element $\tau_2 \in G$ such that all $H\tau_2\sigma, H\tau_1\sigma, \ldots, H\tau_1\sigma^{g_1}$ are distinct and find $g_2 = g(\sigma, \tau_1, \tau_2)$ such that all the sets

$$H\tau_2\sigma, \ldots, H\tau_2\sigma^{g_2}, H\tau_1\sigma, \ldots, H\tau_1\sigma^{g_1}$$

are distinct. Repeating this construction, we finally obtain that G is the disjoint union of the sets $H\tau_n\sigma^{m_n}$, where $1 \leqslant n \leqslant k, 1 \leqslant m_n \leqslant g_n = g(\sigma, \tau_1, \tau_2, \ldots, \tau_n)$. The number g_i can also be determined as the minimal positive integer, for which the element

$$\sigma[\tau_i] = \tau_i \sigma^{g_i} \tau_i^{-1}$$

belongs to H. The definition of Ver shows that in this case

$$\mathrm{Ver}(\sigma \mod G') \equiv \prod_n \sigma[\tau_n] \mod H'.$$

Since the image of $\Upsilon L/F$ is in the abelian group, it defines a homomorphism

$$\Upsilon_{L/F}^{\mathrm{ab}}\colon \mathrm{Gal}(L/F)^{\mathrm{ab}} \longrightarrow A_F/N_{L/F}A_L.$$

PROPOSITION. *Let L/F be a finite Galois extension and let M/F be a subextension in L/F. Then the diagram*

$$\begin{array}{ccc} \mathrm{Gal}(L/F)^{\mathrm{ab}} & \xrightarrow{\Upsilon^{\mathrm{ab}}_{L/F}} & A_F/N_{L/F}A_L \\ \Big\downarrow{\scriptstyle \mathrm{Ver}} & & \Big\downarrow \\ \mathrm{Gal}(L/M)^{\mathrm{ab}} & \xrightarrow{\Upsilon^{\mathrm{ab}}_{L/M}} & A_M/N_{L/M}A_L \end{array}$$

is commutative; here the right vertical homomorphism is induced by the embedding $F \hookrightarrow M$.

Proof. Denote $\tilde{G} = \mathrm{Gal}(\tilde{L}/F), \tilde{H} = \mathrm{Gal}(\tilde{L}/M)$. Let $\sigma \in \mathrm{Gal}(L/F)$, and let $\tilde{\sigma} \in \mathrm{Frob}(L/F)$ be its extension. Let $\tilde{G}$ be the disjoint union of $\tilde{H}\tilde{\tau}_n\tilde{\sigma}^{m_n}$ for $1 \leqslant n \leqslant k, 1 \leqslant m_n \leqslant g_n$, as above. Let $G = \mathrm{Gal}(L/F)$ and $H = \mathrm{Gal}(L/M)$; then G is the disjoint union of $H\tau_n\sigma^{m_n}$ for $\tau_n = \tilde{\tau}_n|_L \in \mathrm{Gal}(L/F)$. This means that

$$\mathrm{Ver}(\sigma \mod G') \equiv \prod_n \sigma[\tau_n] \mod H'.$$

Let S be the subgroup in $\tilde{G}$ generated topologically by $\tilde{\sigma}$ and

$$\tilde{H}_n = \tilde{H} \cap \tilde{\tau}_n S \tilde{\tau}_n^{-1}.$$

Then $\tilde{H}_n$ is a subgroup in $\tilde{H}$, which coincides with the subgroup in $\tilde{H}$ topologically generated by $\tilde{\sigma}[\tilde{\tau}_n]$. Note that $\tilde{\tau}_n S$ is the disjoint union of $\tilde{H}_n\tilde{\tau}_n\tilde{\sigma}^{m_n}$ for $1 \leqslant m_n \leqslant g_n$.

Let $\tilde{H}$ be the disjoint union of $\tilde{\nu}_{n,l}\tilde{H}_n$ for $\tilde{\nu}_{n,l} \in \tilde{H}, 1 \leqslant l \leqslant |\tilde{H} : \tilde{H}_n|$. Then

$$\tilde{G} = \cup\cup\, \tilde{\nu}_{n,l}\tilde{H}_n\tilde{\tau}_n\sigma^{m_n} = \cup\tilde{\nu}_{n,l}\tilde{\tau}_n S.$$

If K is the fixed field of $\tilde{\sigma}$, then it is the fixed field of S, and we obtain that

$$N_{K/F}(\alpha) = \prod_{n,l} \tilde{\nu}_{n,l}\tilde{\tau}_n(\alpha) \qquad \text{for } \alpha \in K.$$

Let K_n be the fixed field of $\tilde{\sigma}[\tilde{\tau}_n] = \tilde{\tau}_n\tilde{\sigma}^{g_n}\tilde{\tau}_n^{-1}$. Then $(\tilde{\tau}_n K)\tilde{F} = \tilde{\tau}_n\tilde{K} = \tilde{\tau}_n\tilde{L} = \tilde{L}$, $\tilde{\tau}_n K \subset K_n$, and $K_n/\tilde{\tau}_n K$ is the unramified extension of degree g_n. Hence, for a prime element π in K, the element $\tilde{\tau}_n(\pi)$ is prime in K_n. Moreover, one can show as before that

$$N_{K_n/M}(\alpha) = \prod_l \tilde{\nu}_{n,l}(\alpha) \qquad \text{for } \alpha \in K_n.$$

We deduce that

$$N_{K/F}(\pi) = \prod_{n,l} \tilde{\nu}_{n,l}\tilde{\tau}_n(\pi) = \prod_n N_{K_n/M}\big(\tilde{\tau}_n(\pi)\big).$$

Since $\tilde{\sigma}[\tilde{\tau}_n] \in \mathrm{Frob}(L/M)$ extends the element $\sigma[\tau_n] \in \mathrm{Gal}(L/M)$, we conclude that

$$\Upsilon^{\mathrm{ab}}_{L/F}(\sigma) = \prod_n \Upsilon^{\mathrm{ab}}_{L/M}(\sigma[\tau_n]) = \Upsilon^{\mathrm{ab}}_{L/M}\big(\prod_n \sigma[\tau_n]\big)$$

and $\Upsilon^{\mathrm{ab}}_{L/F}(\sigma) = \Upsilon^{\mathrm{ab}}_{L/M}\big(\mathrm{Ver}\,(\sigma \bmod \mathrm{Gal}(L/F)')\big)$. □

2.7. In order to prove that $\Upsilon^{\mathrm{ab}}_{L/F}$ is an isomorphism, we need two full axioms for the G_k-module A (axioms of CFT), not just for unramified extensions:

A1. *For any finite separable extension F of k and any cyclic extension L/F of prime degree*

$$\ker N_{L/F} = A_L^{\sigma-1},$$

where σ is any generator of $\mathrm{Gal}(L/F)$.

A2. *For any finite separable extension F of k and any cyclic extension L/F of prime degree*

$$|A_F : N_{L/F}A_L| = |L : F|.$$

Equivalently, $A_F/N_{L/F}A_L \simeq \mathrm{Gal}(L/F)$.

REMARK. Assume that $\Upsilon_{L/M}$ is an isomorphism for a finite abelian extension L/M. Let $\sigma \in G_k$ be such that $\sigma L = L, \sigma M = M$ and $\sigma^*\tau = \tau$ for every $\tau \in \mathrm{Gal}(L/M)$. Then $A_M^{\sigma-1} \subset N_{L/M}A_L$. Indeed, since $\Upsilon_{L/M}$ is an isomorphism, the last Proposition of 2.5 shows that the map $\sigma\colon A_M/N_{L/M}A_L \longrightarrow A_M/N_{L/M}A_L$ is the identity map, i.e. $A_M^{\sigma-1} \subset N_{L/M}A_L$.

THEOREM. *For a finite Galois extension L/F*

$$\Upsilon^{\mathrm{ab}}_{L/F}\colon \mathrm{Gal}(L/F)^{\mathrm{ab}} \longrightarrow A_F/N_{L/F}A_L$$

is an isomorphism.

Proof. First, let L/F be a cyclic extension of prime degree n. If L/F is unramified then $\Upsilon_{L/F}$ is an isomorphism by Lemma 2.5.

If $L \cap \tilde{F} = F$ then, in the notation of Remark 2.4 let σ be a generator of $\mathrm{Gal}(\tilde{L}/\tilde{F})$ and use the same notation for its restriction on L. Let $\varphi = \varphi_L$. Let K be the fixed field of $\sigma\varphi$ with a prime element π_K. Then $K \cap \tilde{F} = F$. Assume that $\Upsilon_{L/F}(\sigma) \equiv N_{K/F}\pi_K \equiv 1 \mod N_{L/F}A_L$ and get a contradiction. Let M be the composite of L and K, it is a subfield of $\tilde{L}$. For a prime element π_L of L there is a unit $\varepsilon \in U_M$ such that $\pi_K = \pi_L\varepsilon$. Using the notation in the proof of Proposition 2.4,

$$\Upsilon_{L/F}(\sigma) \equiv N_{K/F}\pi_K \equiv N_{M/M_0}\varepsilon \mod N_{L/F}A_L.$$

If $N_{M/M_0}\varepsilon \in N_{L/F}A_L$, then since $L \cap \tilde{F} = F$, $N_{M/M_0}\varepsilon = N_{M/M_0}\rho$ for a unit $\rho \in U_L$, and axiom A1 implies $\rho = \varepsilon v^{\sigma-1}$ for some $v \in A_M$. Then

$$(\pi_L\rho)^{\sigma-1} = (\pi_L\rho)^{\sigma\varphi-1} = (\pi_K v^{\sigma-1})^{\sigma\varphi-1} = (v^{\sigma\varphi-1})^{\sigma-1},$$

so $\xi = \pi_L\rho v^{1-\sigma\varphi} \in M_0$. Since $v_M(v^{\sigma\varphi-1}) = 0$, we obtain $1 = v_M(\xi) = n v_{M_0}(\xi)$, a contradiction. Thus, $\Upsilon_{L/F}$ is injective and then by A2 it is also surjective.

Now, for a finite cyclic extension L/F of non-prime degree let M/F be a proper non-trivial subextension of prime degree. By the previous Remark $A_M^{\sigma-1} \subset N_{L/M}A_L$ and therefore $N_{M/F}^*$ in injective in the diagram of Corollary of 2.5. Therefore $\Upsilon_{L/F}$ is injective by induction on the degree. By induction on the degree, A2 and Corollary 2.5, $\Upsilon_{L/F}$ is surjective.

Next, consider the case of a finite abelian extension L/F. Using the commutative diagram in Corollary 2.5, the surjectivity of $\Upsilon_{L/F}$ follows by the induction on the degree, and if $\Upsilon_{L/F}(\sigma) = 1$ then the restriction of σ on every cyclic quotient M/F is trivial, hence $\sigma = 1$.

For a finite Galois extension L/F the same diagram now implies that the kernel of $\Upsilon_{L/F}$ is the commutator subgroup of G. For solvable extensions the surjectivity of $\Upsilon_{L/F}$ follows by induction on the degree. In the general case, the surjectivity follows if the image of $\Upsilon_{L/F}$ includes the p-Sylow subgroup of $A_F/N_{L/F}A_L$ for every prime p. Let M be the fixed field of a p-Sylow subgroup of $\mathrm{Gal}(L/F)$. Then by induction on the degree, $\Upsilon_{L/M}$ is surjective, so the p-Sylow subgroup of $A_M/N_{L/M}A_L$ is in its image. It remains to notice that $N_{M/F}^*$ maps this subgroup isomorphically onto the p-Sylow subgroup of $A_F/N_{L/F}A_L$, since $|M:F|$ is prime to p and the inverse map is induced by the inclusion $A_F \hookrightarrow A_M$. □

COROLLARY. *For every finite cyclic extension L/F with a generator σ properties*

$$\ker N_{L/F} = A_L^{1-\sigma}, \qquad A_F/N_{L/F}A_L \simeq \mathrm{Gal}(L/F).$$

hold.

Proof. The second assertion follows from the isomorphism property of $\Upsilon_{L/F}$.

The first assertion can be proved by induction on the degree of cyclic L/F. Let M/F be a subextension of L/F of prime degree m. Proposition 2.6 for the abelian L/F implies that the homomorphism $j\colon A_F/N_{L/F}A_L \longrightarrow A_M/N_{L/M}A_L$ induced by $A_F \longrightarrow A_M$ corresponds via the reciprocity maps to the homomorphism $\mathrm{Gal}(L/F) \longrightarrow \mathrm{Gal}(L/M)$, $\sigma \mapsto \sigma^m$. For cyclic L/F it is surjective, and hence j is surjective. Therefore, $A_M \subset A_F N_{L/M}A_L$. Now, if $\alpha \in A_L$ is in the kernel of $N_{L/F}$ then by the induction assumption $N_{L/M}\alpha = \beta^{\sigma-1}$ for some $\beta \in A_M$ and

$\sigma \in \mathrm{Gal}(L/F)$. Write $\beta = \gamma N_{L/M}\delta$ with $\gamma \in A_F$ and $\delta \in A_L$. Then $N_{L/M}\alpha = \beta^{\sigma-1} = N_{L/M}\delta^{\sigma-1}$, so $\alpha\delta^{1-\sigma}$ is in the kernel of $N_{L/M}$, and so $\alpha \in C_L^{1-\sigma}$. □

2.8. The inverse of $\Upsilon^{\mathrm{ab}}_{L/F}$ provides the *reciprocity homomorphism*

$$\Psi_{L/F} : A_F \longrightarrow \mathrm{Gal}(L/F)^{\mathrm{ab}},$$

its kernel is $N_{L/F}A_L$.

PROPOSITION. *Let H be a subgroup in $\mathrm{Gal}(L/F)^{\mathrm{ab}}$, and let M be the fixed field of H in $L \cap F^{\mathrm{ab}}$. Then $\Psi^{-1}_{L/F}(H) = N_{M/F}A_M$.*

Let L_1, L_2 be abelian extensions of finite degree over F, and let $L_3 = L_1L_2$, $L_4 = L_1 \cap L_2$. Then

$$N_{L_3/F}A_{L_3} = N_{L_1/F}A_{L_1} \cap N_{L_2/F}A_{L_2}, \quad N_{L_4/F}A_{L_4} = N_{L_1/F}A_{L_1} N_{L_2/F}A_{L_2}.$$

For finite abelian extensions, the field L_1 is a subfield of the field L_2 if and only if $N_{L_2/F}A_{L_2} \subset N_{L_1/F}A_{L_1}$; in particular, $L_1 = L_2$ if and only if $N_{L_1/F}A_{L_1} = N_{L_2/F}A_{L_2}$.

If a subgroup N in A_F contains the norm subgroup $N_{L/F}A_L$ for some finite Galois extension L/F, then N itself is a norm subgroup.

Proof. The first assertion follows immediately from 2.5, 2.7. Let $H_i = \mathrm{Gal}(L_3/L_i)$, $i = 1, 2$. Then

$$\begin{aligned} N_{L_3/F}A_{L_3} &= \Psi^{-1}_{L_3/F}(1) = \Psi^{-1}_{L_3/F}(H_1 \cap H_2) \\ &= \Psi^{-1}_{L_3/F}(H_1) \cap \Psi^{-1}_{L_3/F}(H_2) = N_{L_1/F}A_{L_1} \cap N_{L_2/F}A_{L_2}, \\ N_{L_4/F}A_{L_4} &= \Psi^{-1}_{L_3/F}(H_1H_2) = \Psi^{-1}_{L_3/F}(H_1)\Psi^{-1}_{L_3/F}(H_2) \\ &= N_{L_1/F}A_{L_1} N_{L_2/F}A_{L_2}. \end{aligned}$$

If $L_1 \subset L_2$, then $N_{L_2/F}A_{L_2} \subset N_{L_1/F}A_{L_1}$. Conversely, if $N_{L_2/F}A_{L_2} \subset N_{L_1/F}A_{L_1}$, then $N_{L_1L_2/F}A_{L_1L_2}$ coincides with $N_{L_2/F}A_{L_2}$, and Theorem 2.7 shows that the extension L_1L_2/F is of the same degree as L_2/F, hence $L_1 \subset L_2$.

Finally, if $N \supset N_{L/F}A_L$, then $N = N_{M/F}A_M$, where M is the fixed field of $\Psi_{L/F}(N)$. □

REMARK. The question is how for a specific field k, when the axioms A1 and A2 hold, to characterise norm subgroups $N_{L/F}A_L$ of finite Galois extensions L/F in terms of A_F, e.g. as open subgroup of a certain intrinsic topology of A_F.

2.9. Similarly to 2 of 1.1, passing to the inverse limit for $\Psi_{L/F}$, using 2.5, one gets the *reciprocity map*

$$\Psi_F\colon A_F \longrightarrow \varprojlim A_F/N_{L/F}A_L \longrightarrow \varprojlim \operatorname{Gal}(L/F)^{\mathrm{ab}} = \operatorname{Gal}(F^{\mathrm{ab}}/F)$$

where L runs through all finite Galois (or all finite abelian) extensions of F.

THEOREM. *The reciprocity map is well defined.*

Its image is dense in $\operatorname{Gal}(F^{\mathrm{ab}}/F)$, *and its kernel coincides with the intersection of all norm subgroups* $N_{L/F}A_L$ *in* A_F *for all finite Galois (equivalently, all finite abelian) extensions* L/F.

If L/F *is a finite Galois extension and* $\alpha \in A_F$, *then the automorphism* $\Psi_F(\alpha)$ *acts trivially on* $L \cap F^{\mathrm{ab}}$ *if and only if* $\alpha \in N_{L/F}A_L$.

The restriction of $\Psi_F(\alpha)$ *on* $\tilde{F}$ *coincides with* $\varphi_F^{v_F(\alpha)}$ *for* $\alpha \in A_F$.

Let L *be a finite separable extension of* F, *and let* σ *be an automorphism of* $\operatorname{Gal}(F^{\mathrm{sep}}/F)$. *Then the diagrams*

$$\begin{array}{ccc} A_L & \xrightarrow{\Psi_L} & \operatorname{Gal}(L^{\mathrm{ab}}/L) \\ \downarrow{\scriptstyle\sigma} & & \downarrow{\scriptstyle\sigma^*} \\ A_{\sigma L} & \xrightarrow{\Psi_{\sigma L}} & \operatorname{Gal}\left((\sigma L)^{\mathrm{ab}}/\sigma L\right) \end{array}$$

$$\begin{array}{ccc} A_L & \xrightarrow{\Psi_L} & \operatorname{Gal}(L^{\mathrm{ab}}/L) \\ \downarrow{\scriptstyle N_{L/F}} & & \downarrow \\ A_F & \xrightarrow{\Psi_F} & \operatorname{Gal}\left(F^{\mathrm{ab}}/F\right) \end{array}$$

$$\begin{array}{ccc} A_F & \xrightarrow{\Psi_F} & \operatorname{Gal}(F^{\mathrm{ab}}/F) \\ \downarrow & & \downarrow{\scriptstyle \mathrm{Ver}} \\ A_L & \xrightarrow{\Psi_L} & \operatorname{Gal}\left(L^{\mathrm{ab}}/L\right) \end{array}$$

are commutative, where $\sigma^*(\tau) = \sigma\tau\sigma^{-1}$, *the right vertical homomorphism of the second diagram is the restriction and*

$$\mathrm{Ver}\colon \operatorname{Gal}(F^{\mathrm{sep}}/F)^{\mathrm{ab}} \longrightarrow \operatorname{Gal}(F^{\mathrm{sep}}/L)^{\mathrm{ab}} = \operatorname{Gal}(L^{\mathrm{ab}}/L).$$

Proof. Let $L_1/F, L_2/F$ be finite Galois extensions and $L_1 \subset L_2$. Then the first Proposition 2.5 shows that the restriction of the automorphism

$$\Psi_{L_2/F}(\alpha) \in \operatorname{Gal}(L_2/F)^{\mathrm{ab}}$$

on the field $L_1 \cap F^{\mathrm{ab}}$ coincides with $\Psi_{L_1/F}(\alpha)$ for an element $\alpha \in A_F$. This means that Ψ_F is well defined.

The condition $\alpha \in N_{L/F}A_L$ is equivalent $\Psi_{L/F}(\alpha) = 1$, i.e. $\Psi_F(\alpha)$ acts trivially on $L \cap F^{\mathrm{ab}}$.

Hence, the kernel of Ψ_F is equal to $\bigcap N_{L/F}A_L$, where L runs through all finite Galois extensions of F. Since $\Psi_F(A_F)|_L = \mathrm{Gal}(L/F)$ for a finite abelian extension L/F, we deduce that $\Psi_F(A_F)$ is dense in $\mathrm{Gal}(F^{\mathrm{ab}}/F)$.

Similarly to the proof of Lemma 2.5 we obtain $\Psi_F(\pi_F)|_{\tilde{F}} = \varphi_F$ for a prime element π_F in F. Hence, $\Psi_F(\alpha)|_{\tilde{F}} = \varphi_F^{v_F(\alpha)}$ and $\Psi_F(U_F)|_{\tilde{F}} = 1$.

The commutativity of the diagrams follow from 2.5, 2.6, 2.7. □

3. Local Class Field Theory and Generalisations

In this section k, F, L are a local fields with finite residue field.

3.1. As mentioned in 2.3, for C1 of 2.1, the map $\deg_k\colon G_k \longrightarrow \widehat{\mathbb{Z}}$ is the surjective homomorphism

$$\deg_k\colon G_k \longrightarrow \mathrm{Gal}(k^{\mathrm{ur}}/k) \simeq \widehat{\mathbb{Z}}, \quad \tilde{k} = k^{\mathrm{ur}}.$$

As mentioned in 2.3, $A = k^{\mathrm{sep}\times}$, so $A_F = F^\times$. The map $v\colon A_k \longrightarrow \mathbb{Z}$ is the discrete surjective valuation v_k of k.

A1 of 2.7, i.e. Hilbert Theorem 90, holds by 16.1 of Chapter 2.

A2 of 2.7, the index of the norm group for cyclic extensions of prime degree, holds by 18.5 of Chapter 2.

Thus, for a finite Galois extension L/F we have the homomorphism

$$\Upsilon_{L/F}\colon \mathrm{Gal}(L/F) \longrightarrow F^\times/N_{L/F}L^\times,$$

its kernel is $[\mathrm{Gal}(L/F), \mathrm{Gal}(L/F)]$ and it is surjective, and all the properties proved in Section 2 hold.

We also have the local reciprocity map

$$\Psi_F\colon F^\times \longrightarrow G_F^{\mathrm{ab}}$$

with the properties in 2.8 and 2.9 satisfied.

Its compatibility with 0-dimensional class field theory for finite fields follows from Theorem 2.9.

To check all the properties stated in 1.1, it remains to check that Ψ_F is continuous and injective, its compatibility with ramification theory and Existence theorem.

3.2. Existence Theorem. *The norm groups $N_{L/F}L^\times$ of finite Galois extensions are open of finite index in $F^\times$.*

The reciprocity map Ψ_F is continuous and injective. Its image is dense in $\mathrm{Gal}(F^{\mathrm{ab}}/F)$ and the cokernel is isomorphic to $\widehat{\mathbb{Z}}/\mathbb{Z}$.

The correspondence between open subgroups N of finite index in $F^\times$ and the norm subgroups of finite abelian extensions L/F:

$$N \leftrightarrow L/F,$$

$N = N_{L/F}L^\times = \Psi_F^{-1}(\mathrm{Gal}(F^{\mathrm{ab}}/L))$, *is an order reversing bijection between the lattice of open subgroups of finite index in $F^\times$ (with respect to the intersection $N_1 \cap N_2$ and the product N_1N_2) and the lattice of finite abelian extensions of F (with respect to the compositum L_1L_2 and intersection $L_1 \cap L_2$).*

Proof. To show that the norm group $N_{L/F}L^\times$ is an open subgroup of $F^\times$, note that the norm map for cyclic extensions of prime degree maps open subgroups of the group of units to open subgroups, this follows from the explicit description of the norm map in Section 13 of Chapter 2. Hence by induction on the degree we deduce that the norm map $N_{L/F}$ is open. In particular, $N_{L/F}L^\times$ is open. By Theorem 2.7 it is of finite index.

The preimage $\Psi_F^{-1}(\mathrm{Gal}(F^{\mathrm{ab}}/L))$ of an open subgroup $\mathrm{Gal}(F^{\mathrm{ab}}/L)$ of the Galois group of F^{ab}/F is the norm group $N_{L/F}L^\times$ by Theorem 2.9, hence Ψ_F is continuous.

Since U_F is compact, its image with respect to Ψ_F is closed, hence equals $\mathrm{Gal}(F^{\mathrm{ab}}/F^{\mathrm{ur}})$, so the cokernel of Ψ_F is isomorphic to $\mathrm{Gal}(F^{\mathrm{ur}}/F)/\varphi_F^{\mathbb{Z}} \simeq \widehat{\mathbb{Z}}/\mathbb{Z}$.

We will verify that an open subgroup N of finite index in $F^\times$ coincides with the norm subgroup $N_{L/F}L^\times$ of some finite abelian extension L/F. It suffices to verify that N contains the norm subgroup $N_{M/F}M^\times$ of some finite separable extension M/F. Indeed, in this case N contains $N_{E/F}E^\times$, where E/F is a finite Galois extension, $E \supset M$. Then by Proposition 2.8 we deduce that $N = N_{M/F}M^\times$, where M is the fixed field of $\Psi_{E/F}(N)$ and M/F is abelian.

Denote by n the index of N in $F^\times$. First, assume that n is not divisible by characteristic of F. If roots μ_n of order dividing n are in F, then consider the Kummer extension $L = F(\sqrt[n]{F^\times})$. By Kummer theory $\mathrm{Hom}(\mathrm{Gal}(L/F), \mu_n) \simeq F^\times/F^{\times n}$. Since the latter is finite by Proposition 4.9, L/F is an abelian extension of exponent n. The index of its norm group in $F^\times$ is the order of $\mathrm{Gal}(L/F)$ equal to the index of $F^{\times n}$, and the latter is included in the former, hence they are equal. Thus, in this case N contains the norm group $N_{L/F}L^\times$. If μ_n is not in $F^\times$, then put $F_1 = F(\mu_n)$. By the same arguments, $F_1^{\times n} = N_{L_1/F_1}L^\times$ for the finite abelian extension L_1/F_1. Then $N_{L_1/F}L_1^\times \subset F^{\times n} \subset N$.

Assume now that $\mathrm{char}(F) = p$. We will show by induction on $m \geqslant 1$ that any open subgroup N of index p^m in $F^\times$ contains a norm group. Let $m = 1$. If $N \supset U_F$, then N is the norm group of the unramified extension of degree p. If $N \not\supset U_F$, then it is the norm group by 18.5 of Chapter 2. Let $m > 1$, and let N_1 be an open subgroup of index p^{m-1} in $F^\times$ such that $N \subset N_1$. By the induction assumption, $N_1 \supset N_{L_1/F}L_1^\times$. The subgroup $N \cap N_{L_1/F}L_1^\times$ is of index 1 or p in $N_{L_1/F}L_1^\times$. In the first case $N \supset N_{L_1/F}L_1^\times$, and in the second case let L/L_1 be a finite separable extension with $N_{L_1/F}^{-1}(N \cap N_{L_1/F}L_1^\times) \supset N_{L/L_1}L^\times$, then $N \supset N_{L/F}L^\times$. For an open subgroup N of index np^m in $F^\times$ with $p \nmid n$ we now take open subgroups N_1 and N_2 of indices n and p^m, respectively, in $F^\times$ such that $N \subset N_i$. Then $N = N_1 \cap N_2 \supset N_{L_1/F}L_1^\times \cap N_{L_2/F}L_2^\times \supset N_{L_1L_2/F}(L_1L_2)^\times$ and we have proved the desired assertion for N.

The kernel of Ψ_F is the intersection of all norm groups $N_{L/F}L^\times$ equal to the intersection of all open subgroups of $F^\times$, hence Ψ_F is injective.

Everything else follows from Proposition 2.8. □

3.3. Theorem. *Every finite abelian extension of $\mathbb{Q}_p$ is contained in an appropriate finite cyclotomic extension $\mathbb{Q}_p^{(n)} = \mathbb{Q}_p(\zeta_n)$ where ζ_n is a primitive nth root of unity. Hence*

$$\mathbb{Q}_p^{\mathrm{ab}} = \mathbb{Q}_p^{\mathrm{cycl}} = \varinjlim \mathbb{Q}_p^{(n)}$$

and

$$\mathrm{Gal}(\mathbb{Q}_p^{\mathrm{ab}}/\mathbb{Q}_p) = \varprojlim \mathrm{Gal}(\mathbb{Q}_p^{(n)}/\mathbb{Q}_p) \simeq \widehat{\mathbb{Z}} \times \mathbb{Z}_p^\times.$$

Proof. Let's look at the extension $M = \mathbb{Q}_p^{(p^m)}$, $p^m > 2$. We have $v_M(\zeta_{p^m}) = 0$, so $\zeta_{p^m} \in \mathcal{O}_M$. Let

$$f_m(X) = \frac{X^{p^m} - 1}{X^{p^{m-1}} - 1} = X^{(p-1)p^{m-1}} + X^{(p-2)p^{m-1}} + \cdots + 1.$$

Then ζ_{p^m} is a root of $f_m(X)$, and hence $|M : \mathbb{Q}_p| \leqslant (p-1)p^{m-1}$. The elements $\zeta_{p^m}^i$, $0 < i < p^m, p \nmid i$, are roots of $f_m(X)$. Hence

$$f_m(X) = \prod_{\substack{p \nmid i \\ 0<i<p^m}} (X - \zeta_{p^m}^i) \quad \text{and} \quad p = f_m(1) = \prod_{\substack{p \nmid i \\ 0<i<p^m}} (1 - \zeta_{p^m}^i).$$

Also,

$$(1 - \zeta_{p^m}^i)(1 - \zeta_{p^m})^{-1} = 1 + \zeta_{p^m} + \cdots + \zeta_{p^m}^{i-1}$$

belongs to the ring of integers of M. When i is prime to p, for the same reason $(1 - \zeta_{p^m})(1 - \zeta_{p^m}^i)^{-1}$ belongs to the ring of integers of M, so $(1 - \zeta_{p^m}^i)(1 - \zeta_{p^m})^{-1}$ is a unit and $p = (1 - \zeta_{p^m})^{p^{m-1}(p-1)}\varepsilon$ for some unit ε. Therefore, $e(M|\mathbb{Q}_p) \geqslant (p-1)p^{m-1}$, and M is a cyclic totally ramified extension with the prime element $1 - \zeta_{p^m}$, and of degree $(p-1)p^{m-1}$ over $\mathbb{Q}_p$. Thus, polynomial $f_m(X)$ is the monic irreducible polynomial of ζ_{p^m} over $\mathbb{Q}_p$, and $p = N_{M/\mathbb{Q}_p}(1-\zeta_{p^m})$. If p is odd then $U_{m,\mathbb{Q}_p} = U_{\mathbb{Q}_p}^{(p-1)p^{m-1}}$ so it is $\subset N_{M/\mathbb{Q}_p} U_M$. If $p=2, m>1$ then $U_{m,\mathbb{Q}_2} = U_{2,\mathbb{Q}_2}^{2^{m-2}} = U_{\mathbb{Q}_2}^{2\cdot 2^{m-2}} \cup 5^{2^{m-2}} U_{\mathbb{Q}_2}^{2\cdot 2^{m-2}} \subset N_{M/\mathbb{Q}_2} U_M$, as $5 = N_{\mathbb{Q}_2^{(4)}/\mathbb{Q}_2}(2+\zeta_4)$. Since the index of the norm group equals to the index of $U_{m,\mathbb{Q}_2}$, they are equal. Thus, $N_{M/\mathbb{Q}_p} M^\times = \langle p \rangle \times U_{m,\mathbb{Q}_p}$.

Let $L/\mathbb{Q}_p$ be a finite abelian extension and N its norm group. Then $\langle p^r \rangle \times U_{\mathbb{Q}_p} \cap \langle p \rangle \times U_{m,\mathbb{Q}_p}$ is in N for some r and m. The first group on the left is the norm group of $\mathbb{Q}_p(\mu_{p^r-1})/\mathbb{Q}_p$, the second group is the norm group of the extension $\mathbb{Q}_p(\mu_{p^m})/\mathbb{Q}_p$. Hence by Theorem 3.2 L is a subfield of $\mathbb{Q}_p(\mu_{(p^r-1)p^m})$.

We also obtain $\mathrm{Gal}(\mathbb{Q}_p^{(p^m)}/\mathbb{Q}_p) \simeq (\mathbb{Z}_p/p^m\mathbb{Z}_p)^\times$ and hence the Galois group of the extension of $\mathbb{Q}_p$ generated by all roots of order a power of p is isomorphic to $\mathbb{Z}_p^\times$. Of course, the extension of $\mathbb{Q}_p$ generated by all roots of order prime to p is $\mathbb{Q}_p^{\mathrm{ur}}$. Hence $\mathrm{Gal}(\mathbb{Q}_p^{\mathrm{ab}}/\mathbb{Q}_p) \simeq \widehat{\mathbb{Z}} \times \mathbb{Z}_p^\times$. □

Proposition. *Let $M = \mathbb{Q}_p^{(p^m)}$, $p^m > 2$. Let $\alpha = u p^{v_p(\alpha)} \in \mathbb{Q}_p^\times$, $u \in \mathbb{Z}_p^\times$. Then*

$$\zeta_{p^m}^{\Psi_{M/\mathbb{Q}_p}(\alpha)} = \zeta_{p^m}^{u^{-1}}.$$

Proof. By the previous Theorem $\Psi_{M/\mathbb{Q}_p}(\alpha) = \Psi_{M/\mathbb{Q}_p}(u)$. Also, by the previous Theorem only u mod p^m matters for the statement of this Proposition, so we can assume that u is a positive integer.

Denote by Q the completion of the maximal unramified extension of $\mathbb{Q}_p$ and let ϕ be the continuous extension of $\varphi_{\mathbb{Q}_p}$ on Q. Denote by T be completion of the maximal unramified extension of M and let φ be the continuous extension of φ_M on T. Then $\varphi|_Q = \phi$.

Let $\sigma \in \mathrm{Gal}(M/\mathbb{Q}_p)$ be such that $\zeta_{p^m}^\sigma = \zeta_{p^m}^{u^{-1}}$. Let K be the fixed field of $\sigma\varphi$. Similarly to the proof of Proposition 2.2, K is a finite extension of $\mathbb{Q}_p$ and hence it is the fixed field of $\sigma\varphi_M$ in the maximal unramified extension of M. Put $\pi_M = \zeta_{p^m} - 1$. We would like to find a certain prime element π_K of K in the form $\theta(\pi_M)$ where θ is a power series in $X\mathcal{O}_Q[[X]]$ satisfying an equation $\theta^\phi = \theta((1+X)^u - 1)$. Using this equation, we will show $\Upsilon_{M/\mathbb{Q}_p}(\sigma) \equiv N_{K/\mathbb{Q}_p}\pi_K \equiv u \mod N_{M/\mathbb{Q}_p} M^\times$ and prove the Proposition.

Proposition 18.2 of Chapter 2 implies $t^\phi \equiv t^p \mod p$ for every $t \in \mathcal{O}_Q$. Extend the action of ϕ to power series in $\mathcal{O}_Q[[X]]$ by acting on their coefficients and sending X to X. Then $h(X)^p \equiv h^\phi(X^p) \mod p$ for any $h \in \mathcal{O}_Q[[X]]$.

Denote the set of multiplicative representatives in Q by R. We claim that the equation $a^{\phi-1} = b$ with $b \in \mathcal{O}_Q$ has a solution $a \in \mathcal{O}_Q$. Indeed, write $b = \sum_{i \geqslant 0} b_i p^i$

with $b_i \in R$, and find coefficients of $a = \sum_{i\geqslant 0} a_i p^i$, $a_i \in R$, inductively. The equation $a_0^{p-1} = a_0^{\phi-1} \equiv b_0 \bmod p$ has a solution in R by the Henselian property. If $(\sum_{i\geqslant 0}^n a_i p^i)^\phi \equiv (\sum_{i\geqslant 0}^n a_i p^i) b \bmod p^{n+1}$ then a_{n+1} is a solution in R of $a_{n+1}^p - a_{n+1} b_0 \equiv \sum_{i=0}^n a_i b_{n+1-i} \bmod p$ by the Henselian property. If $b \in \mathcal{O}_Q^\times$ then $a \in \mathcal{O}_Q^\times$.

Define

$$g_u(X) = upX + X^p, \quad f_n(X) = (1+X)^n - 1$$

for a positive integer n. We now prove that there is a power series $\theta(X) \in X\mathcal{O}_Q[[X]]$ such that

$$g_u \circ \theta = \theta^\phi \circ f_p$$

and $\theta(X)$ is uniquely determined by its first coefficient. We will find coefficients of $\theta(X) = \sum_{i\geqslant 1} t_i X^i$, $t_1 \in \mathcal{O}_Q^\times$, inductively. The first coefficient is a solution of $t_1^{\phi-1} = u$, it exists by the previous paragraph. If $g_u \circ \theta_n \equiv \theta_n^\phi \circ f_p \bmod \deg(n+1)$ with $\theta_n = \sum_{i=1}^n t_i X^i$ then $\theta_{n+1} = \theta_n + t_{n+1} X^{n+1}$ where $p^{n+1} t_{n+1}^\phi - upt_{n+1} = b$ where b is the coefficient of X^{n+1} of $g_u \circ \theta_n - \theta_n^\phi \circ f_p$, note that the latter $\equiv \theta_n(X)^p - \theta_n^\phi(X^p) \equiv 0 \bmod p$, so $b \in p\mathcal{O}_Q$. Rewrite the equation for t_{n+1} as $t_{n+1} - \beta t_{n+1}^\phi = \gamma$ with $\beta \in \mathcal{M}_Q$, $\gamma \in \mathcal{O}_Q$. Then $t_{n+1} = \gamma + \beta\gamma^\phi + \beta^{1+\phi}\gamma^{\phi^2} + \cdots$. The uniqueness of t_{n+1} follows, since the only solution of $c = \beta c^\phi$ is 0.

Then

$$g_u \circ (\theta^{\phi^{-1}} \circ f_u) = (g_u \circ \theta)^{\phi^{-1}} \circ f_u = (\theta^\phi \circ f_p)^{\phi^{-1}} \circ f_u = \theta \circ f_{up} = (\theta^{\phi^{-1}} \circ f_u)^\phi \circ f_p,$$

since $f_{up} = f_u \circ f_p$. The series θ and $\theta^{\phi^{-1}} \circ f_u$ have the same first coefficient, so the uniqueness of θ with fixed first coefficient implies $\theta = \theta^{\phi^{-1}} \circ f_u$. Thus, we obtain the equation

$$\theta^\phi = \theta \circ f_u.$$

We deduce $f_u(\pi_M^\sigma) = (1+\pi_M^\sigma)^u - 1 = \zeta_{p^m}^{\sigma u} - 1 = \zeta_{p^m} - 1 = \pi_M$ and

$$\pi_K^{\sigma\varphi} = \theta^\phi(\pi_M^\sigma) = \theta(f_u(\pi_M^\sigma)) = \theta(\pi_M) = \pi_K,$$

so π_K belongs to K and it is its prime element. Hence

$$\Upsilon_{M/\mathbb{Q}_p}(\sigma) \equiv N_{K/\mathbb{Q}_p}\pi_K \mod N_{M/\mathbb{Q}_p} M^\times.$$

For a polynomial h define $h^{(n)}$ as the composite of n copies of h. Then

$$g_u^{(n)}(\pi_K) = g_u^{(n)}(\theta(\pi_M)) = \theta^{\phi^n}(f_p^{(n)}(\pi_M)) = \theta^{\phi^n}(\zeta_{p^m}^{p^n} - 1)$$

is zero if $n = m$. It is non-zero if $n = m-1$. Indeed, if p is odd then $|K:\mathbb{Q}_p| = |M:\mathbb{Q}_p| = (p-1)p^{m-1} > \deg g_u^{(m-1)} = p^{m-1}$; if $p = 2$ then $\theta^{\phi^{m-1}}(\zeta_{p^m}^{p^{m-1}} - 1) = \theta^{\phi^{m-1}}(-2) \equiv -t_1^{\phi^{m-1}} 2 \bmod 4$. Hence π_K is a root of the polynomial

$g(X) = g_u^{(m)}(X)/g_u^{(m-1)}(X) = g_u^{(m-1)}(X)^{p-1} + up \equiv X^{p^{m-1}(p-1)} \bmod p$, and g is irreducible over $\mathbb{Q}_p$ by Eisenstein's criterion. Finally, $N_{K/\mathbb{Q}_p}\pi_K = (-1)^{|M:\mathbb{Q}_p|}g(0) = (-1)^{|M:\mathbb{Q}_p|}pu$, $p = (-1)^{|M:\mathbb{Q}_p|}N_{M/\mathbb{Q}_p}\pi_M$, so

$$N_{K/\mathbb{Q}_p}\pi_K \equiv u \mod N_{M/\mathbb{Q}_p}M^\times,$$

and $\Psi_{M/\mathbb{Q}_p}(\alpha) = \Psi_{M/\mathbb{Q}_p}(u) = \sigma$ as desired. □

The next Theorem includes another proof of the Hasse–Arf Theorem by using class field theory.

3.4. THEOREM. *Let L/F be a finite abelian extension, $G = \mathrm{Gal}(L/F)$. Denote by h the Hasse–Herbrand function $h_{L/F}$. Put $U_{0,F} = U_F$. Then for every non-negative integer n the reciprocity map $\Psi_{L/F}$ maps the quotient group $U_{n,F}N_{L/F}L^\times/N_{L/F}L^\times$ isomorphically onto the ramification group $G(n) = G_{h(n)}$ and $U_{n,F}N_{L/F}L^\times/U_{n+1,F}N_{L/F}L^\times$ isomorphically onto $G_{h(n)}/G_{h(n)+1}$. Therefore*

$$G_{h(n)+1} = G_{h(n+1)},$$

i.e., upper ramification jumps of L/F are integers.

Proof. Let L_0 be the maximal unramified extension of F in L. We know that $h_{L/F} = h_{L/L_0}$, and the norm $N_{L_0/F}$ maps U_{n,L_0} onto $U_{n,F}$ for $n \geqslant 0$. Using the first Proposition of 2.5 (for $E = L, M = F, L = L_0$) we can therefore assume that $L \cap \tilde{F} = F$.

By Remark 2.4 and using its notation

$$\Upsilon_{L/F}(\sigma) \equiv N_{M/M_0}\,\varepsilon \mod N_{L/F}L^\times, \quad \varepsilon^{1-\varphi} = \pi_M^{\sigma-1},$$

where $M_0 = M \cap F^{\mathrm{ur}}$. If $\sigma \in G_{h(n)}$, then $\pi_M^{1-\sigma}$ belongs to $U_{h(n),M}$. Writing $\varepsilon = \prod(1+\theta_i\pi^i)$ with a prime element π of L, one immediately deduces that $\varepsilon \in U_{h(n),M}U_L$. Hence

$$N_{M/M_0}\,\varepsilon \in N_{M/M_0}(U_{h(n),M}U_L) \cap U_F \subset U_{n,F}N_{L/F}U_L.$$

So $\Upsilon(G_{h(n)}) \subset U_{n,F}N_{L/F}L^\times$. Similarly, $\Upsilon(G_{h(n)+1}) \subset U_{n+1,F}N_{L/F}L^\times$.

In the rest of the proof we will show that $\Upsilon(G_{h(n)}) \supset U_{n,F}N_{L/F}L^\times$. Then $\Upsilon(G_{h(n)}) = U_{n,F}N_{L/F}L^\times$, and therefore $\Upsilon_{L/F}(G_{h(n)+1}) = \Upsilon_{L/F}(G_{h(n+1)})$, $G_{h(n)+1} = G_{h(n+1)}$.

Let R/F be a subextension of L/F such that L/R is of prime degree l and its ramification jump s is such that $G_{s+1} = \{1\}$.

If $h(n) > s$ then $G_{h(n)} = \{1\}$. Let's show in this case, by induction on the degree, that $U_{n,F} \subset N_{L/F}U_{h(n),L}$. The inequality $h(n) > s$ and the description of the Hasse–Herbrand function for cyclic extensions of prime degree implies that $h_{R/F}(n) > s$. By induction $U_{n,F} \subset N_{R/F}U_{h_{R/F}(n),R}$. Since every unit in $U_{h_{R/F}(n),R}$ is the image with respect to $N_{L/R}$ of a unit in $U_{h(n),L}$, we deduce the claim. Thus, if $h(n) > s$ then $N_{L/F}L^\times = \Upsilon(\{1\}) = \Upsilon(G_{h(n)}) \supset U_{n,F}N_{L/F}L^\times = N_{L/F}L^\times$.

Let $h(n) \leqslant s$. If $s = 0$ there is nothing to prove, so let $s > 0$ and hence L/R is of degree p. Then $h_{R/F}(n) = h(n) \leqslant s$. Let's show by induction on the degree that

$$\Psi_{L/F}(U_{n,F}N_{L/F}L^\times/N_{L/F}L^\times) \subset G_{h(n)}.$$

Assume that this inclusion is not true for L/F. Then, using the previous notation, there is a $\sigma \in G_j \setminus G_{j+1}$, $j < h(n)$ such that $\pi_M^{\sigma-1} = \varepsilon^{1-\varphi}$ and $N_{M/M_0}\varepsilon \in U_{n,F}N_{L/F}U_L$. Denote by E the composite of R and M_0. Applying the norm map $N_{M/E}$, since $j < s$ we deduce that $\sigma|_R \in \mathrm{Gal}(R/F)_j \setminus \mathrm{Gal}(R/F)_{j+1}$, and $(N_{M/E}\pi_M)^{\sigma-1} = (N_{M/E}\varepsilon)^{1-\varphi}$, $N_{E/M_0}(N_{M/E}\varepsilon) \in U_{n,F}N_{L/F}U_L$, which contradicts the induction assumption. □

COROLLARY.

For $n \geqslant 0$ the reciprocity map Ψ_F maps $U_{n,F}$ isomorphically onto $G(n)$, where $G = \mathrm{Gal}(F^{\mathrm{ab}}/F)$.

Every abelian extension with finite residue field extension is arithmetically profinite.

Every abelian extension has integer upper ramification jumps.

Proof. By the previous Theorem $\Psi_{L/F}(U_{n,F}N_{L/F}L^\times) = \mathrm{Gal}(L/F)(n)$ for every finite abelian extension L/F. We deduce that $\Psi_F(U_{n,F})$ is a dense subset of $G(n)$. Since $U_{n,F}$ is compact when the residue field is finite and Ψ_F is continuous, $\Psi_F(U_{n,F})$ is closed and we conclude that $\Psi_F(U_{n,F}) = G(n)$.

For every abelian extension L/F the group $\mathrm{Gal}(L/F)(n)$ is the image of $G(n)$ in $\mathrm{Gal}(L/F)$. Since every group of principal units of F has finite index in U_F, the previous paragraph implies that $G(n)$ has finite index in $G(0)$ and so $\mathrm{Gal}(L/F)(x)$ for every x has finite index in $\mathrm{Gal}(L/F)$. Thus, L/F is arithmetically profinite.

For an upper ramification jump x of L/F, the group $\mathrm{Gal}(L/F)(x+1)$ is an open subgroup of $\mathrm{Gal}(L/F)$. Therefore, the fixed field E of $\mathrm{Gal}(L/F)(x+1)$ is a finite abelian extension of F. The jump x corresponds to the jump x of $\mathrm{Gal}(E/F)$ and therefore is integer by the previous Theorem. □

3.5. The Hilbert symbol plays a prominent role in class field theory and its applications.

Let the set μ_n of all nth roots of unity in the separable closure F^{sep} be contained in F and let $p \nmid n$ if $\mathrm{char}(F) = p$.

The *norm residue symbol* or *Hilbert symbol* or *Hilbert pairing*

$$(\cdot,\cdot)_n \colon F^\times \times F^\times \longrightarrow \mu_n$$

is defined by the formula

$$(\alpha,\beta)_n = \gamma^{-1}\Psi_F(\alpha)(\gamma), \quad \text{where } \gamma^n = \beta, \gamma \in F^{\text{sep}}.$$

If $\gamma' \in F^{\text{sep}}$ is another element with $\gamma'^n = \beta$, then $\gamma^{-1}\gamma' \in \mu_n$ and

$$\gamma'^{-1}\Psi_F(\alpha)(\gamma') = \gamma^{-1}\Psi_F(\alpha)(\gamma),$$

so the Hilbert symbol is well defined.

Kummer theory asserts that abelian extensions L/F of exponent n ($\mu_n \subset F^\times, p \nmid n$ if $\text{char}(F) = p$) are in one-to-one correspondence with subgroups $\mathrm{B}_L \subset F^\times$, such that $\mathrm{B}_L \supset F^{\times n}$,

$$L = F(\sqrt[n]{\mathrm{B}_L}) = F(\gamma_i : \gamma_i^n \in \mathrm{B}_L)$$

and the group $\mathrm{B}_L/F^{\times n}$ is dual to $\text{Gal}(L/F)$.

PROPOSITION. *The Hilbert symbol possesses the following properties:*

(1) $(\cdot,\cdot)_n$ *is bilinear;*
(2) $(1-\alpha,\alpha)_n = 1$ *for* $\alpha \in F^\times, \alpha \neq 1$ *(Steinberg property);*
(3) $(-\alpha,\alpha)_n = 1$ *for* $\alpha \in F^\times$;
(4) $(\alpha,\beta)_n = (\beta,\alpha)_n^{-1}$;
(5) $(\alpha,\beta)_n = 1$ *if and only if* $\alpha \in N_{F(\sqrt[n]{\beta})/F}F(\sqrt[n]{\beta})^\times$ *and if and only if* $\beta \in N_{F(\sqrt[n]{\alpha})/F}F(\sqrt[n]{\alpha})^\times$;
(6) $(\alpha,\beta)_n = 1$ *for all* $\beta \in F^\times$ *if and only if* $\alpha \in F^{\times n}$, $(\alpha,\beta)_n = 1$ *for all* $\alpha \in F^\times$ *if and only if* $\beta \in F^{\times n}$;
(7) $(\alpha,\beta)_{nm}^m = (\alpha,\beta)_n$ *for* $m \geqslant 1, \mu_{nm} \subset F^\times$;
(8) $(\alpha,\beta)_{n,L} = (N_{L/F}\alpha,\beta)_{n,F}$ *for* $\alpha \in L^\times, \beta \in F^\times$, *where* $(\cdot,\cdot)_{n,L}$ *is the Hilbert symbol in L,* $(\cdot,\cdot)_{n,F}$ *is the Hilbert symbol in F, and L is a finite separable extension of F;*
(9) $(\sigma\alpha,\sigma\beta)_{n,\sigma L} = \sigma(\alpha,\beta)_{n,L}$, *where L is a finite separable extension of F,* $\sigma \in \text{Gal}(F^{\text{sep}}/F)$, *and* $\mu_n \subset L^\times$ *but not necessarily* $\mu_n \subset F^\times$.

Proof.
(1): For $\gamma \in F^{\text{sep}}, \gamma^n = \beta$ we get

$$\begin{aligned}\gamma^{-1}\Psi_F(\alpha_1\alpha_2)(\gamma) &= \Psi_F(\alpha_1)\big(\gamma^{-1}\Psi_F(\alpha_2)(\gamma)\big)\cdot\big(\gamma^{-1}\Psi_F(\alpha_1)(\gamma)\big)\\ &= \big(\gamma^{-1}\Psi_F(\alpha_2)(\gamma)\big)\big(\gamma^{-1}\Psi_F(\alpha_1)(\gamma)\big),\end{aligned}$$

since $\Psi_F(\alpha_1)$ acts trivially on $(\alpha_2,\beta)_n \in \mu_n$. We also obtain

$$\begin{aligned}(\alpha,\beta_1\beta_2)_n = \gamma_1^{-1}\gamma_2^{-1}\Psi_F(\alpha)(\gamma_1\gamma_2) &= \big(\gamma_1^{-1}\Psi_F(\alpha)(\gamma_1)\big)\big(\gamma_2^{-1}\Psi_F(\alpha)(\gamma_2)\big)\\ &= (\alpha,\beta_1)_n(\alpha,\beta_2)_n.\end{aligned}$$

for $\gamma_1, \gamma_2 \in F^{\text{sep}}$, $\gamma_1^n = \beta_1, \gamma_2^n = \beta_2$.

(5),(2),(3),(4): $(\alpha,\beta)_n = 1$ if and only if $\Psi_F(\alpha)$ acts trivially on $F(\sqrt[n]{\beta})$ and if and only if $\alpha \in N_{F(\sqrt[n]{\beta})/F}F(\sqrt[n]{\beta}))^\times$ by Theorem 2.9.

Let d be the order of α in $F^\times/F^{\times n}$. Let $\delta \in F^{\mathrm{sep}}, \delta^n = \alpha$. By Kummer theory, d is the degree of the cyclic extension $F(\delta)/F$. Then

$$1-\alpha = \prod_{i=1}^{n}(1-\zeta_n^i\delta) = \prod_{i=1}^{nd^{-1}}\prod_{j=1}^{d}\left(1-\zeta_n^i\zeta_d^j\delta\right)$$

$$= N_{F(\delta)/F}\prod_{i=1}^{nd^{-1}}\left(1-\zeta_n^i\delta\right) \in N_{F(\delta)/F}F(\delta)^\times.$$

Hence, $(1-\alpha,\alpha)_n = 1$. Further, $-\alpha = (1-\alpha)(1-\alpha^{-1})^{-1}$ for $\alpha \neq 0, \alpha \neq 1$. This means that $(-\alpha,\alpha)_n = (1-\alpha,\alpha)_n(1-\alpha^{-1},\alpha^{-1})_n^{-1} = 1$. Moreover,

$$1 = (-\alpha\beta,\alpha\beta)_n = (-\alpha,\alpha)_n(\alpha,\beta)_n(\beta,\alpha)_n(-\beta,\beta)_n = (\alpha,\beta)_n(\beta,\alpha)_n,$$

i.e., $(\alpha,\beta)_n = (\beta,\alpha)_n^{-1}$.

Finally, if $(\alpha,\beta)_n = 1$, then $(\beta,\alpha)_n = 1$, which is equivalent to

$$\beta \in N_{F(\sqrt[n]{\alpha})/F}F(\sqrt[n]{\alpha})^\times.$$

(6): Let $\beta \in F^{\times n}$; then $(\alpha,\beta)_n = 1$ for all $\alpha \in F^\times$. Let $\beta \notin F^{\times n}$, then $L = F(\sqrt[n]{\beta}) \neq F$, and L/F is a non-trivial abelian extension. By Theorem 2.9 the subgroup $N_{L/F}L^\times$ does not coincide with $F^\times$. If we take an element $\alpha \in F^\times$ such that $\alpha \notin N_{L/F}L^\times$ then, by property (5), we get $(\alpha,\beta)_n \neq 1$.
(7): For $\gamma \in F^{\mathrm{sep}}, \gamma^{nm} = \beta$, one has

$$(\alpha,\beta)_{nm}^m = (\gamma^{-1}\Psi_F(\alpha)(\gamma))^m = (\gamma^{-m}\Psi_F(\alpha)(\gamma^m)) = (\alpha,\beta)_n,$$

because $(\gamma^m)^n = \beta$.
(8): Theorem 2.9 shows that

$$(\alpha,\beta)_{n,L} = \gamma^{-1}\Psi_L(\alpha)(\gamma) = \gamma^{-1}\Psi_F\left(N_{L/F}(\alpha)\right)(\gamma) = \left(N_{L/F}\alpha,\beta\right)_{n,F},$$

where $\gamma \in F^{\mathrm{sep}}, \gamma^n = \beta$.
(9): Theorem 2.9 shows that for $\gamma \in F^{\mathrm{sep}}, \gamma^n = \beta$,

$$(\sigma\alpha,\sigma\beta)_{n,\sigma L} = \sigma\left(\gamma^{-1}\Psi_L(\alpha)(\gamma)\right) = \sigma(\alpha,\beta)_{n,L}.$$

□

COROLLARY. *The Hilbert symbol induces the non-degenerate pairing*

$$(\cdot,\cdot)_n\colon F^\times/F^{\times n} \times F^\times/F^{\times n} \longrightarrow \mu_n.$$

THEOREM. *Let* $\mu_n \subset F^\times, p \nmid n$, *if* $\operatorname{char}(F) = p$. *Let* A *be a subgroup in* $F^\times$ *such that* $F^{\times n} \subset \mathrm{A}$. *Denote its orthogonal complement with respect to the Hilbert symbol* $(\cdot,\cdot)_n$ *by* $\mathrm{B} = \mathrm{A}^\perp$, *i.e.,*

$$\mathrm{B} = \{\beta \in F^\times : (\alpha,\beta)_n = 1 \quad \textit{for all } \alpha \in \mathrm{A}\}.$$

Then $\mathrm{A} = N_{L/F}L^\times$, *where* $L = F(\sqrt[n]{\mathrm{B}})$ *and* $A = B^\perp$.

Proof. We first recall that $F^{\times n}$ is of finite index in $F^\times$ by Proposition 4.9.

Let B be a subgroup in $F^\times$ with $F^{\times n} \subset \mathrm{B}$ and $|\mathrm{B} : F^{\times n}| = m$. Let $\mathrm{A} = \mathrm{B}^\perp$. Then $\Psi_F(\alpha)$, for $\alpha \in \mathrm{A}$, acts trivially on $F(\sqrt[n]{\beta})$ for $\beta \in \mathrm{B}$. This means that $\Psi_F(\alpha)$ acts trivially on $L = F(\sqrt[n]{\mathrm{B}})$ and, by Theorem 2.9, $\alpha \in N_{L/F}L^\times$. Hence

$$\mathrm{A} \subset N_{L/F}L^\times.$$

Conversely, if $\alpha \in N_{L/F}L^\times$, then $\Psi_F(\alpha)$ acts trivially on $F(\sqrt[n]{\beta}) \subset L$ and

$$\alpha \in N_{F(\sqrt[n]{\beta})/F}F(\sqrt[n]{\beta})^\times$$

for every $\beta \in \mathrm{B}$. Property (5) of the previous Proposition shows that $(\alpha,\beta)_n = 1$ and hence $N_{L/F}L^\times \subset \mathrm{A}$. Thus, $\mathrm{A} = N_{L/F}L^\times$.

Furthermore, to complete the proof it suffices to verify that a subgroup A in $F^\times$ with $F^{*n} \subset \mathrm{A}$ coincides with $(\mathrm{A}^\perp)^\perp$. Restricting the Hilbert symbol on $\mathrm{A} \times F^\times$ we obtain that it induces the non-degenerate pairing $\mathrm{A}/F^{*n} \times F^\times/\mathrm{A}^\perp \longrightarrow \mu_n$. The order of A/F^{*n} coincides with the order of $F^\times/\mathrm{A}^\perp$. Similarly, one can verify that the order of $\mathrm{A}^\perp/F^{\times n}$ is the same as that of $F^\times/(\mathrm{A}^\perp)^\perp$, and hence the order of $F^\times/\mathrm{A}^\perp$ equals the order of $(\mathrm{A}^\perp)^\perp/F^{\times n}$. From $\mathrm{A} \subset (\mathrm{A}^\perp)^\perp$ we deduce that $\mathrm{A} = (\mathrm{A}^\perp)^\perp$. □

Explicit formulas for the norm residue symbol play important role in local number theory and applications in arithmetic geometry. In the case under consideration the challenge is to find a formula for the Hilbert symbol $(\alpha,\beta)_n$ in terms of the elements α,β of the field F. This problem is complicated when $p|n$. There is a simple (essentially linear algebra) answer when $p \nmid n$.

PROPOSITION. *Let* n *be relatively prime with* p *and* $\mu_n \subset F^\times$. *Then*

$$(\alpha,\beta)_n = c(\alpha,\beta)^{(q-1)/n},$$

where q *is the cardinality of the residue field* $\overline{F}$ *and*

$$c\colon F^\times \times F^\times \longrightarrow \mu_{q-1}$$

is the tame symbol defined by the formula

$$c(\alpha,\beta) = \operatorname{pr}\left(\beta^{v_F(\alpha)}\alpha^{-v_F(\beta)}(-1)^{v_F(\alpha)v_F(\beta)}\right),$$

with the projection pr: $U_F \longrightarrow \mu_{q-1}$ *induced by the decomposition* $U_F \simeq \mu_{q-1} \times U_{1,F}$, *i.e.,* $\mathrm{pr}(u)$ *is the multiplicative representative of* $\overline{u} \in \overline{F}$.

Proof. Note that the elements of the group μ_n, for $p \nmid n$, are isomorphically mapped onto the subgroup in the multiplicative group $\mathbb{F}_q^\times$. Hence, $n|(q-1)$. Note that the prime elements generate $F^\times$. Indeed, if $\alpha = \pi^a\varepsilon$ with $\varepsilon \in U_F$, then $\alpha = \pi_1\pi^{a-1}$ for the prime element $\pi_1 = \pi\varepsilon$, when $a \neq 1$, and $\alpha = \pi_2$ for the prime element $\pi_2 = \pi\varepsilon$, when $a = 1$. Using properties (1) and (7) of the Hilbert symbol it suffices to verify that $c(\pi,\beta) = (\pi,\beta)_{q-1}$ for $\beta \in F^\times$.

Let $\beta = (-\pi)^a\theta\varepsilon$ with $a = v_F(\beta), \theta \in \mu_{q-1}, \varepsilon \in U_{1,F}$. Then $c(\pi,-\pi) = 1$. Since $\varepsilon = \varepsilon_1^{q-1}$ for some $\varepsilon_1 \in U_{1,F}$ due to $(q-1)$-divisibility of $U_{1,F}$, we obtain $c(\pi,\varepsilon) = 1$. Hence $c(\pi,\beta) = c(\pi,\theta) = \theta$. Property (3) of the Hilbert symbol shows that $(\pi,-\pi)_{q-1} = 1$. Since the group $U_{1,F}$ is $(q-1)$-divisible, $(\pi,\varepsilon)_{q-1} = 1$. Finally, since the extension $F(\sqrt[n]{\theta})/F$ is unramified, for $\eta \in F^{\mathrm{sep}}, \eta^{q-1} = \theta$ we have

$$(\pi,\theta)_{q-1} = \eta^{-1}\Psi_F(\pi)(\eta) = \eta^{-1}\varphi_F(\eta) = \eta^{q-1} = \theta.$$

We conclude that $(\pi,\beta)_{q-1} = \theta = c(\pi,\beta)$. □

REMARK. There are two types of explicit formulas for the p^rth Hilbert symbol: explicit formulas of Shafarevich, Vostokov, Kato type and explicit formulas of Eisenstein, Kummer, Artin–Hasse, Iwasawa, Sen, Coates–Wiles, Kato–Kurihara . type, [**36**]. See Exercises 3.4, 3.5 for several Artin–Hasse formulas.

Here is the *Brückner–Vostokov explicit formula* for the (p^nth) Hilbert pairing. It is a better version of the preceding Shafarevich explicit formula, [**32**], [**35**]. Let F contain a primitive p^nth root ζ_{p^n} of unity, $p > 2, n \geqslant 1$. Choose a prime element π of F. Let $\mathcal{O}_0$ be the ring of integers of the inertia subfield $F_0 = F \cap \mathbb{Q}_p^{\mathrm{ur}}$ of F. Let $\mathrm{Tr} = \mathrm{Tr}_{\mathcal{O}_0/\mathbb{Z}_p}$ and let φ be the Frobenius automorphism of $\mathbb{Q}_p$. Then for $\alpha,\beta \in F^\times$

$$(\alpha,\beta)_{p^n} = \zeta_{p^n}^{\mathrm{Tr}\,\mathrm{res}\,\Phi(A,B)\,(1/S+1/2)},$$

$$\text{where } \Phi(A,B) = l(B)dA/A - l(A)\frac{1}{p}dB^{\triangle}/B^{\triangle},$$

where $A,B \in \mathcal{O}_0((X))^\times$ are any series such that

$$A(\pi) = \alpha, \quad B(\pi) = \beta,$$

$S = S_1^{p^n} - 1$ where the series $S_1 \in 1 + X\mathcal{O}_0[[X]]$ is any series such that $S_1(\pi) = \zeta_{p^n}$,

$$l(A) = \log(A^p/A^{\triangle})/p, \quad (\sum a_iX^i)^{\triangle} = \sum \varphi(a_i)X^{pi}, \quad \mathrm{res}(\sum a_iX^i dX) = a_{-1}.$$

Thus, this formula for the Hilbert pairing involves indeterminacies in relation to the choice of π, A, B, S_1.

The right hand side of the previous displayed formula is defined independently of class field theory, it is called the *Vostokov symbol*. Its arithmetical importance is much higher than that of the (more of linear algebra type) tame symbol in the previous Proposition. The Vostokov symbol can be used to provide an alternative presentation of class field theory for Kummer extensions of local fields without using the local reciprocity map.

For the case $p=2$ see [**2**] and Section 2 Chapter VII of [**12**].

3.6. In positive characteristic the p-primary pairing is the *Artin–Schreier pairing*.

Abelian extensions of exponent p of a field F of characteristic p are described by Artin–Schreier theory. The polynomial $\wp(X) = X^p - X$ is additive. Abelian extensions L/F of exponent p are in one-to-one correspondence with subgroups $\mathrm{B}_L \subset F$ such that $\wp(F) \subset \mathrm{B}_L$. The quotient group $\mathrm{B}_L/\wp(F)$ is dual to $\mathrm{Gal}(L/F)$, where

$$L = F\big(\wp^{-1}(\mathrm{B}_L)\big) = F\big(\gamma\colon \wp(\gamma) \in \mathrm{B}_L\big).$$

For a complete discrete valuation field F of characteristic p with a finite residue field we define the map

$$(\cdot,\cdot]\colon F^\times \times F \longrightarrow \mathbb{F}_p$$

by the formula

$$(\alpha,\beta] = \Psi_F(\alpha)(\gamma) - \gamma,$$

where γ is a root of the polynomial $X^p - X - \beta$. All the roots of this polynomial are $\gamma + c$ where c runs through $\mathbb{F}_p$, so the pairing $(\cdot,\cdot]$ is well defined.

PROPOSITION. *The map $(\cdot,\cdot]$ has the following properties:*

(1) $(\alpha_1\alpha_2,\beta] = (\alpha_1,\beta] + (\alpha_2,\beta]$, $(\alpha,\beta_1+\beta_2] = (\alpha,\beta_1] + (\alpha,\beta_2]$;

(2) $(-\alpha,\alpha] = 0$ *for* $\alpha \in F^\times$;

(3) $(\alpha,\beta] = 0$ *if and only if* $\alpha \in N_{F(\gamma)/F}F(\gamma)^\times$, *where* $\gamma^p - \gamma = \beta$;

(4) $(\alpha,\beta] = 0$ *for all* $\alpha \in F^\times$ *if and only if* $\beta \in \wp(F)$;

(5) $(\alpha,\beta] = 0$ *for all* $\beta \in F$ *if and only if* $\alpha \in F^{\times p}$;

(6) $(\pi,\beta] = \mathrm{Tr}_{\mathbb{F}_q/\mathbb{F}_p}\theta_0$, *where π is a prime element in F and* $\beta = \sum_{i\geqslant a}\theta_i\pi^i$ *with* $\theta_i \in \mathbb{F}_q$.

Proof.
(1): One has

$$\begin{aligned}\Psi_F(\alpha_1\alpha_2)(\gamma) - \gamma &= \Psi_F(\alpha_1)\big(\Psi_F(\alpha_2)(\gamma) - \gamma\big) + \Psi_F(\alpha_1)(\gamma) - \gamma\\ &= \Psi_F(\alpha_1)(\gamma) - \gamma + \Psi(\alpha_2)(\gamma) - \gamma,\end{aligned}$$

since $\Psi_F(\alpha_2)(\gamma)-\gamma\in F$. One also has

$$\Psi_F(\alpha)(\gamma_1+\gamma_2)-(\gamma_1+\gamma_2)=\Psi_F(\alpha)(\gamma_1)-\gamma_1+\Psi_F(\alpha)(\gamma_2)-\gamma_2.$$

(3): $(\alpha,\beta]=0$ if and only if $\Psi_F(\alpha)$ acts trivially on $F(\gamma)$, where $\gamma^p-\gamma=\beta$. Theorem 2.9 shows that this is equivalent to $\alpha\in N_{F(\gamma)/F}F(\gamma)^\times$.
(2): If $\alpha\in\wp(F)$, then $(-\alpha,\alpha]=0$ by property (3). If a root γ of the polynomial $X^p-X-\alpha$ does not belong to F, then $-\alpha=N_{F(\gamma)/F}(-\gamma)$ and property (3) shows that $(-\alpha,\alpha]=0$.
(4): If $\beta\notin\wp(F)$, then $L=F(\gamma)\neq F$ for a root γ of the polynomial $X^p-X-\beta$; L/F is an abelian extension of degree p, and hence $N_{L/F}L^\times\neq F^\times$. For an element $\alpha\in F^\times$, such that $\alpha\notin N_{L/F}L^\times$, we deduce by Theorem 2.9 that $\Psi_F(\alpha)$ acts non-trivially on L, i.e., $\Psi_F(\alpha)(\gamma)\neq\gamma$ and $(\alpha,\beta]\neq 0$.
(5): Let A denote the set of those $\alpha\in F^\times$, for which $(\alpha,\beta]=0$ for all $\beta\in F$. Note that for $\alpha,\beta\in F^\times$ properties (1) and (2) imply

$$(-\beta,\alpha\beta]=(-\alpha\beta,\alpha\beta]-(\alpha,\alpha\beta]=-(\alpha,\alpha\beta].$$

The condition $\alpha\in\mathrm{A}$ is equivalent to $(\alpha,\alpha\beta]=0$ for all $\beta\in F^\times$ and hence to $(-\beta,\alpha\beta]=0$ for all $\beta\in F^\times$. Then, if $\alpha_1,\alpha_2\in\mathrm{A}$ we get $(-\beta,(\alpha_1+\alpha_2)\beta]=(-\beta,\alpha_1\beta]+(-\beta,\alpha_2\beta]=0$, and $(-\beta,-\alpha_1\beta]=-(-\beta,\alpha_1\beta]=0$. This means that $\alpha_1+\alpha_2,-\alpha_1\in\mathrm{A}$. Obviously, $\alpha_1\alpha_2\in\mathrm{A},\alpha_1^{-1}\in\mathrm{A}$. Therefore, the set $\mathrm{A}\cup\{0\}$ is a subfield in F. Further, $F^p\subset\mathrm{A}\cup\{0\}$ by property (1), and we obtain $F^p\subset\mathrm{A}\cup\{0\}\subset F$.

One can identify the field F with $\mathbb{F}_q((\pi))$. Then the field F^p is identified with the field $\mathbb{F}_q((\pi^p))$ and we obtain that the extension $\mathbb{F}_q((\pi))/\mathbb{F}_q((\pi^p))$ is of degree p. Hence, $\mathrm{A}\cup\{0\}=F^p$ or $\mathrm{A}\cup\{0\}=F$. Since $\wp(F)\neq F$, property (4) shows that $(\alpha,\beta]\neq 0$ for some $\beta\in F,\alpha\in F^\times$. Thus, $\mathrm{A}\cup\{0\}\neq F$, i.e., $\mathrm{A}=F^{\times p}$.
(6): If $\theta\in\mathbb{F}_q$ and $\gamma\in F^{\mathrm{sep}}$, $\gamma^p-\gamma=\theta$, then $F(\gamma)=F$ or $F(\gamma)/F$ is the unramified extension of degree p. Theorem 2.9 implies

$$(\pi,\theta]=\varphi_F(\gamma)-\gamma=\gamma^q-\gamma=\theta^{q/p}+\theta^{q/p^2}+\cdots+\theta=\mathrm{Tr}_{\mathbb{F}_q/\mathbb{F}_p}\theta.$$

Let a be a positive integer and $\theta\in\mathbb{F}_q^\times$. Then

$$a(\pi,\theta\pi^a]=(\pi^a,\theta\pi^a]=(\theta\pi^a,\theta\pi^a]=(-1,\theta\pi^a]=0,$$

since the group $\mathbb{F}_q^\times$ is p-divisible and $-1\in\mathbb{F}_q^p$. Hence $(\pi,\theta\pi^a]=0$ for $p\nmid a$. Finally, let $a=p^sb$, where $s>0$ and $p\nmid b,b>0$. Then

$$\theta\pi^a=(\theta_1\pi^{p^{s-1}b})^p-\theta_1\pi^{p^{s-1}b}+\theta_1\pi^{p^{s-1}b}\ \in\ \theta_1\pi^{p^{s-1}b}+\wp(F),$$

where $\theta_1^p=\theta$. Continuing in this way we deduce that $\theta\pi^a=\theta_s\pi^b+\wp(\lambda)$, where $\theta_s^{p^s}=\theta$ and $\lambda\in F$. Then $(\pi,\theta\pi^a]=(\pi,\theta_s\pi^b]=0$. We obtain property (6) and complete the proof. □

COROLLARY. *The pairing* $(\cdot,\cdot]$ *determines the non-degenerate pairing*

$$F^\times/F^{\times p}\times F/\wp(F)\longrightarrow \mathbb{F}_p$$

To obtain an explicit formula for $(\cdot,\cdot]$, introduce a map d_π as follows.

Let π be a prime element of a complete residue field F of characteristic p with the residue field $\mathbb{F}_q$. Then an element $\alpha\in F$ can be uniquely expanded as

$$\alpha=\sum_{i\geqslant a}\theta_i\pi^i,\quad \theta_i\in\mathbb{F}_q.$$

Put

$$d_\pi\alpha=\sum_{i\geqslant a}i\theta_i\pi^{i-1}d\pi,\quad \operatorname{res}_\pi\left(\sum\eta_i\pi^i d\pi\right)=\eta_{-1}.$$

Define the *Artin–Schreier pairing*

$$D_\pi\colon F^\times\times F\longrightarrow\mathbb{F}_p,\qquad D_\pi(\alpha,\beta)=\operatorname{Tr}_{\mathbb{F}_q/\mathbb{F}_p}\operatorname{res}_\pi(\beta d_\pi\alpha/\alpha).$$

PROPOSITION. *The map* D_π *possesses the following properties:*

(1) *linearity*

$$D_\pi(\alpha_1\alpha_2,\beta)=D_\pi(\alpha_1,\beta)+D_\pi(\alpha_2,\beta),$$
$$D_\pi(\alpha,\beta_1+\beta_2)=D_\pi(\alpha,\beta_1)+D_\pi(\alpha,\beta_2);$$

(2) *if* π_1 *is a prime element in* F, *then*

$$D_\pi(\pi_1,\beta)=D_{\pi_1}(\pi_1,\beta)=\operatorname{Tr}_{\mathbb{F}_q/\mathbb{F}_p}\theta_0,$$

where $\beta=\sum_{i\geqslant a}\theta_i\pi_1^i,\theta_i\in\mathbb{F}_q$;

(3) *let* $\theta,\eta\in\mathbb{F}_q^\times$. *If* $i>-j$, $i>0$, *then* $D_\pi(1+\theta\pi^i,\eta\pi^j)=0$. *If* $i=-j>0$, *then* $D_\pi(1+\theta\pi^i,\eta\pi^j)=\operatorname{Tr}_{\mathbb{F}_q/\mathbb{F}_p}(\theta\eta)$.

Proof.

(1): We have

$$\frac{d_\pi(\alpha_1\alpha_2)}{\alpha_1\alpha_2}=\frac{d_\pi\alpha_1}{\alpha_1}+\frac{d_\pi\alpha_2}{\alpha_2},$$

since $d_\pi\alpha$ can be treated as a formal differential $d\alpha(X)|_{X=\pi}$ for the series $\alpha(X)=\sum a_iX^i$. Hence, we get $D_\pi(\alpha_1\alpha_2,\beta)=D_\pi(\alpha_1,\beta)+D_\pi(\alpha_2,\beta)$.

The other formula follows immediately.

(2): Let $C=\mathbb{Z}[X_1,X_2,\dots]$, where $X_1,X_2,\dots$ are independent indeterminates. Let X be an indeterminate over C. Put

$$\alpha(X)=X_1X+X_2X^2+X_3X^3+\cdots\in C[[X]].$$

For an element $\sum_{j\geqslant a}\kappa_jX^j\in C[[X]]$, $\kappa_i\in C$, we put

$$d\Big(\sum_{j\geqslant a}\kappa_jX^j\Big)=\sum_{j\geqslant a}j\kappa_jX^{j-1}dX,\quad \operatorname{res}_X\Big(\sum_{j\geqslant a}\kappa_jX^jdX\Big)=\kappa_{-1}.$$

Note that

$$\operatorname{res}_X d\big(\sum_{j\geqslant a} \kappa_j X^j\big) = 0.$$

Hence, for $i \neq 0$ we get

$$\operatorname{res}_X \left(\alpha(X)^{i-1} d\alpha(X)\right) = \operatorname{res}_X \left(\frac{1}{i} d\left(\alpha(X)^i\right)\right) = 0.$$

One can define a ring-homomorphism $C[[X]] \longrightarrow F$ as follows: $X_i \in C \to \eta_i \in \mathbb{F}_q, X \to \pi$. The series $\alpha(X)$ is mapped to $\alpha(\pi) = \eta_1\pi + \eta_2\pi^2 + \cdots \in F$, and we conclude that

$$\operatorname{res}_\pi \left(\alpha(\pi)^{i-1} d_\pi \alpha(\pi)\right) = 0 \qquad \text{if} \quad i \neq 0.$$

Now let $\beta = \sum_{i\geqslant a} \theta_i \pi_1^i, \theta_i \in \mathbb{F}_q$. The definition of D_{π_1} shows that

$$D_{\pi_1}(\pi_1, \beta) = \operatorname{Tr}_{\mathbb{F}_q/\mathbb{F}_p} \theta_0.$$

Writing $\pi_1 = \eta_1\pi + \eta_2\pi^2 + \cdots = \alpha(\pi)$ with $\eta_i \in \mathbb{F}_q$, we get

$$D_\pi(\pi_1, \theta_i\pi_1^i) = \operatorname{res}_\pi \left(\theta_i \pi_1^{i-1} d_\pi \pi_1\right) = \operatorname{res}_\pi \left(\theta_i \alpha(\pi)^{i-1} d_\pi \alpha(\pi)\right) = 0 \quad \text{if } i \neq 0,$$

and

$$D_\pi(\pi_1, \theta_0) = \operatorname{res}_\pi \left(\theta_0 \alpha(\pi)^{-1} d_\pi \alpha(\pi)\right) = \operatorname{res}_\pi((\theta_0\pi^{-1} + \delta)d\pi) = \operatorname{Tr}_{\mathbb{F}_q/\mathbb{F}_p} \theta_0$$

where $\delta \in \mathcal{O}_F$. Thus $D_{\pi_1}(\pi_1, \beta) = D_\pi(\pi_1, \beta) = \operatorname{Tr}_{\mathbb{F}_q/\mathbb{F}_p} \theta_0$, as desired.

(3) follows immediately from the definitions. □

PROPOSITION. *Let F be a complete discrete valuation field of characteristic p with the residue field $\mathbb{F}_q$. Then the pairing $(\cdot,\cdot]$ coincides with D_π. In particular, the pairing D_π does not depend on the choice of the prime element π.*

Proof. As the prime elements generate $F^\times$, it suffices to show, using property (1) of $(\cdot,\cdot]$ and property (1) of D_π, that for a prime element π_1 in F the following equality holds:

$$(\pi_1, \beta] = D_\pi(\pi_1, \beta), \quad \beta \in F.$$

Let $\beta = \sum_{i\geqslant a} \theta_i \pi_1^i$. Then property (6) of $(\cdot,\cdot]$ and property (2) of d_π imply that

$$(\pi_1, \beta] = D_\pi(\pi_1, \beta) = \operatorname{Tr}_{\mathbb{F}_q/\mathbb{F}_p} \theta_0,$$

as desired. □

REMARKS.

1. One can prove directly, without using class field theory, that D_π induces a continuous perfect pairing $F^\times/F^{\times p} \times F/\wp(F) \longrightarrow \mathbb{F}_p$, using explicit computations of D_π in the Proposition preceding the previous one. Using Artin–Schreier theory, this gives an algebraic and topological isomorphism $F^\times/F^{\times p} \xrightarrow{\sim} \mathrm{Gal}(F_p/F)$ where F_p is the composite of all cyclic extensions of degree p of F.

2. Similar to the study of the Hilbert symbol, one can prove that for an open subgroup A in $F^\times$ such that $F^{\times p} \subset \mathrm{A}$, its orthogonal complement B with respect to the Artin–Schreier pairing $(\ ,\ ]$ produces an abelian extension $L = F(\wp^{-1}(\mathrm{B}))$ of F such that $\mathrm{A} = N_{L/F}L^\times$. In particular, every open subgroup A of index p in $F^\times$ is the norm group $N_{L/F}L^\times$ of $L = F(\wp^{-1}(\beta))$ where $\beta \notin \wp(F)$ satisfies $(A,\beta] = 0$.

3. Using Witt vectors over F one can extended the previous theory to the Artin–Schreier–Witt pairing. A map defined by

$$(\cdot,\cdot]_n \colon F^\times \times W_n(F) \longrightarrow W_n(\mathbb{F}_p) \simeq \mathbb{Z}/p^n\mathbb{Z}$$

by the formula

$$(\alpha,x]_n = \Psi_F(\alpha)(z) - z,$$

where $z \in W_n(F^{\mathrm{sep}})$ and $z^p - z = x$, produces a non-degenerate pairing

$$F^\times/F^{\times p^n} \times W_n(F)/\wp W_n(F) \longrightarrow W_n(\mathbb{F}_p) \simeq \mathbb{Z}/p^n\mathbb{Z}.$$

Similar to the previous material, there is an explicit formula for it.

3.7. FURTHER REMARKS.

1. Let L be an infinite arithmetically profinite extension of a local number field F, and let E/L be a finite Galois extension. If L is the union of finite field extensions L_i of F and $E = L(\alpha)$, then E is the union of $E_i = L_i(\alpha)$ and E_i/L_i is Galois extension with the Galois group isomorphic to $\mathrm{Gal}(E/L)$ for all sufficiently large i. Define

$$\Upsilon_{E/L} \colon \mathrm{Gal}(E/L) \longrightarrow N(L|F)^\times/N_{N(E|F)/N(L|F)}N(E|F)^\times$$

as the inverse limit (see 5.1 of Chapter 1) of $\Upsilon_{E_i/L_i} \colon \mathrm{Gal}(E/L) \xrightarrow{\sim} \mathrm{Gal}(E_i/L_i) \longrightarrow L_i^\times/N_{E_i/L_i}E_i^\times$ with respect to the norm maps. Then $\Upsilon_{E/L}$ equals the composition of

$$\mathrm{Gal}(E/L) \xrightarrow{\sim} \mathrm{Gal}(N(E|F)/N(L|F))$$

and the homomorphism

$$\Upsilon_{N(E|F)/N(L|F)} \colon \mathrm{Gal}(N(E|F)/N(L|F)) \longrightarrow N(L|F)^\times/N_{N(E|F)/N(L|F)}N(E|F)^\times.$$

Thus, the reciprocity map in characteristic 0 or zero is connected with the reciprocity map in characteristic p.

Using this observation and the explicit formula for the Artin–Schereir pairing and its generalisation, the Artin-Scheirer–Witt pairing, and field of norms of a local number field containing μ_{p^r} and its appropriate arithmetically profinite extension L/F, one can obtain new proofs of explicit formulas for the p^rth Hilbert symbol. Using the arithmetically profinite extension described in Remark 4 of 17.1 of Chapter 2 one obtains explicit formulas of the Shafarevich–Vostokov type. Using the arithmetically profinite extension generated by all roots of order a power of p one obtains explicit formulas of the Kummer–Artin–Hasse–Iwasawa type.

An open question is whether there are other classes of arithmetically profinite extensions that can lead to new types of useful explicit formulas for the Hilbert symbol.

2. Let π be a prime element in F and $\Psi_F(\pi) = \varphi$. Then $\varphi|_{F^{\mathrm{ur}}} = \varphi_F$, and for the fixed field F_π of φ we get

$$F_\pi \cap F^{\mathrm{ur}} = F, \quad F_\pi F^{\mathrm{ur}} = F^{\mathrm{ab}}.$$

The prime element π belongs to the norm group of every finite subextension L/F of F_π/F. The group $\mathrm{Gal}(F^{\mathrm{ab}}/F_\pi)$ is mapped isomorphically onto $\mathrm{Gal}(F^{\mathrm{ur}}/F)$ and the group $\mathrm{Gal}(F_\pi/F)$ is isomorphic to $\mathrm{Gal}(F^{\mathrm{ab}}/F^{\mathrm{ur}})$, the *inertia subgroup* of $G_F^{\mathrm{ab}} = \mathrm{Gal}(F^{\mathrm{ab}}/F)$.

We have

$$\mathrm{Gal}(F^{\mathrm{ab}}/F) \simeq \mathrm{Gal}(F_\pi/F) \times \mathrm{Gal}(F^{\mathrm{ur}}/F), \quad \mathrm{Gal}(F_\pi/F) \simeq U_F, \quad \mathrm{Gal}(F^{\mathrm{ur}}/F) \simeq \widehat{\mathbb{Z}}$$

and

$$\Psi_F(F^\times) = \langle\varphi\rangle \times \mathrm{Gal}(F^{\mathrm{ab}}/F^{\mathrm{ur}}),$$

where $\langle\varphi\rangle$ is the cyclic group generated by φ.

Using formal Lubin–Tate groups, one gets a generalisation of Proposition 3.3 to arbitrary local fields with finite residue field. The theory explicitly describes the field F_π as generated by roots of iterated powers of the isogeny of a formal Lubin–Tate group associated to π. This theory can be used to provide an explicit cohomology-free approach to class field theory of local fields with finite residue fields, [**20**]. However, this theory is not known to have generalisations to local fields with infinite residue field.

3. In addition to Neukirch's approach and Lubin–Tate formal groups approach, other approaches to class field theory of local fields with finite residue field include:

– historically the first one, by Hasse–Schmidt–Chevalley, using the computation of the Brauer group of the field to define a canonical pairing of the group of characters of the field k with $k^\times$ and use its properties to derive the reciprocity map, see [**15**]

– historically the second one, using group cohomology, as described in the first edition of Artin–Tate's book, [**1**]

– in positive characteristic, Kawada–Satake's cohomology-free approach that uses Artin–Schreier–Witt theory and explicit pairings, [**23**]

– explicit cohomology-free approach of Hazewinkel (it is the inverse to the Neukirch approach in the local field case), [**16**]

– using ϕ-γ-modules theory, [**17**].

Hazewinkel's approach to local class field theory constructs

$$\Psi_{L/F}\colon F^\times/N_{L/F}L^\times \longrightarrow \mathrm{Gal}(L/F)$$

for a finite totally ramified abelian extension L/F by sending $\alpha \in U_L$ to the unique $\sigma \in \mathrm{Gal}(L/F)$ that satisfies the congruence $\pi_L^{1-\sigma} \equiv \beta^{\varphi-1}$ modulo the subgroup generated by $u^{1-\tau}$, $\tau \in \mathrm{Gal}(L/F)$, $u \in U_{\mathscr{L}}$, where $\mathscr{L} = L\mathscr{F}$, $\mathscr{F}$ is the completion of F^{ur} and $\beta \in U_{\mathscr{L}}$ is any such that $N_{\mathscr{L}/\mathscr{F}}\beta = \alpha$, [**16**].

4. In the context of Neukirch's abstract class field theory mechanism, it is an open question whether there is another local class field theory with different deg that uses another $\widehat{\mathbb{Z}}$-quotient of the maximal abelian extension of $\mathbb{Q}_p$.

5. For odd p, two absolute Galois groups G_F and G_L of finite extensions F, L of $\mathbb{Q}_p$ are topologically isomorphic if and only if $|F : \mathbb{Q}_p| = |L : \mathbb{Q}_p|$ and the maximal abelian subextension of $\mathbb{Q}_p$ in F coincides with the maximal abelian subextension of $\mathbb{Q}_p$ in L, [**31**], [**21**]. Since there are non-isomorphic fields F, L satisfying these two conditions, one cannot in general restore a non-trivial finite extension of $\mathbb{Q}_p$ from its absolute Galois group.

On the other hand, if two absolute Galois groups G_F and G_L of finite extensions F, L of $\mathbb{Q}_p$ are topologically isomorphic such that this isomorphism maps the upper ramification filtration of G_F to the upper ramification filtration of G_L, then F and L are isomorphic, [**28**].

6. A more recent theory is arithmetic non-abelian class field theory for local fields with finite residue field, [**8**], [**18**]. Let φ in the absolute Galois group G_F of F be an extension of the Frobenius automorphism φ_F. Let F_φ be the fixed field of φ. It is a totally ramified extension of F and its compositum with F^{ur} coincides with the maximal separable extension of F. For every finite subextension E/F

of F_φ/F put $\pi_E = \Upsilon_E(\varphi|_{E^{ab}})$. Then π_E is a prime element of E and from functorial properties of the reciprocity maps we deduce that $\pi_M = N_{E/M}\pi_E$ for every subextension M/F of E/F.

Let $L \subset F_\varphi$ be an infinite Galois totally ramified arithmetically profinite extension of F. Then the prime elements (π_E) in finite subextensions E of F_φ/F supply the sequence of norm-compatible prime elements (π_E) in finite subextensions of L/F and therefore by the theory of fields of norms a prime element X of the local field $N = N(L|F)$. Denote by φ the automorphism of N^{ur} and of its completion $\widehat{N^{\mathrm{ur}}} \simeq N(\widehat{L^{\mathrm{ur}}}/\widehat{F^{\mathrm{ur}}})$ corresponding to φ. Note that N and $\widehat{N^{\mathrm{ur}}}$ are G_F-modules.

Define a *noncommutative local reciprocity map*

$$\Theta_{L/F}\colon \operatorname{Gal}(L/F) \longrightarrow U_{\widehat{N^{\mathrm{ur}}}}/U_N$$

by

$$\Theta_{L/F}(\sigma) = U \mod U_N,$$

where $U \in U_{\widehat{N^{\mathrm{ur}}}}$ satisfies the equation

$$U^{\varphi-1} = X^{1-\sigma}.$$

The element U exists by the properties of local fields with separably closed residue field. Compare this equation with that in Remark 2.4.

The ground component $u_{\widehat{F^{\mathrm{ur}}}}$ of $U = (u_{\widehat{M^{\mathrm{ur}}}})$ belongs to F. We have compatibility with the usual local class field theory at the lowest component:

$$\Theta_{L/F}(\sigma)_{\widehat{F^{\mathrm{ur}}}} = u_{\widehat{F^{\mathrm{ur}}}} = \Upsilon_F(\sigma) \mod N_{L/F}U_L.$$

The reciprocity map $\Theta_{L/F}$ is injective and satisfies the 1-cocyle relation:

$$\Theta_{L/F}(\sigma\tau) = \Theta_{L/F}(\sigma)\,\sigma(\Theta_{L/F}(\tau)).$$

For arithmetically profinite extensions whose Galois group is n-nilpotent, this noncommutative reciprocity map implies explicit Koch–de Shalit–Gurevich class field theory [**24**].

7. Generalisation of class field theory to local fields with quasi-finite residue field $\overline{F}$, i.e. $G_{\overline{F}} \simeq \widehat{\mathbb{Z}}$, using $A_F = F^\times$ can be produced by checking axioms A1 and A2. When the residue field is infinite, existence theorem becomes more complicated.

There is a generalisation of class field theory to local fields with perfect residue field $\overline{F}$ of characteristic p such that $\overline{F} \neq \wp(\overline{F})$, i.e. the field $\overline{F}$ is not separably p-closed, i.e., it has non-trivial separable extensions of degree p, [**4**].

Let F^{abur} denote the maximal abelian unramified p-extension of F and let L/F be a finite Galois totally ramified p-extension. In this theory, its reciprocity map $\Upsilon_{L/F}$ induces an isomorphism

$$\mathrm{Hom}_{\mathbb{Z}_p}\big(\mathrm{Gal}(F^{\mathrm{abur}}/F),\mathrm{Gal}(L/F)^{\mathrm{ab}}\big) \xrightarrow{\sim} U_{1,F}/N_{L/F}U_{1,L},$$

where $\mathrm{Hom}_{\mathbb{Z}_p}$ denotes continuous $\mathbb{Z}_p$-homomorphisms from the profinite group $\mathrm{Gal}(F^{\mathrm{abur}}/F)$ endowed with the topology of profinite group to the discrete finite group $\mathrm{Gal}(L/F)^{\mathrm{ab}}$.

The group $U_{1,F}/N_{L/F}U_{1,L}$ is no longer finite if the residue field if not quasi-finite, so the numerical property in A2 has to be replaced with the isomorphism property $U_{1,F}/N_{L/F}U_{1,L} \xrightarrow{\sim} \mathrm{Hom}_{\mathbb{Z}_p}\big(\mathrm{Gal}(F^{\mathrm{abur}}/F),\mathrm{Gal}(L/F)^{\mathrm{ab}}\big)$ for cyclic totally ramified extensions L/F of degree p.

Existence theorem in this theory implies the following property: let π be a prime element in F and let F_π be the compositum of all finite abelian extensions L of F such that $\pi \in N_{L/F}L^\times$. Then F_π is a maximal abelian totally ramified p-extension of F and the maximal abelian p-extension F^{abp} of F is the compositum of linearly disjoint extensions F_π and F^{abur}. No explicit construction of F_π is known unless the residue field is finite.

8. There is even a generalisation of class field theory to some partial class field theory of complete discrete valuation fields with general (i.e. possibly imperfect) residue field $\overline{F}$ of characteristic p such that $\overline{F} \neq \wp(\overline{F})$, [**5**]. Unlike other local class field theories, there is no induction on the degree in this theory.

9. *Higher class field theory* of an n-dimensional local field F, see 3.5 of Chapter 2, with last finite residue field describes abelian extensions of F by using the Milnor $K_n(F)$-group of F, and induction on the degree works fine there. This theory works with $K_n(F)$ with appropriate definitions of v and deg. However, for $K_n(F)$ there is in general no Galois descent, i.e. for a finite Galois extension L/F the map $K_n(F) \longrightarrow K_n(L)^{\mathrm{Gal}(L/F)}$ is not an isomorphism, so one needs to modify the abstract class field theory to be applicable here. In C2 of 2.3 one needs to take A_F as $K_n(F)$ and not as the G_F-fixed elements of the inductive limit of K_n-groups. Axiom A1 is satisfied; this is much more nontrivial than for usual local fields. A version of axiom A2, enough for the abstract class field theory mechanism, is satisfied as well. Explicit higher local class field theory constructs the higher local reciprocity map

$$K_n(F) \longrightarrow \mathrm{Gal}(F^{\mathrm{ab}}/F)$$

with everywhere dense image and with the kernel $\cap_{m\geqslant 1} mK_n(F)$, such that all the properties in Section 2 hold, [**3**]. For Kato's cohomological approach to higher local class field theory see [**22**], [**26**]. There are more sophisticated approaches by Koya and by Spieß. Kawada–Satake's cohomology-free approach mentioned in 3 above is generalisable to higher local fields of positive characteristic, using generalisations of Artin–Schreier–Witt theory and relevant explicit pairings, [**11**].

4. Adeles of Global Fields

4.1. A *global field* F is either a number field, i.e. a finite extension of $\mathbb{Q}$, or a global function field, i.e. a finite separable extension of $\mathbb{F}_p(t)$.

The largest finite subfield of a global function field is called its *constant field* or *field of constants*.

Note that every finitely generated extension F of $\mathbb{F}_p$ of transcendence degree 1 over $\mathbb{F}_p$ is a global function field. Indeed, if $F = \mathbb{F}_p(a_1, \dots, a_n)$ with a_1 transcendental over $\mathbb{F}_p$, then by induction one can assume that $\mathbb{F}_p(a_2, \dots, a_n)$ is a finite separable extension of $\mathbb{F}_p(a_2)$, so F is a finite separable extension of $\mathbb{F}_p(a_1, a_2)$. Find a non-zero irreducible polynomial $f(X_1, X_2)$ over $\mathbb{F}_p$ such that $f(a_1, a_2) = 0$, it contains a term in which the degree of X_i is prime to p for i equal 1 or 2, and then F is separable over $\mathbb{F}_p(a_j)$, where $\{i, j\} = \{1, 2\}$.

Many results of algebraic number theory hold for global function fields, with $\mathbb{Z}$ replaced by $\mathbb{F}_p[t]$. The ring of integers $\mathcal{O}_F$ of a global field is a Dedekind ring, therefore every non-zero proper ideal uniquely, up to permutation of the factors, factorises into the product of maximal ideals. The norm $N(I)$ of ideals is a multiplicative function and the maximal ideals of $\mathcal{O}_L$ lying over maximal ideals of $\mathcal{O}_F$ are described similarly to the number field case. In positive characteristic, instead of working with the ideal class group of the ring of integers $\mathcal{O}_F$ it is better to work with the Picard group, as we will see later in this section.

DEFINITION. A *completion* F_v of F is $\mathbb{R}$ or $\mathbb{C}$ or a local field with finite residue field, such that there is a ring isomorphism ξ between F and its dense subfield.

Two completions F_v, $F_{v'}$ of F are called equivalent if there is a ring isomorphism $\tau\colon F_v \longrightarrow F_{v'}$ such that $\xi = \xi' \circ \tau$.

A *place* of F is an equivalence class of completions of F.

A place is called real, resp. complex, if the completion is isomorphic to $\mathbb{R}$, resp. $\mathbb{C}$; and then the place is also called archimedean or infinite. Their set is denoted S_∞, in positive characteristic $S_\infty = \emptyset$. The rest of the places is called non-archimedean or finite.

The residue field of a non-archimedean place v will be denote $k(v)$ in the rest of this chapter.

EXAMPLES.

1. Finite places of $\mathbb{Q}$ correspond to positive primes, and there is one infinite real place.

2. A complex place has two representatives, a complex embedding and its composite with complex conjugation.

3. All places of $\mathbb{F}_q(t)$ are finite, they correspond to monic irreducible polynomials over $\mathbb{F}_q$ or to $-\deg$, see Example 2 of 1.3 of Chapter 2.

For a non-zero element $\alpha \in \mathcal{O}_F$ and a maximal ideal P of $\mathcal{O}_F$ the valuation $v_P(\alpha)$ is the power of P participating in the factorisation of the principal ideal $\alpha\mathcal{O}_F$ into the product of maximal ideals. This immediately extends by multiplicativity to the discrete valuation v_P of F and its completion F_{v_P}. In turn, every finite place v of F, different from those lying over $-\deg$ place of $\mathbb{F}_p(t)$, produces a maximal ideal P of $\mathcal{O}_F$ which is the preimage with respect to v of positive integers. The correspondence $v \leftrightarrow P$ is a one-to-one correspondence between finite places, different from those lying over $-\deg$ place of $\mathbb{F}_p(t)$, and maximal ideals of $\mathcal{O}_F$.

Similarly to Section 9 of Chapter 2,

DEFINITION. For a finite separable extension L/F of global fields a place w of L is said to lie over a place v of F, we write $w|v$, if L_w/F_v is a finite extension of complete fields.

Due to 9.6 and Remark 1 of 9.7 of Chapter 2, for a finite separable extension L/F of global fields and a place v of F places w of L over v are determined by the right hand side of the isomorphism

$$L \otimes_F F_v \simeq \oplus_{w|v} L_w$$

(the same argument as in 9.7 of Chapter 2 works for infinite places as well).

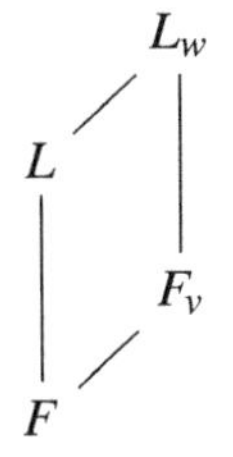

So

$$\sum_{w|v} |L_w : F_v| = |L : F|$$

and

$$\mathrm{Tr}_{L/F}(\alpha) = \sum_{w|v} \mathrm{Tr}_{L_w/F_v}(\alpha_w), \quad N_{L/F}(\alpha) = \prod_{w|v} N_{L_w/F_v}(\alpha_w)$$

where (α_w) is the image of an element α of L in $\oplus_{w|v} L_w$. For a finite place v we have due to 9.4 of Chapter 2

$$\sum_{w|v} e(w|v) f(w|v) = \sum_{w|v} |L_w : F_v| = |L : F|.$$

DEFINITION. Let L/F be a finite Galois extension. Let a place w of L lie over a place v of F. The group $\mathrm{Gal}(L/F)$ acts on the set of w over v, and this action is transitive by 9.7 of Chapter 2. An automorphism $\sigma \in \mathrm{Gal}(L/F)$ induces an isomorphism $L_w \simeq L_{\sigma w}$.

The decomposition group $\mathrm{Gal}(L/F)_w$ of w in L/F is the subgroup $\{\sigma \in \mathrm{Gal}(L/F) : w \circ \sigma = w\}$ of $\mathrm{Gal}(L/F)$. These are exactly elements of $\mathrm{Gal}(L/F)$ continuous with respect to the discrete topology of w: indeed, if the topologies associated to the valuations w and $w \circ \sigma$ are equivalent and then $w \circ \sigma = w$ by 2.7.

Each $\sigma \in \mathrm{Gal}(L/F)_w$ induces a K_v-automorphism of L_w, which is the continuous extension of σ from L to L_w. The restriction of automorphisms gives the injective map $i_w\colon \mathrm{Gal}(L_w/F_v) \longrightarrow \mathrm{Gal}(L/F)$ whose image is $\mathrm{Gal}(L/F)_w$.

4.2. The name 'adele' in number theory is the evolution of 'ideal number' to 'ideal' to 'idèle' to 'additive idèle' to 'adèle'.

DEFINITION. For a global field F its *ring of adeles* ring of *adeles* A_F is the restricted product of all its non-equivalent completions

$$\mathrm{A}_F = \prod{}' F_v = \{\alpha = (\alpha_v) : \alpha_v \in F_v, \alpha_v \in \mathcal{O}_v \text{ for almost all } v\}$$

where v runs through all places of F, and $\mathcal{O}_v$ is the ring of integers of F_v for finite v. Due to 'almost all' we do not need to know what the rings of integers of $\mathbb{R}$ and $\mathbb{C}$ are.

Equivalently,

$$\mathrm{A}_F = \varinjlim_S \mathrm{A}_F(S), \quad \mathrm{A}_F(S) = \prod_{v \in S} F_v \times \prod_{v \notin S} \mathcal{O}_v$$

with S running through all finite subsets of places of F containing S_∞.

The addition and multiplication on $\mathrm{A}_F(S)$ are component-wise.

For example, $\mathrm{A}_{\mathbb{Q}} = \widehat{\mathbb{Z}} \otimes \mathbb{Q} \oplus \mathbb{R}$.

DEFINITION. Define the translation invariant topology on $\mathrm{A}_F(S)$ as the product topology of the topology of F_v for $v \in S$ and the topology of $\mathcal{O}_v$ for $v \notin S$.

Define the topology of A_F is the direct limit topology $\varinjlim \mathrm{A}_F(S)$. So $\mathrm{A}_F(S)$ are open subrings of A_F and a subset of A_F is open if and only if it is an open subset of some $\mathrm{A}_F(S)$. This topology is translation invariant. A base of neighbourhoods of zero in A_F is formed by $\prod_{v \in S} W_v \times \prod_{v \notin S} \mathcal{O}_v$ where W_v are open neighbourhoods of zero in F_v. Due to the local results, one can choose a countable set of neighbourhoods of zero as a fundamental system.

Then the ring $\mathbb{A}_F$ is a topological ring.

LEMMA. *$\mathbb{A}_F(S)$ and $\mathbb{A}_F$ are locally compact complete topological spaces.*

Proof. Since $\mathcal{O}_v$ are compact and F_v are locally compact (for finite v see 18.1 of Chapter 2), $\mathbb{A}_F(S)$ is locally compact. A sequence (α_n) of adeles in $\mathbb{A}_F(S)$ is a fundamental sequence if and only if it is a fundamental sequence at each place. Since each F_v and $\mathcal{O}_v$ are complete topological spaces (for finite v see 18.1 of Chapter 2), $\mathbb{A}_F(S)$ is a complete topological space.

Let's show that for every fundamental sequence (α_n) of adeles there is a finite set S such that $\alpha_n - \alpha_m \in \mathbb{A}_F(S)$ for all n, m; then $\mathbb{A}_F$ is complete. If this property is not true then one can choose an increasing sequence of finite sets S_i containing v_j for $j < i$, such that there is a place $v_i \notin S_i$ and $m_i > n_i$ such that the v_i-component of $\alpha_{m_i} - \alpha_{n_i}$ is not in $\mathcal{O}_{v_i}$, contradicting to (α_n) being a fundamental sequence.

Since $\mathbb{A}_F(S)$ are locally compact, $\mathbb{A}_F$ is locally compact. □

We have the canonical diagonal injective homomorphism

$$F \longrightarrow \mathbb{A}_F, \quad a \mapsto (a).$$

We will identify F with its image in $\mathbb{A}_F$.

Due to the relation between completions in finite field extensions, for a finite separable extension L/F of global fields we immediately deduce

$$\mathbb{A}_L \simeq \mathbb{A}_F \otimes_F L.$$

If L/F is of degree n, then $\mathbb{A}_L$ is a free $\mathbb{A}_F$-module of rank n. It is easy to see that the topology of $\mathbb{A}_L$ is equivalent to the topology of a free module of rank n over topological ring $\mathbb{A}_F$.

We also have the trace and norm maps $\mathrm{Tr}_{L/F}, N_{L/F} \colon \mathbb{A}_L \longrightarrow \mathbb{A}_F$. Using 4.1 we deduce

$$(\mathrm{Tr}_{L/F}(\alpha_w))_v = \sum_{w|v} \mathrm{Tr}_{L_w/F_v} \alpha_w, \quad (N_{L/F}(\alpha_w))_v = \prod_{w|v} N_{L_w/F_v} \alpha_w.$$

The definitions imply that these maps are continuous. If M/F is a finite Galois extension with $M \supset L$, and σ runs through representatives of $\mathrm{Gal}(M/F)/\mathrm{Gal}(M/L)$, then

$$\mathrm{Tr}_{L/F}(\alpha) = \sum \sigma\alpha, \quad N_{L/F}(\alpha) = \prod \sigma\alpha.$$

4.3. Recall that a character of a topological group is a continuous homomorphism from the group to the unit complex circle S^1 (with respect to multiplication). Let G be a topological group with the property that it has some open subgroups forming a base of neighbourhoods of its identity element. By Lemma 18.1 of Chapter 2 the additive and multiplicative groups of a local field with finite residue field have this property. Hence the subgroup of A_F with zero component at all infinite places and the subgroup of $\mathrm{A}_F^\times$ (its topology is defined in 4.5) with component 1 at all infinite places also have this property. Choose an open neighbourhood U of $1 \in S^1$ which contains no non-trivial subgroups of S^1. For every character χ of G, the preimage $\chi^{-1}(U)$ must contain an open subgroup V of G, which is therefore mapped to 1; thus the kernel of χ is an open subgroup. All topological groups whose characters we will study will have this property. If, in addition, G is compact, then V is of finite index and so the character χ is of finite order.

THEOREM. *The topological additive group of a completion F_v of a global field is topologically self-dual: it is non-canonically isomorphic to its character group $X(F_v)$.*

The topological additive group of A_F is topologically self-dual: it is non-canonically isomorphic to its character group $X(\mathrm{A}_F)$.

F is discrete and closed in A_F, and A_F/F is compact.

Proof. Let $k = \mathbb{Q}$ in characteristic zero and $k = \mathbb{F}_p(t)$ in positive characteristic and a global field F be a finite separable extension of k.

The additive group $\mathbb{Q}_p$ is $\mathbb{Z}_p + A_p$ where $A_p = \{a/p^n : a \in \mathbb{Z}, n \geqslant 0\}$, and $\mathbb{Z}_p \cap A_p = \mathbb{Z}$, so we get a continuous additive homomorphism

$$\omega_p \colon \mathbb{Q}_p \longrightarrow \mathbb{Q}_p/\mathbb{Z}_p \xrightarrow{\sim} A_p/\mathbb{Z} \longrightarrow \mathbb{R}/\mathbb{Z}$$

by sending $z + a$ to $a \mod \mathbb{Z}$, $z \in \mathbb{Z}_p, a \in A_p$. We have $\omega_p(\mathbb{Z}_p) = 0$.

For a discrete valuation v of $k = \mathbb{F}_p(t)$ corresponding to a monic irreducible polynomial with a root α_v and residue field $\mathbb{F}_p(v) = \mathbb{F}_p(\alpha_v)$ and the completion k_v with prime element $p(t)$, the field $k_v(\alpha_v)$ is a totally ramified extension of k_v with a prime element $t_v = t - \alpha_v$. Define a continuous additive homomorphism

$$\omega_v \colon k_v \to \mathbb{R}/\mathbb{Z}$$

which sends f in k_v to $\mathfrak{c} \circ \mathrm{Tr}_{\mathbb{F}_p(v)/\mathbb{F}_p} \circ \mathrm{res}_{t_v}(f\,dt_v)$ with $f\,dt_v$ in the differential module of $k_v(\alpha_v)$, where $\mathfrak{c}$ is a homomorphism which sends $1 \in \mathbb{F}_p$ to $1/p \mod \mathbb{Z}$ and res is as in 3.6. We have $\omega_v(\mathcal{O}_v) = 0$.

For a discrete valuation $v = -\deg$ of k define define a continuous additive homomorphism

$$\omega_v \colon k_v \to \mathbb{R}/\mathbb{Z}$$

which sends f in $k_v = \mathbb{F}_p((t^{-1}))$ with prime element t^{-1} to $\mathfrak{c} \circ \mathrm{res}_{t^{-1}}(f\,dt^{-1})$. Note that we have $\mathrm{res}_{t^{-1}}(\sum a_i t^{-i} dt^{-1}) = -a_1$.

Using these homomorphisms, define their analogs for completions of F: For a non-archimedean completion F_v in characteristic zero denote by ψ_v^0 its character

$$\alpha \mapsto \exp(2\pi i \omega_p \circ \mathrm{Tr}_{F_v/\mathbb{Q}_p}(\alpha)).$$

For an archimedean completion F_v denote by ψ_v^0 its character

$$\alpha \mapsto \exp(-2\pi i \,\mathrm{Tr}_{F_v/\mathbb{R}}(\alpha)).$$

In positive characteristic p for a non-archimedean completion F_w of F with respect to a discrete valuation w lying over a discrete valuation v of k denote by ψ_w^0 its character

$$\alpha \mapsto \exp(2\pi i \omega_v \circ \mathrm{Tr}_{F_w/k_v}(\alpha)).$$

Since the trace sends integral elements to integral elements, we get $\psi_v^0(\mathscr{O}_v) = 1$ for all finite places v.

Denote the character $\alpha \mapsto \psi_v^0(\alpha\gamma)$ by ${}_\gamma\psi_v^0$. For every character ψ_v of F_v one can find $\gamma \in F_v$ such that $\psi_v = {}_\gamma\psi_v^0$. If v is infinite, this is well known. If v is finite, then γ can be found by choosing its successive coefficients of powers of a prime element appropriately. Indeed, by the first paragraph of this subsection, the kernel of ψ_v is an open subgroup of F_v. So there is integer m such that $\psi_v(\mathscr{M}_v^m) = 1$, $\psi_v(\mathscr{M}_v^{m-1}) \neq 1$, and there is a similar m_0 for ψ_v^0. If π_v is a prime element of F_v, let $\gamma = \theta_{m_0-m}\pi_v^{m_0-m} + \dots$ with a non-zero multiplicative representative $\theta_{m_0-m} \in \mathscr{O}_v^\times$ such that the induced by ψ_v character of the finite field $k(v) = \mathscr{M}_v^{m-1}/\mathscr{M}_v^m$ coincides with the character induced by ${}_\gamma\psi_v^0$. Then ${}_\gamma\psi_v^0\psi_v^{-1}$ vanishes on $\mathscr{M}_v^{m-1}$. Repeat the procedure to get $\gamma = \theta_{m_0-m}\pi_v^{m_0-m} + \theta_{1+m_0-m}\pi_v^{1+m_0-m} + \dots \in F_v$, etc. Thus, $X(F_v) = \{{}_\gamma\psi_v^0 : \gamma \in F_v\} \leftrightarrows F_v$.

Open neighbourhoods in $X(F_v)$ of the character ψ^1, $\psi^1(F_v) = 1$, are $W(U) = \{\psi \in X(F_v) : \psi(B_v) \subset U\}$ where U runs through open neighbourhoods of 1 of the complex unit circle and B_v is some fixed non-trivial closed ball of F_v. The set $\{\gamma \in F_v : {}_\gamma\psi_v^0 \in W(U)\}$ equals $W = \{\gamma \in F_v : \psi_v^0(\gamma B_v) \subset U\}$ which is open in F_v. Conversely, for any non-empty open subset V of F_v the set $\{{}_\gamma\psi_v^0 : \gamma \in V\}$ is open in $X(F_v)$ since V contains an open set $\{\gamma \in V : \psi_v^0(\gamma B_v) \subset U\}$ for some open U and hence $\{{}_\gamma\psi_v^0 : \gamma \in V\}$ contains $W(U)$.

Then the pairing $F_v \times F_v \longrightarrow \mathbb{R}/\mathbb{Z}$, $(\alpha,\beta) \mapsto \psi_v^0(\alpha\beta)$ induces an algebraic and topological isomorphism of the additive group F_v and its group of characters $X(F_v)$. This pairing is a perfect pairing.

A character ψ of the additive group of A_F induces a character of $\mathrm{A}_F(S)$ and it induces as character ψ_v on F_v via the embedding of F_v into A_F where the components of the image at $v' \neq v$ are 1. The induced characters ψ_v are trivial on almost all $\mathscr{O}_v$, so $\psi((\alpha)_v) = \prod \psi_v(\alpha_v)$ on $\mathrm{A}_F(S)$ and hence on A_F. Conversely, if ψ_v are characters of F_v trivial on almost all $\mathscr{O}_v$, then $(\alpha_v) \mapsto \prod \psi_v(\alpha_v)$ is a character of A_F.

Using the previously defined local characters Ψ_v^0 we get the character

$$\psi^0 = \psi^0_{\mathrm{A}_F} = \prod_v \psi_v^0.$$

The definitions imply that for a finite separable extension of global fields we have

$$\psi^0_{\mathrm{A}_L} = \psi^0_{\mathrm{A}_F} \circ \mathrm{Tr}_{L/F}.$$

The definitions also imply, using the local results, that the pairing $\mathrm{A}_F \times \mathrm{A}_F \longrightarrow \mathbb{R}/\mathbb{Z}$, $(\alpha,\beta) \mapsto \psi^0(\alpha\beta)$ induces an (algebraic and topological) isomorphism of A_F with its group of characters. This pairing is a perfect pairing.

Due to the formula $\mathrm{A}_L = \mathrm{A}_F \otimes_F L$ for a finite separable field extension L/F, A_L/L is a free A_F/F-module of finite rank. It suffices to show the last claim of the Proposition for k.

In characteristic zero, using the second paragraph of the proof, $\mathrm{A}_k = k + \mathrm{A}_k(\infty)$, where $\mathrm{A}_k(\infty) = \prod \mathbb{Z}_p \times \mathbb{R}$, and $k \cap \mathrm{A}_k(\infty) = \mathbb{Z}$. Hence we have a homeomorphism $\mathrm{A}_k/k \rightrightarrows \mathrm{A}_k(\infty)/\mathbb{Z}$. The group $\mathbb{Z}$ is discrete in $\mathrm{A}_k(\infty)$ as one immediately sees looking at the real component, hence $\mathbb{Q}$ is discrete in $\mathrm{A}_\mathbb{Q}$. Since $\mathbb{Z}$ is the intersection of its open neighbourhoods in each of $\mathbb{Z}_p$ and in $\mathbb{R}$, it is the intersection of its open (hence closed) neighbourhoods in $\mathrm{A}_k(\infty)$, and therefore it is closed in $\mathrm{A}_k(\infty)$, so k is closed in A_k.

Also, $\mathrm{A}_\mathbb{Q} = \mathbb{Q} + \prod \mathbb{Z}_p \times [-1/2,1/2]'$ where $[-1/2,1/2]'$ with glued $-1/2$ and $1/2$ is isomorphic to the complex unit circle with respect to $\alpha \mapsto \exp(2\pi i\alpha)$. Then $\mathbb{Q} \cap \prod \mathbb{Z}_p \times [-1/2,1/2]' = \{0\}$. We obtain a homeomorphism of $\mathrm{A}_\mathbb{Q}/\mathbb{Q}$ with the compact set $\prod \mathbb{Z}_p \times [-1/2,1/2]'$.

In positive characteristic, using $k_v = \mathscr{O}_v + k \cap \mathrm{A}_k(\{v\})$ for every place v, we deduce $\mathrm{A}_k = \mathrm{A}_k(\emptyset) + k$. Since $k \cap \mathrm{A}_k(\emptyset) = \mathbb{F}_p$, we get a homeomorphism $\mathrm{A}_k/k \rightrightarrows \prod_v \mathscr{O}_v/\mathbb{F}_p$ (diagonal image of $\mathbb{F}_p$), so k is discrete and closed in A_k and A_k/k is compact. Due to $\mathrm{A}_k = \mathrm{A}_k(-\deg) + k$ and $k \cap \mathrm{A}_k(-\deg) = \mathbb{F}_p[t] = \mathscr{O}_k$, we also have a homeomorphism of A_k/k with $\mathrm{A}_k(-\deg)/\mathscr{O}_k$. □

So F sits in $\mathbb{A}_F$ similarly to how a full lattice of n-dimensional real space sits in it: discrete and cocompact.

COROLLARY 1. *For the standard character ψ^0 we have $\psi^0(F) = 1$.*

In characteristic zero, the induced character of $\mathbb{A}_{\mathbb{Q}}/\mathbb{Q}$ via its isomorphism with $\mathbb{A}_{\mathbb{Q}}(\infty)/\mathbb{Z}$ comes from the character of the latter which sends $\mathbb{Z}_p$ to 1 and is ψ_v^0 for the real place v.

In positive characteristic, the character of $\mathbb{A}_k/k$ via its isomorphism with $\mathbb{A}_k(-\deg)/\mathcal{O}_k$ comes from the character of the latter which sends $\mathcal{O}_v$ to 1 for $v \neq -\deg$ and is $\psi^0_{-\deg}$ for $v = -\deg$.

Proof. Due to the definitions, it suffices to check that $\psi^0(k) = 1$.

In characteristic zero this follows from $-\alpha + \sum_p \omega_p(\alpha) \in \mathbb{Z}$ for $\alpha \in \mathbb{Q}$, since $v_q(\omega_p(\alpha)) \geqslant 0$ if $p \neq q$ and $v_p(\omega_p(\alpha) - \alpha) \geqslant 0$. Moreover, the character of $\mathbb{A}_{\mathbb{Q}}(\infty)/\mathbb{Z}$ described in the statement of the corollary, corresponds to the character of $\mathbb{A}_{\mathbb{Q}}$ whose restriction on $\mathbb{Q}_p$ on an element $\alpha \in A_p$ equals $\psi_v^0(\alpha)^{-1}$ for the real place v, hence it is $\psi_p^0(\alpha)$.

In positive characteristic, clearly $\psi^0(\mathcal{O}_k) = 1$. Then it is sufficient to check for a rational function $f(t) = g(t)/h(t)^n \in \mathbb{F}_p(t)$ where $h(t)$ is an irreducible monic polynomial over $\mathbb{F}_p$ corresponding to a discrete valuation v and $\deg(g) < n\deg(h)$. We have $\omega_{-\deg}(f) = \mathfrak{c} \circ \operatorname{res}_{t^{-1}}(f(t)dt^{-1}) = \mathfrak{c}(-a)$ where a is the coefficient of degree $n\deg(h) - 1$ of g. If α is a root of $h(t)$, then $h(t) = \prod(t - \sigma_i\alpha)$ with σ_i running through the Galois group of $\mathbb{F}_p(\alpha)/\mathbb{F}_p$. Write
$f(t) = \sum_i \sum_{m \geqslant -n} a_m^{(i)} (t - \sigma_i\alpha)^m$. Then a, the t^{-1} coefficient of $f(t)$ is

$$\sum_i a_{-1}^{(i)} = \sum_i \operatorname{res}_{t-\sigma_i(\alpha)}(f(t)dt) = \operatorname{Tr}_{\mathbb{F}_p(\alpha)/\mathbb{F}_p} \operatorname{res}_{t-\alpha}(f(t)dt).$$

Hence $\omega_v(f) = \mathfrak{c} \circ \operatorname{Tr}_{\mathbb{F}_p(\alpha)/\mathbb{F}_p} \circ \operatorname{res}_{t-\alpha}(f(t)dt) = \mathfrak{c}(a)$ and $\psi^0(f) = 1$. Therefore, $\psi^0(F) = 1$.

The remaining assertion is checked in the same way as in characteristic zero. □

COROLLARY 2. *The orthogonal complement $F^\perp$ of F with respect to ψ^0 is F.*

Proof. Indeed, this complement is isomorphic to the group of characters of the compact group $\mathbb{A}_F/F$, hence it is a discrete subgroup of $\mathbb{A}_F$. Hence $F^\perp/F$ is a discrete subgroup of the compact $\mathbb{A}_F/F$, so it is finite. Therefore, since $F^\perp$ is an F-vector space, it coincides with F. □

4.4. Adeles in the function field case and the Riemann–Roch formula.

DEFINITION. For a global field F positive characteristic, denote by $\mathrm{Div}(F)$ the free abelian group generated by classes of discrete valuations of F, i.e.

$$\mathrm{Div}(F) = \{\sum_v n_v[v] : n_v \in \mathbb{Z},\ n_v = 0 \text{ for almost all } v.\}$$

Endow it with the discrete topology.

Every divisor $\sum_v n_v[v]$ has degree $\sum_v n_v \log_q |k(v)|$ where q is the cardinality of the constant subfield of F. Divisors of degree zero form a subgroup $\mathrm{Div}^0(F)$.

The image of elements of $F^\times$ in $\mathrm{Div}(F)$ is denoted $\mathrm{PDiv}(F)$, it consists of principal divisors $\{\sum_v v(\alpha)[v] : \alpha \in F^\times\}$. The form a subgroup of $\mathrm{Div}^0(F)$.

The quotient group $\mathrm{Div}(F)/\mathrm{PDiv}(F)$ is denoted $\mathrm{Pic}(F)$ and is called the Picard group of F. The quotient group $\mathrm{Div}^0(F)/\mathrm{PDiv}(F)$ is denoted $\mathrm{Pic}^0(F)$ and is called the Picard group of degree zero of F.

If $\mathscr{C}$ is a smooth proper geometrically irreducible curve over a finite field $\mathbb{F}_q$ with the function field F, then the divisor group, principal divisor group, Picard group of $\mathscr{C}$ are equal to the divisor group, principal divisor group, Picard group of F.

For a divisor $d = \sum v(d)[v]$ of F define

$$\mathrm{A}_F(d) = \{\alpha = (\alpha_v) \in \mathrm{A}_F : v(\alpha_v) \geqslant -v(d) \text{ for all } v\}$$

In particular, $\mathrm{A}_F(0) = \mathrm{A}_F(\emptyset)$. We have an adelic complex

$$\mathscr{A}_F(d) : \ F \oplus \mathrm{A}_F(d) \longrightarrow \mathrm{A}_F, \ (a,b) \mapsto a - b,$$

and

$$H^0(\mathscr{A}_F(d)) = F \cap \mathrm{A}_F(d), \quad H^1(\mathscr{A}_F(d)) = \mathrm{A}_F/(F + \mathrm{A}_F(d)).$$

By definition, the space of 1-differentials $\Omega^1_{F/\mathbb{F}_q}$ is an F-module generated by symbols df, $f \in F$, such that $d(af) = adf$, $d(f+g) = df + dg$, $d(fg) = fdg + gdf$ for $a \in \mathbb{F}_q$, $f, g \in F$. It is a free module of rank 1 over F, since F is of transcendence degree 1 over $\mathbb{F}_q$. For $\omega \in \Omega^1_{F/\mathbb{F}_q}$ and $\alpha_v \in F_v$, the residue $\mathrm{res}_v(\alpha_v\omega)$ at v is $\mathrm{res}_v(\beta_v d\pi)$ for a prime element π of F_v as in 3.6, where $\alpha_v\omega = \beta_v d\pi$ for a prime element π of F_v and $\beta_v \in F_v$. As explained in 3.6, this does not depend on the choice of π.

For a non-zero differential form $\omega \in \Omega^1_{F/\mathbb{F}_q}$ define a continuous $\mathbb{F}_q$-linear map

$$d_\omega : \mathrm{A}_F \longrightarrow \mathbb{F}_q, \quad (\alpha_v) \mapsto \sum_v \mathrm{Tr}_{k(v)/\mathbb{F}_q} \mathrm{res}_v(\alpha_v\omega),$$

where $k(v)$ is the residue field of F_v. Finite fields are endowed with the discrete topology. There are only finitely many non-zero terms in the sum, since almost all $\alpha_v \in \mathscr{O}_v$ and ω has poles at finitely many places.

By the classical result on differential forms, $d_\omega(F) = 0$.

In characteristic p, characters of A_F are in one-to-one correspondence with continuous linear maps from A_F to $\mathbb{F}_p$. The composite of the map $\mathrm{Tr}_{\mathbb{F}_q/\mathbb{F}_p} \circ d_\omega$ with an isomorphism from $\mathbb{F}_p$ to the cyclic group of order p on the unit circle is a non-trivial character of A_F. By Corollary 2 in the previous subsection, the F-space of continuous linear maps from A_F to $\mathbb{F}_p$ is of dimension 1, which is the dimension of the F-space $\Omega^1_{F/\mathbb{F}_p}$. Since the F-linear map $\omega \mapsto d_\omega$, with 0 sent to 0, is non-zero as one can see by choosing adeles with only one non-zero component, it is an isomorphism of $\Omega^1_{F/\mathbb{F}_p}$ with the space of continuous $\mathbb{F}_p$-linear maps from A_F to $\mathbb{F}_p$ vanishing on F. This also implies an isomorphism of F-spaces $\Omega^1_{F/\mathbb{F}_q}$ and the space of continuous $\mathbb{F}_q$-linear maps from A_F to $\mathbb{F}_q$ vanishing on F.

Composing with the multiplication $\mathrm{A}_F \times \mathrm{A}_F \longrightarrow \mathrm{A}_F$ we get the differential pairing

$$\mathrm{A}_F \times \mathrm{A}_F \longrightarrow \mathbb{F}_q, \qquad (\alpha,\beta) \mapsto \sum_v \mathrm{Tr}_{k(v)/\mathbb{F}_q} \mathrm{res}_v(\alpha_v \beta_v \omega).$$

Due to the relation between $\Omega^1_{F/\mathbb{F}_q}$ and the space of continuous $\mathbb{F}_q$-linear maps from A_F/F to $\mathbb{F}_q$, and the proof of Theorem 4.3, this pairing is perfect.

For a subspace H denote $H^\perp = \{\beta \in \mathrm{A}_F : (H,\beta) = 0\}$. The complement $\mathrm{A}_F(0)^\perp$ of $\mathrm{A}_F(0)$ with respect to the pairing is $\mathrm{A}_F(\kappa)$, κ is the divisor of ω and is called a canonical divisor of F (and of $\mathscr{C}$). It is uniquely determined modulo principal divisors. We get $\mathrm{A}_F(d)^\perp = \mathrm{A}_F(\kappa - d)$, hence the space of continuous linear maps from $H^0(\mathscr{A}_F(d))$ to $\mathbb{F}_q$ is isomorphic to $\mathrm{A}_F/H^0(\mathscr{A}_F(d))^\perp$, i.e. to $H^1(\mathscr{A}_F(\kappa - d))$. The space $\mathrm{A}_F(0)$ and hence $\mathrm{A}_F(d)$ are compact, and their intersection with F is discrete, which implies that $H^0(\mathscr{A}_F(d))$ is of finite $\mathbb{F}_q$-dimension and so is $H^1(\mathscr{A}_F(d))$. We now obtain

$$\dim_{\mathbb{F}_q} H^0(\mathscr{A}_F(d)) = \dim_{\mathbb{F}_q} H^1(\mathscr{A}_F(\kappa - d))$$

and

$$\chi_{\mathscr{A}_F}(d) := \dim_{\mathbb{F}_q} H^0(\mathscr{A}_F(d)) - \dim_{\mathbb{F}_q} H^1(\mathscr{A}_F(d)) = \chi_{\mathscr{A}_F}(\kappa - d).$$

We will use the virtual dimension of two $\mathbb{F}_q$-commensurable spaces G, H (i.e $G \cap H$ is of $\mathbb{F}_q$-finite codimension in each of them), $\dim_{\mathbb{F}_q}(G:H) := \dim_{\mathbb{F}_q} G/(G \cap H) - \dim_{\mathbb{F}_q} H/(G \cap H)$. Noting it is additive on short exact sequences and comparing $\mathscr{A}_F(d)$ and $\mathscr{A}_F(0)$, we obtain

$$\deg_{\mathbb{F}_q} d = \dim_{\mathbb{F}_q}(\mathrm{A}_F(d) : \mathrm{A}_F(0)) = \chi_{\mathscr{A}_F}(d) - \chi_{\mathscr{A}_F}(0).$$

Then

$$-\deg_{\mathbb{F}_q} d = \dim_{\mathbb{F}_q} H^0(\mathscr{A}_F(0)) - \dim_{\mathbb{F}_q} H^0(\mathscr{A}_F(\kappa)) - \dim_{\mathbb{F}_q} H^0(\mathscr{A}_F(d)) + \dim_{\mathbb{F}_q} H^0(\mathscr{A}_F(\kappa - d)),$$

and we obtain

$$\dim_{\mathbb{F}_q} H^0(\mathscr{A}_F(d)) = \dim_{\mathbb{F}_q} H^0(\mathscr{A}_F(\kappa - d)) + \deg_{\mathbb{F}_q} d + \chi_{\mathscr{A}_F}(0),$$

the *adelic Riemann–Roch formula.* Since we assume that $\mathscr{C}$ is geometrically irreducible, we obtain $\dim_{\mathbb{F}_q} H^0(\mathscr{A}_F(0)) = 1$ and $\chi_{\mathscr{A}_F}(0) = 1 - g$ where $g = \dim_{\mathbb{F}_q} H^1(\mathscr{A}_F(0))$, g is called the genus of $\mathscr{C}$.

REMARKS. 1. This adelic proof is extendable to any (not necessarily smooth) proper irreducible curve over a perfect field (in particular, $\mathbb{C}$) by working with its adelic space and complex.

2. One can prove that adelic $H^i(\mathscr{A}_F(d))$ is isomorphic to the usual $H^i(\mathscr{C}, d)$ and hence obtain the usual Riemann–Roch theorem in algebraic geometry.

4.5. The group of *ideles* is the multiplicative group of the ring of adeles A_F:

$$J_F = \mathrm{A}_F^\times = \prod{}' F_v^\times = \{\alpha = (\alpha_v) : \alpha_v \in F_v^\times, \alpha_v \in U_v \text{ for almost all } v\}$$

where v runs through all places of F, $U_v = \mathscr{O}_v^\times$. Thus, J_F consists of adeles (α_v) such that $v(\alpha_v) = 0$ for almost all v and $\alpha_v \neq 0$ for all v.

For $\mathrm{A}_F^\times$ to become a topological group, its topology cannot be the induced topology from A_F. In the induced topology from A_F the inversion is not continuous, for example, the sequence of adeles α_v with one finite v-component equal to a prime element of F_v and all other components equal to 1 converges to 1, but the inverse sequence does not converge.

By definition, the topology of $\mathrm{A}_F^\times$ is the induced topology from $\mathrm{A}_F \times \mathrm{A}_F$ in which $\mathrm{A}_F^\times$ is viewed with respect to the embedding

$$\mathrm{A}_F^\times \longrightarrow \mathrm{A}_F \times \mathrm{A}_F, \quad \alpha \mapsto (\alpha, \alpha^{-1}).$$

Then J_F is a topological group. Note that the topology of the multiplicative group of a complete discrete valuation field F_v is the induced topology from $F_v \times F_v$ in which $F_v^\times$ is viewed with respect to the embedding $\alpha \mapsto (\alpha, \alpha^{-1})$, see 18.1 of Chapter 2; hence the topology of J_F induces the usual topology on each local multiplicative group $F_v^\times$.

For a finite set S of places containing all archimedean ones in characteristic zero denote

$$J_F(S) = \mathrm{A}_F(S)^\times = \prod_{v \in S} F_v^\times \times \prod_{v \notin S} U_v.$$

Then

$$J_F = \varinjlim J_F(S).$$

Define the translation invariant topology on $J_F(S)$ as the product topology of the topology of $F_v^\times$ for $v \in S$ and the topology of $\mathcal{O}_v^\times$ for $v \notin S$. Since $\mathcal{O}_v^\times$ are compact and $F_v^\times$ are locally compact, $J_F(S)$ is locally compact. The direct limit topology of J_F is equivalent to the previously defined topology. Then J_F is locally compact.

We have the diagonal injective homomorphism

$$F^\times \longrightarrow \mathbb{A}_F^\times.$$

We will identify $F^\times$ with its image in $\mathbb{A}_F^\times$.

The quotient

$$C_F = \mathbb{A}_F^\times / F^\times$$

is called the *idele class group*. It plays a fundamental central role in global class field theory, similarly to the role of the multiplicative group of a local field in local class field theory.

DEFINITION. For a local field with (surjective) discrete valuation v and finite residue field define the normalised absolute value $|\alpha|_v = |k(v)|^{-v(\alpha)}$ where $|k(v)|$ is the cardinality of the residue field $k(v)$. For a field isomorphic to $\mathbb{R}$ define its absolute value as the usual absolute value, for a field isomorphic to $\mathbb{C}$ define its absolute value is the *square* of the usual complex norm/module. Note that the triangle inequality does not hold for this absolute value on $\mathbb{C}$.

Due to Theorem 9.5 of Chapter 2 for an extension L_w/F_v of complete discrete valuation fields, the normalised absolute values are related by the formula

$$|\alpha|_w = |N_{L_w/F_v}\alpha|_v, \quad \text{for } \alpha \in L_w^\times,$$

since $w = f(w|v)^{-1} v \circ N_{L_w/F_v}$, $|k(w)| = |k(v)|^{f(w|v)}$. Also, for the extension of archimedean completions L_w/F_v we have the same formula $|\alpha|_w = |N_{L_w/F_v}\alpha|_v$ as easily checked from the definitions.

When k is $\mathbb{Q}$ or $\mathbb{F}_p(t)$ we have the property $\prod_v |\alpha|_v = 1$ for a non-zero $\alpha \in k$, where v runs through all places of k. Hence, for a global field F and $\alpha \in F^\times$ we obtain the *product formula*

$$\prod_w |\alpha|_w = \prod_v \prod_{w|v} |N_{F_w/k_v}\alpha|_v = \prod_v |N_{F/k}\alpha|_v = 1.$$

REMARK. Approximation Theorem 2.8 for discrete valuations can be rewritten as a statement about non-equivalent absolute values $|\ |_v$, and also including archimedean absolute values, with exactly the same proof. Thus,

for any $\varepsilon > 0$ and finitely many distinct places v_i and elements $\alpha_i \in F_{v_i}$ there is an element $a \in F$ such that $|a - \alpha_i|_{v_i} < \varepsilon$ for all i.

In particular, for any $\alpha \in \mathbb{A}_F^\times$ and $\varepsilon_i > 0$ there is $a \in F^\times$ such that $|a - \alpha_{v_i}^{-1}|_{v_i} < \varepsilon_i$. Therefore, given positive integer n_{v_i} and choosing $\varepsilon_i = |\alpha_{v_i}|_{v_i}^{-1} |k(v)|^{-n_{v_i}}$ and $\varepsilon_i = |\alpha_{v_i}|_{v_i}^{-1}$ at real v_i, there is $a \in F^\times$ such that $a\alpha_{v_i} \in U_{F_{v_i}, n_{v_i}}$ for finite v_i and $a\alpha_{v_i} > 0$ for real v_i.

The *adelic module* (absolute value) is the product of local modules (absolute values)

$$| \; | = \prod | \; |_v \colon \mathbb{A}_F \longrightarrow [0, \infty).$$

It induces a continuous homomorphism

$$| \; | = | \; |_{J_F} = \prod | \; |_v \colon J_F \longrightarrow \mathbb{R}_{>0}^\times.$$

Its image is $\mathbb{R}_{>0}^\times$ in the number field case: look, for example, at the image of ideles with only one infinite place component different from 1. Its image is an infinite cyclic group in positive characteristic: for each completion the image of the local absolute value is a non-trivial subgroup of $q^{\mathbb{Z}}$, where q is the cardinality of the largest finite subfield of F; hence the image of the adelic module is its non-trivial subgroup as well.

Its kernel J_F^1 is a closed subgroup of J_F.

LEMMA. *The topology of J_F^1 induced by the topology of J_F is equivalent to the topology induced by the topology of* $\mathbb{A}_F$.

Proof. If $1 \in V \cap J_F^1$ for an $\mathbb{A}_F$-neighbourhood V of 1 of the type $|\beta_v - 1|_v < \varepsilon$ for $v \in S$ and $|\beta_v|_v \leqslant 1$ for $v \notin S$, for a finite set S, then $V \cap J_F^1 \supset W \cap J_F^1$ with a J_F-neighbourhood W for which $\leqslant$ is replaced with $=$ for $v \notin S$. If $1 \in W \cap J_F^1$ for an J_F-neighbourhood W of 1 of the type $|\beta_v - 1|_v < \varepsilon$ for $v \in S$ and $|\beta_v|_v = 1$ for $v \notin S$, for a finite set S containing all infinite places, we can assume ε is small enough so that $\prod_{v \in S} |\beta_v|_v < 2$ for all $\beta \in W$. Since the nearest to and smaller than 1 element of $|F_v^\times|_v$ is $p^{-1} \leqslant 1/2$, we deduce that $W \cap J_F^1 = V \cap J_F^1$ with an $\mathbb{A}_F$-neighbourhood V for which $=$ is replaced with $\leqslant$ for $v \notin S$. □

Due to the product formula, $F^\times$ is a subgroup of J_F^1. Since F is discrete in $\mathbb{A}_F$ by Theorem 4.3, $F^\times$ is a discrete subgroup of J_F^1 by the previous Lemma and an observation that for $\alpha \in J_F$ and an open neighbourhood V of 0 in $\mathbb{A}_F$ there is an open neighbourhood W of 1 in J_F such that $\alpha W \subset \alpha + V$. Since F is closed in $\mathbb{A}_F$ by Theorem 4.3, $F \times F$ is closed in $\mathbb{A}_F \times \mathbb{A}_F$ and hence $F^\times$ is closed in $\mathbb{A}_F^\times$. Since J_F^1 is a closed subgroup of J_F, $C_F^1 = J_F^1 / F^\times$ is a closed subgroup of C_F.

DEFINITION. Let S be a finite set, containing S_∞ in the number field case. The intersection

$$F^\times(S) = F^\times \cap J_F(S) = \{\alpha \in F^\times : |\alpha|_v = 1 \text{ for all } v \notin S\}$$

is called the *group of S-units of F*.

In particular, $F^\times(S_\infty) = F^\times \cap \prod_{v\in S_\infty} F_v^\times \times \prod_{v\notin S_\infty} \mathcal{O}_v^\times$ is the group of units $\mathcal{O}_F^\times$ of $\mathcal{O}_F$.

The quotient

$$C_F(S) = J_F(S)/F^\times(S)$$

is called the group of *S-idele classes of F*.

Put

$$C_F^1(S) = J_F^1(S)/F^\times(S).$$

4.6. Let L/F be a finite Galois extension of global fields, $G = \mathrm{Gal}(L/F)$. The group G acts on A_L, $\sigma(\alpha_w) = (\sigma\alpha_w)_{\sigma w}$. We have $\sigma w = w$ if and only if σ belongs to the decomposition subgroup $G_w \simeq \mathrm{Gal}(L_w/F_v)$ where v is the place of F under w.

The G-fixed elements are $\mathrm{A}_L^G = \mathrm{A}_F$, $J_L^G = J_F$. Indeed, if $(\alpha_w) = \sigma(\alpha_w)$, then $\alpha_w = (\sigma\alpha_w)_{\sigma w}$. When σ runs through all elements of the subgroup G_w, $\sigma w = w$, so $\alpha_w \in F_v$ for all $w|v$. Then $\alpha_w = \alpha_{\sigma w}$ for all $\sigma \in G$ and so (α_w) is in the image of A_F in A_L.

LEMMA. *For a finite separable extension L/F the map $C_F \longrightarrow C_L$ induced by $J_F \longrightarrow J_L$ is injective. For a finite Galois extension L/F with Galois group G we have $C_L^G = C_F$.*

Proof. To check the assertions, we can assume L/F is a finite Galois extension with Galois group G. Then $J_F \cap L^\times \subset (J_F \cap L^\times)^G = J_F \cap F^\times$.

For the second assertion, we only need to show the surjectivity of $J_L^G = J_F \longrightarrow C_L^G$. Let $\alpha \in J_L$, $\sigma \in G$, and $\sigma(\alpha L^\times) = \alpha L^\times$. Then $\sigma\alpha = \alpha\beta_\sigma$ for some $\beta_\sigma \in L^\times$. We have the relation $\beta_{\sigma\tau} = \beta_\sigma\beta_\tau^\sigma$ for all $\sigma, \tau \in G$. Since automorphisms $\sigma \in G$ are linearly independent as F- linear operators, see Proposition 2.3.5 and its proof in Chapter 1, there is $\delta \in L^\times$ such that $\sum_{\tau\in G}\beta_\tau\tau(\delta) \neq 0$. Denote it by $\gamma^{-1} \in L^\times$. Then $\gamma^{-\sigma} = \sum_\tau \beta_\tau^\sigma \delta^{\sigma\tau} = \beta_\sigma^{-1}\gamma^{-1}$, so $\alpha^{\sigma-1} = \beta_\sigma = \gamma^{\sigma-1}$ for all σ, hence $\alpha\gamma^{-1} \in J_F$ and $\alpha L^\times = (\alpha\gamma^{-1})L^\times$. □

For a finite separable extension L/F of global fields, the local formulas in 4.5 imply

$$|\alpha|_{J_L} = |N_{L/F}\alpha|_{J_F}, \quad \alpha \in J_L.$$

The norm map induces the norm map

$$N_{L/F}\colon C_L \longrightarrow C_F.$$

Our next main aims, towards checking the axioms of class field theory for global fields, are to show that

(a) *$N_{L/F}C_L$ is open in C_F,*
(b) *$N_{L/F}C_L$ is of finite index in C_F,*
(c) *for a cyclic extension L/F of prime degree the index of $N_{L/F}C_L$ in C_F equals the degree and the kernel of $N_{L/F}$ consists of elements $\alpha^{\sigma-1}$, $\alpha \in C_L$, σ in the Galois group of L/F.*

For (a) see Corollary 4.6, for (b) see Corollary 2 of 4.7 and for (c) see Corollary 1 of 4.9 and Theorem 5.7.

PROPOSITION. *In a finite separable extension L/F only finitely many places v of F have at least one ramification index $e(w|v) > 1$.*

Proof. Let $\alpha \in \mathcal{O}_L$ such that $L = F(\alpha)$. The $\mathcal{O}_F$-module $\mathcal{O}_F[\alpha]$ is in general not an $\mathcal{O}_L$-module. Denote by K the largest ideal of $\mathcal{O}_L$ which is contained in the $\mathcal{O}_F$-module $\mathcal{O}_F[\alpha]$ of $\mathcal{O}_L$. Every maximal ideal Q of $\mathcal{O}_L$ not dividing K satisfies $Q + K = \mathcal{O}_L$ and hence $Q^n + K = \mathcal{O}_L$ by Lemma 3.3.9 in Chapter 1. Therefore for every maximal ideal P of $\mathcal{O}_F$ such that $P\mathcal{O}_L = \prod Q_i^{e_i}$ with maximal ideals Q_i not dividing K, we have $P\mathcal{O}_L + K = \mathcal{O}_L$ and $P\mathcal{O}_L + \mathcal{O}_F[\alpha] = \mathcal{O}_L$. Then $\mathcal{O}_F[\alpha] \cap P\mathcal{O}_L = (P\mathcal{O}_L + \mathcal{O}_F[\alpha])(\mathcal{O}_F[\alpha] \cap P\mathcal{O}_L) \subset P\mathcal{O}_F[\alpha]$ and so $\mathcal{O}_F[\alpha] \cap P\mathcal{O}_L = P\mathcal{O}_F[\alpha]$. Therefore,

$$\mathcal{O}_L/P\mathcal{O}_L = (P\mathcal{O}_L + \mathcal{O}_F[\alpha])/P\mathcal{O}_L \simeq \mathcal{O}_F[\alpha]/P\mathcal{O}_F[\alpha] \simeq (\mathcal{O}_F/P)[X]/(\overline{f})$$

where $f \in \mathcal{O}_F[X]$ is the monic irreducible polynomial of α over F and $\overline{f}$ is its image in $\mathcal{O}_F/P$. Then the factorisation $\overline{f} = \prod \overline{f_i}^{e_i}$ into powers of irreducible polynomials $\overline{f_i}$ over $\mathcal{O}_F/P$ corresponds to the factorisation of $P\mathcal{O}_L = \prod Q_i^{e_i}$ into the product of maximal ideals Q_i of $\mathcal{O}_L$ and $Q_i = P\mathcal{O}_L + f_i(\alpha)\mathcal{O}_L$, the proof is entirely similar to the proof of Theorem 3.5.8 in Chapter 1. The product $\prod_i e_i = 1$ if and only if $\overline{f}$ has no multiple roots if and only if the discriminant of f is not in P. Thus, every maximal ideal P of $\mathcal{O}_F$ which is relatively prime to the ideal generated by the discriminant of f and to every maximal ideal of $\mathcal{O}_F$ lying below a maximal ideal of $\mathcal{O}_L$ dividing K has ramification index $e(Q|P) = 1$ for every maximal ideal Q over $\mathcal{O}_L$ over P. □

REMARK. One can show that a maximal ideal P of $\mathcal{O}_F$ ramifies in L/F if and only if it divides the discriminant of $\mathcal{O}_L$ over $\mathcal{O}_F$.

COROLLARY. *For a finite Galois extension L/F the norm group $N_{L/F}C_L$ is an open subgroup of C_F.*

Proof. By the previous Lemma almost all places v of F are unramified in L/F. The norm map in finite unramified extensions sends the group of units surjectively on the group of units. For the remaining finitely many places the local norm is open, see the proof of Theorem 3.2 in the case of finite places and the case of infinite places is obvious. Open neighbourhoods of 1 in J_F contain the product of the group of local units for almost all places. Thus, we deduce that $N_{L/F}\colon J_L \longrightarrow J_F$ is open. Hence for a finite Galois extension L/F the norm group $N_{L/F}C_L$ is an open subgroup of C_F. □

4.7. In the number field case let I_F be the group of fractional ideals of $\mathcal{O}_F$ generated by maximal ideals P of $\mathcal{O}_F$; endowed it with the discrete topology. We have a surjective continuous homomorphism

$$\rho\colon J_F \longrightarrow I_F, \quad \rho((\alpha_v)) = \prod_P P^{v_P(\alpha_{v_P})},$$

sending components at infinite places to 1. The kernel of ρ is $J_F(S_\infty)$. The image $\rho(F^\times)$ is the group P_F of principal fractional ideals, and we have the induced isomorphism

$$J_F/(F^\times J_F(S_\infty)) \xrightarrow{\sim} I_F/P_F$$

and the right hand side is the class group Cl_F of $\mathcal{O}_F$, defined in 3.6.1 of Chapter 1.

Adjusting archimedean components, we see that ρ induces a surjective homomorphism $J_F^1 \longrightarrow I_F$, so ρ induces a surjective continuous homomorphism

$$\overline{\rho}\colon C_F^1 \longrightarrow I_F/P_F.$$

For a global field F of positive characteristic we have a surjective continuous homomorphism

$$\rho\colon J_F \longrightarrow \mathrm{Div}(F), \quad \rho((\alpha_v)) = \sum v(\alpha_v)[v].$$

The kernel of ρ is $J_F(\emptyset)$, the image $\rho(F^\times)$ is the group of principal divisors $\mathrm{PDiv}(F)$. Hence we obtain an induced isomorphism

$$J_F/(F^\times J_F(\emptyset)) \xrightarrow{\sim} \mathrm{Pic}(F).$$

The map ρ induces the surjective continuous homomorphism

$$\overline{\rho}\colon J_F^1/(F^\times J_F(\emptyset)) \longrightarrow \mathrm{Pic}^0(F).$$

Also, forgetting the components of ideles for valuations lying over $-\deg$, we have, similar to the number field case, a continuous homomorphism

$$\rho\colon J_F \longrightarrow I_F, \quad \rho((\alpha_v)) = \prod_P P^{v_P(\alpha_v)},$$

where P runs through maximal ideals of $\mathcal{O}_F$.

PROPOSITION. *C_F^1 and $C_F^1(S)$ are compact. C_F and $C_F(S)$ are locally compact.*

Proof. By 4.3 A_F/F is a compact abelian group, let μ_0 be its probability measure and let μ be the translation invariant measure on A_F whose quotient on A_F/F is μ_0. Then the measure μ of a compact set $\{\gamma = (\gamma_v) \in \mathrm{A}_F : |\gamma_v|_v \leqslant 1 \text{ for all } v\}$ is positive, denote it by c^{-1}.

Suppose that $|\alpha| > c$ for an idele α. Then for the compact set

$$L = \{(\delta_v) \colon |\delta_v|_v \leqslant |\alpha_v|_v \text{ for all } v\}$$

we have $\mu(L) > 1$. Hence there are two distinct elements λ_i of L which have the same image in A_F/F, so their difference $a = \lambda_1 - \lambda_2 \in F^\times$ and $|a|_v \leqslant |\alpha_v|_v$ for all finite v, $|a|_v \leqslant 2|\alpha_v|_v$ for real v and $|a|_v \leqslant 4|\alpha_v|_v$ for complex v. Denote by m_v the coefficient in front of $|\alpha_v|_v$.

Let ν be an idele with $|\nu| > c$. The previous paragraph implies that for every $\gamma \in J_F^1$ there is an $a \in F^\times$ such that $|a|_v \leqslant m_v|\gamma_v^{-1}\nu_v|_v$ for all v. Consider $K = \{(\beta_v) : |\beta_v|_v \leqslant m_v|\nu_v|_v\}$, a compact subset of A_F. We deduce $\gamma a \in K \cap J_F^1$. Thus, $J_F^1 = (K \cap J_F^1)F^\times$. The set $K \cap J_F^1$ is compact in A_F and Lemma 4.5 implies it is compact in J_F^1. Hence $J_F^1/F^\times$ is compact.

Since $C_F^1(S)$ is a closed subgroup of C_F^1, it is compact.

The last sentence of the Proposition follows from the description in 4.5 of the quotient C_F/C_F^1 using the adelic module. □

COROLLARY 1. *In the number field case the class group* $\mathrm{Cl}_F = I_F/P_F$ *is finite.*

In the global function field case the group $\mathrm{Pic}^0(F)$ *is finite.*

For sufficiently large finite sets S including S_∞ we have $J_F = F^\times J_F(S)$.

Proof. Since C_F^1 is compact, its $\overline{\rho}$-image is compact. Now the discreteness of the class group I_F/P_F and of $\mathrm{Pic}^0(F)$ implies their finiteness.

Since the class group and $\mathrm{Pic}^0(F)$ are finite, enlarging the set S_∞ (or the empty set in the global function field case) to a finite non-empty set S to include in it places corresponding to finitely many maximal ideals that generate the class group or $\mathrm{Pic}^0(F)$, we have $J_F^1 = F^\times J_F^1(S)$. In characteristic zero $|J_F| = |J_F(S)|$, hence we deduce $J_F = F^\times J_F(S)$. In positive characteristic enlarge S to include places at which components of ideles whose adelic modules generates $|J_F|$ are not units, then $|J_F(S)| = |J_F|$ and hence $J_F = F^\times J_F(S)$. □

COROLLARY 2. *For a finite Galois extension the norm group $N_{L/F}C_L$ is an open subgroup of finite index in C_F.*

Proof. From Corollary 4.6 we know that $N_{L/F}C_L$ is an open subgroup of C_F. Hence $N_{L/F}C_L^1$ is an open subgroup of compact C_F^1 and so it is of finite index in C_F^1. In the number field case, the adelic module of the image with respect to $N_{L/F}$ of the subgroup of ideles where all components except at one infinite place are 1 and at that infinite place the component runs through all elements of the corresponding completion is $\mathbb{R}^\times_{>0}$. In the global function field case, the adelic module of the image with respect to $N_{L/F}$ of the subgroup of ideles where all components except at one place are 1 and at that place the component runs through all elements of the corresponding completion is a subgroup of finite index in $|J_F|$. Hence $N_{L/F}C_L$ is a subgroup of finite index in C_F. □

REMARKS.

1. This gives a new proof of the finiteness of the class group, using the compactness of C_F^1. In turn, using the finiteness of the class group and of the zero part of the Picard group, one can deduce the compactness property of C_F^1.

2. The arguments in the first paragraph of the proof of the Proposition can be used for an adelic proof of Minkowski's bound Theorem 3.6.6 of Chapter 1.

3. An alternative independent and very different proof of the compactness of C_F^1 will be obtained later, see Remark 2 of 5.6.

4.8. For a finite S with $s > 0$ elements and containing S_∞ in the number field case we have a homomorphism

$$\mathrm{Log}_S\colon J_F(S) \longrightarrow \mathbb{R}^s, \quad (\alpha_v) \mapsto (\log|\alpha_v|_v).$$

Due to the product formula, it sends $J_F^1(S)$ to the hyperplane $H_s = \{(x_1,\dots,x_s) \in \mathbb{R}^s : x_1 + \dots + x_s = 0\}$ of $\mathbb{R}^s$. The homomorphism Log_S induces the homomorphism

$$\log_S\colon F^\times(S) \longrightarrow H_s.$$

The following Proposition gives an adelic proof of a generalisation of Dirichlet's unit theorem stated in 3.8.1 of Chapter 1.

PROPOSITION. *Let S be a finite non-empty set of places containing S_∞ in the number field case. The kernel of* $\log_S$ *is μ_F, the image is a discrete subgroup of* *rank $s-1$ of H_s, i.e. a complete lattice of H_s, $s=|S|$. Hence the group of units $F^\times(S)$ is isomorphic to the direct sum of its torsion part and a free group of rank $s-1$. In particular, the group of units $\mathcal{O}_F^\times$ is isomorphic to the direct sum of its torsion part and a free group of rank r_1+r_2-1, where r_1,r_2 are the same as in 3.6.5 of Chapter 1.*

Proof. The kernel of Log_S is UJ_F where $UJ_F=\prod_v S_v^1$ and $S_v^1=\{\alpha_v\in F_v : |\alpha_v|_v = 1\}$ for all v, so UJ_F is a compact subgroup of $J_F(S)$. The kernel of $\log_S$ is the intersection of the discrete set $F^\times(S)$ in J_F with the compact subgroup UJ_F, hence it is a finite group. Hence the kernel consists of all roots of unity in F.

Denote $T=\log_S(F^\times(S))$, it is isomorphic to the quotient $F^\times(S)/(UJ_F\cap F^\times(S))$ of the discrete group $F^\times(S)$, hence T is discrete in H_s. Via Log_S, the quotient H_s/T is isomorphic to $J_F^1(S)/(F^\times(S)UJ_F)$ which is the quotient of the compact $J_F^1(S)/F^\times(S)$ by its closed subgroup $(F^\times(S)UJ_F)/F^\times(S)$. Thus, T is co-compact and discrete in $H_s\simeq\mathbb{R}^{s-1}$. Hence $T\simeq\mathbb{Z}^{s-1}$ is a full lattice in $\mathbb{R}^{s-1}$. □

4.9. Let A be an abelian group written additively and let $f,g\colon A\longrightarrow A$ be group homomorphisms such that $f\circ g=g\circ f=0$. Denote by A_f the kernel of f and by A^f the image of f. Define the *Herbrand quotient*

$$Q_{f,g}(A)=\frac{|A_f:A^g|}{|A_g:A^f|}$$

when both the numerator and denominator are finite.

LEMMA. *$Q_{f,g}(A)=1$ for a finite group A.*

If B is a subgroup of A such that $f(B),g(B)\subset B$, then f,G induce homomorphisms $B\longrightarrow B$ and $A/B\longrightarrow A/B$, denote them by the same f,g. Then

$$Q_{f,g}(A)=Q_{f,g}(B)Q_{f,g}(A/B)$$

when any two of the factors are defined.

Proof. For the first property, consider finite groups $A\supset A_g\supset A^f\supset 0\subset A^g\subset A_f\subset A$ in which the index for the first inclusion equals the index for the fourth inclusion, the index for the third inclusion equals the index for the sixth inclusion. Hence the index for the second inclusion equals the index for the fifth inclusion.

For the second property, denote $C=A/B$. We have an exact sequence of homomorphisms

$$B_f/B^g\longrightarrow A_f/A^g\longrightarrow C_f/C^g\longrightarrow B_g/B^f\longrightarrow A_g/A^f\longrightarrow C_g/C^f\longrightarrow B_f/B^g$$

in which the first, second, fourth, fifth maps are induced by $B\longrightarrow A$ and $A\longrightarrow C$. To define the third map, take $c\in C$ such that $f(c)=0$, take any $a\in A$ such that

$a+B=c$, then c is sent to $f(a) \in B_g$; this is well defined. Similarly one defines the sixth map. The exactness is immediate. The order of each object is the product of the order of the image of the map to that object and the order of the image of the map from that object. One deduces $Q_{f,g}(A) = Q_{f,g}(B)Q_{f,g}(A/B)$. □

We will use $Q_{f,g}$ in the situation when a cyclic group G of order n with a generator σ acts on an abelian group A, $f = 1-\sigma$ and $g = \mathrm{Tr}_G = \sum_{i=0}^{n-1} \sigma^i$, so $A_f = A^G$, $A^f = I_G A = \{a^{\sigma-1} : a \in A\}$, $A_g = \ker \mathrm{Tr}_G$, $A^g = \mathrm{Tr}_G(A)$. In this situation denote $Q(G,A) = Q_{f,g}(A)$.

EXAMPLES.

1. If the action on an infinite cyclic group $A \simeq \mathbb{Z}$ is trivial, then $Q(G,A) = n$.

2. If A is a G-module $\oplus_{\sigma \in G} B\sigma$, B an abelian group, with the action

$$\tau(\sum b_\sigma \sigma) = \sum b_\sigma \tau\sigma, \quad b_\sigma \in B,$$

then $Q(G,A) = 1$.

3. Let L/F be a cyclic extension of local fields with finite residue field, $G = \mathrm{Gal}(L/F)$ of order n. Then

$$Q(G,L^\times) = \frac{|F^\times : N_{L/F}L^\times|}{|\ker N_{L/F} : {L^\times}^{1-\sigma}|} = n$$

by local class field theory and Hilbert 90 Theorem. We also have

$$Q(G,U_L) = 1$$

due to $L^\times/U_L \simeq \mathbb{Z}$, Lemma and Example 1.

When L/F be a cyclic extension of archimedean completions, then the only non-trivial case is $G = \mathrm{Gal}(\mathbb{C}/\mathbb{R})$, and $Q(G,\mathbb{C}^\times) = |\mathbb{R}^\times : \mathrm{N}_{\mathbb{C}/\mathbb{R}}\mathbb{C}^\times| = 2$.

4. Let L/F be a cyclic extension of global fields of prime degree n. Let v be a place of F, w be a fixed place of L over v, and let $G_v = \mathrm{Gal}(L_w/F_v)$ be the decomposition group of v of order n_v dividing n. Then

$$Q(G, \oplus_{\sigma \in G/G_v} \sigma L_w^\times) = n_v = Q(G, \oplus_{\sigma \in G/G_v} \mathbb{Z}\sigma).$$

Indeed, since the order of G is prime, either $G_v = 1$ or $G_v = G$. In the first case, using Example 2, we deduce $Q(G, \oplus_{\sigma \in G/G_v} \sigma L_w^\times) = 1 = n_v = Q(G, \oplus_{\sigma \in G/G_v} \mathbb{Z}\sigma)$; in the second case, using Examples 3 and 1, $Q(G, \oplus_{\sigma \in G/G_v} \sigma L_w^\times) = Q(G_v, L_w^\times) = n_v = Q(G, \oplus_{\sigma \in G/G_v} \mathbb{Z}\sigma)$.

THEOREM. *Let L/F be a cyclic extension of global fields with Galois group G of prime order n. Then $Q(G,C_L) = n$.*

Proof. For a finite place v of F and a place w of L, $w|v$, the preceding Examples imply $Q(G,L_w^\times) = |L_w : F_v|$ and $Q(G,U_{L_w}) = 1$.

In positive characteristic we have

$$Q(G,C_L) = Q(G,J_L/J_L^1)Q(G,J_L^1/L^\times J_L(\emptyset))Q(G,L^\times J_L(\emptyset)/L^\times),$$

and $Q(G,J_L/J_L^1) = Q(G,\mathbb{Z}) = n$, $Q(G,J_L^1/(L^\times J_L(\emptyset))) = 1$ since $J_L^1/(L^\times J_L(\emptyset))$ is isomorphic to finite $\mathrm{Pic}^0(F)$, see 4.7. Also,

$$Q(G,L^\times J_L(\emptyset)/L^\times) = Q(G,J_L(\emptyset))Q(G,L^\times(\emptyset))^{-1} = Q(G,J_L(\emptyset)),$$

since $L^\times(\emptyset)$ is the multiplicative group of the finite field of constants of L. Using $Q(G,J_L(\emptyset)) = \prod_v Q(G,U_{L_v}) = 1$, we conclude $Q(G,C_L) = n$.

For number fields the proof is longer. Choose a finite set S of places of L, which is invariant under the action of G and which contains all archimedean places and is sufficiently large so that $J_L = L^\times J_L(S)$ in accordance with Corollary 1 of 4.7. Denote by s its cardinality. Denote by S_0 the set of places of F under the places in S. Then $C_L = J_L/L^\times = (L^\times J_L(S))/L^\times \simeq J_L(S)/L^\times(S)$ and

$$Q(G,C_L) = Q(G,J_L(S))Q(G,L^\times(S))^{-1}.$$

Using Examples 3 and 4, we get

$$Q(G,J_L(S)) = \prod_{v\in S_0} Q(G,\oplus_{\sigma\in G/G_v}\sigma L_w^\times) = \prod_{v\in S_0} n_v.$$

To complete the proof, it remains to show that

$$Q(G,L^\times(S)) = n^{-1}\prod_{v\in S_0} n_v.$$

In order to achieve that, we use the map $\log_S : L^\times(S) \longrightarrow \mathbb{R}^s$. Let $\{e_w : w\in S\}$ be the standard basis of $V = \mathbb{R}^s$. Let the group G act $\mathbb{R}$-linearly on V by $\sigma e_w = e_{\sigma w}$. Then $\log_S(\sigma a) = \sum_{w\in S}\log|\sigma a|_w e_w = \sigma\sum_{w\in S}\log|a|_{\sigma^{-1}w}e_{\sigma^{-1}w} = \sigma\log_S(a)$. By Proposition 4.8, $T = \log_S(L^\times(S))$ together with $e' = \sum_{w\in S} e_w$ generate a complete lattice M in V. We deduce that M is a G-invariant, i.e. $\sigma M\subset M$ for every $\sigma\in G$; and $\sigma e' = e'$ for every $\sigma\in G$. We have $M/\mathbb{Z}e'\simeq T$. Since the kernel of $\log_S$ is finite, by the previous Lemma we obtain

$$Q(G,L^\times(S)) = Q(G,T) = Q(G,\mathbb{Z})^{-1}Q(G,M) = n^{-1}Q(G,M).$$

It remains to show $Q(G,M) = \prod_{v\in S_0} n_v$.

Note that $M\otimes\mathbb{R}\simeq\mathbb{R}^s\simeq(\oplus\mathbb{Z}e_w)\otimes\mathbb{R}$ as $\mathbb{R}[G]$-modules. In this situation it can be shown that $Q(G,M) = Q(G,\oplus\mathbb{Z}e_w)$, while the latter, similarly to the argument below for N is $\prod_{v\in S_0} n_v$. We will use a different more explicit argument though. We will find a complete sublattice $N = \sum\mathbb{Z}z_w$ of M such that $\sigma z_w = z_{\sigma w}$ for all $\sigma\in G$. Then $N = \sum\mathbb{Z}z_w$ is a sublattice of M of finite index, and it is a complete G-invariant lattice of $\mathbb{R}^s$, and $\sigma z_w = z_{\sigma w}$. So $N = \oplus_{v \in S_0} N_v$ where $N_v = \oplus_{\sigma \in G/G_v} \mathbb{Z}\sigma z_{w_v}$. Hence, $Q(G,M) = Q(G,N)$ and

$$Q(G,N) = \prod_{v \in S_0} Q(G, \oplus_{\sigma \in G/G_v} \mathbb{Z}\sigma z_{w_v}) = \prod_{v \in S_0} Q(G, \oplus_{\sigma \in G/G_v} \mathbb{Z}\sigma) = \prod_{v \in S_0} n_v,$$

using Example 4, and the proof will be completed.

Denote by $|\ |$ the sup-norm with respect to the coordinates of the basis e_w of V. The group G maps this basis into itself, hence $|\sigma\gamma| = |\gamma|$ for every $\nu \in V$. Since M is a lattice, there is $c > 0$ such that for every vector in V there is an element of M such that the distance between them is $< c$. For every $v \in S_0$ choose $w_v \in S$ such that $w_v | v$. Let $t = ncs$. Then there is $m_v \in M$ such that for $x_v = te_{w_v} - m_v$ we have $|x_v| < c$. Hence $|\sigma x_v| < c$ for every $\sigma \in G$.

For $w \in S$, $w|v$ define

$$z_w = \sum_{\sigma : \sigma w_v = w} \sigma m_v \in M,$$

the number of terms is n_v. Then $\tau z_w = \sum_{\sigma : \sigma w_v = w} \tau\sigma m_v = \sum_{\rho : \rho w_v = \tau w} \rho m_v = z_{\tau w}$ for every $\tau \in G$. We have

$$z_w = \sum_{\sigma : \sigma w_v = w} \sigma m_v = t \sum_{\sigma : \sigma w_v = w} e_{w_v} - y_w = tn_v e_{w_v} - y_w, \quad \text{where } y_w = \sum_{\sigma : \sigma w_v = w} \sigma x_v.$$

Hence $|y_w| < n_v c$.

Write $y_{w'} = \sum_{w \in S} d_{w'}^{w} e_w$ with real $d_{w'}^{w}$, then $|d_{w'}^{w}| < n_{v'} c$ when $w' | v'$. Let $\sum_{w \in S} c_w z_w = 0$ with real c_w. Then

$$t \sum_{v \in S_0} n_v \sum_{w|v} c_w e_{w_v} = \sum_{w' \in S} y_{w'} c_{w'} = \sum_{w \in S} \sum_{w' \in S} c_{w'} d_{w'}^{w} e_w.$$

We deduce

$$tn_v c_w = \sum_{w' \in S} d_{w'}^{w_v} c_{w'} \quad \text{for } w|v$$

and

$$\begin{aligned} n_v ncs|c_w| = |tn_v c_w| &= \Big| \sum_{v' \in S_0} \sum_{w'|v'} d_{w'}^{w_v} c_{w'} \Big| \\ &< \sum_{v' \in S_0} cn_{v'} nn_{v'}^{-1} \max\{|c_{w'}| : w'|v'\} \leqslant cns \max\{|c_{w'}|\} \end{aligned}$$

when $w|v$. Here we use $|\{w : w|v\}| = nn_v^{-1}$. If not all c_w are zero, choose w such that $0 < |c_w| = \max\{|c_{w'}|\}$ to get a contradiction. Hence the vectors z_w, $w \in S$, are linearly independent. □

COROLLARY 1. *$|C_F : N_{L/F} C_L| = |J_F : F^\times N_{L/F} J_L|$ is divisible by n for cyclic extensions of prime degree n.*

Proof. $Q(G, C_L) = \dfrac{|C_F : N_{L/F} C_L|}{|\ker N_{L/F} : C_L^{1-\sigma}|} = n.$ □

DEFINITION. A place v of F is said to place split completely *split completely* or *totally decomposed* in L/F if $L_w = F_v$ for every place $w|v$ of L. In other words, due to the formula $|L : F| = \sum_{w|v} e(w|v) f(w|v)$, there are exactly $|L : F|$ distinct places w of L over the place v and for each of them $e(w|v) = f(w|v) = 1$.

COROLLARY 2. *Let L/F be a non-trivial finite Galois extension. Then there are infinitely many places of F which do not split completely in L.*

Proof. Take any cyclic subgroup of prime order of $\mathrm{Gal}(L/F)$ and consider its fixed field E, then L/E is cyclic of prime order. If $L_w = F_v$ for almost all places v of F and $w|v$, then $L_w = E_u$ for almost all places u of E and $w|u$. Let $\alpha \in J_E$. Denote by S the finite set of places of E where $L_w \neq E_u$. Using Remark 4.5 find an element $a \in E^\times$ such that αa^{-1} is a local norm at every $u \in S$. Then $\alpha a^{-1} \in N_{L/E} J_L$, so $C_E / N_{L/E} C_L = 1$, a contradiction. □

Compare the statement of Corollary 2 with that of Proposition 2.6.

COROLLARY 3. *Let F be a global field whose field of constants is $\mathbb{F}_q$. Then for the adelic module $|J_F| = q^{\mathbb{Z}}$.*

Proof. Let q^d be the greatest common divisor of the cardinalities of the residue fields of places of F, and let $F' = F\mathbb{F}_{q^d}$. Since for every place v the residue field of F_v contains $\mathbb{F}_{q^d}$, $F_v = F'_w$ for $w|v$. Hence $F = F'$ by Corollary 2 and $d = 1$. □

5. Zeta Functions and Zeta Integrals

5.1. Zeta functions is one of the key objects of number theory. Class field theory can be presented without the use of material of this section. However, there are numerous important links between the two, and the use of zeta integrals and higher zeta integrals plays a key role in generalisations of class field theory such as the Langlands program and higher adelic theory.

DEFINITION. The *zeta function* of a scheme X of finite type over $\mathrm{Spec}(\mathbb{Z})$ is

$$\zeta_X(s) = \prod_{x \in X_0} (1 - |k(x)|^{-s})^{-1}$$

where x runs through closed points of X, $k(x)$ is the finite residue field of x.

EXAMPLES.

1. When $X = \mathrm{Spec}(\mathbb{Z})$, this is the Euler–Riemann zeta function

$$\zeta_{\mathrm{Spec}(\mathbb{Z})}(s) = \zeta_{\mathbb{Q}}(s) = \prod_p (1-p^{-s})^{-1},$$

where p runs through all positive primes. On $\Re e(s) > 1$ we have

$$\zeta_{\mathrm{Spec}(\mathbb{Z})}(s) = \sum_{n\geqslant 1} \frac{1}{n^s}.$$

2. When $X = \mathrm{Spec}(\mathscr{O}_F)$, $\mathscr{O}_F$ is the ring of integers of an algebraic number field, this is the Dedekind zeta function

$$\zeta_{\mathrm{Spec}(\mathscr{O}_F)}(s) = \zeta_F(s) = \prod_v (1-|k(v)|^{-s})^{-1} = \prod_P (1-N(P)^{-s})^{-1}$$

where v runs through all finite places of F, P runs through maximal ideals of $\mathscr{O}_F$. When $P = P_v$ corresponds to v, the norm $N(P)$ of P equals to $|k(v)|$.

3. When $X = \mathscr{C}$ with a smooth proper irreducible curve $\mathscr{C}$ over a finite field $\mathbb{F}_q$ with function field F and F is a finite separable extension of $\mathbb{F}_q(t)$, we have

$$\begin{aligned}\zeta_{\mathscr{C}}(s) = \zeta_F(s) &= \prod_{x\in\mathscr{C}_0} (1-|k(x)|^{-s})^{-1} = \prod_v (1-|k(v)|^{-s})^{-1} \\ &= \prod_{v|-\deg} (1-|k(v)|^{-s})^{-1} \prod_P (1-N(P)^{-s})^{-1}\end{aligned}$$

where v runs through all places of F, P runs through maximal ideals of $\mathscr{O}_F$, $-\deg$ is the discrete valuation of $\mathbb{F}_q(t)$ associated to the minus degree of polynomials. Note that each Euler factor $(1-|k(x)|^{-s})^{-1}$ absolutely and uniformly converges for $\Re e(s) > 0$ and meromorphically extends to the complex plane with the only pole at $s = 0$.

Recall that if a Dirichlet series $\sum_{n\geqslant 1} a_n/n^s$ converges at real s_0, then it converges absolutely and uniformly on compact subsets for $\Re e(s) > s_0$. Indeed, consider partial sums $q_r(s) = \sum_{n=1}^{n=r} a_n/n^s$. Then

$$\begin{aligned}q_r(s) - q_m(s) &= \sum_{n=m+1}^{n=r} a_n/n^s \\ &= \sum_{n=m}^{n=r-1} q_n\left(1/n^{s-s_0} - 1/(n+1)^{s-s_0}\right) - q_m/m^{s-s_0} + q_{r-1}/r^{s-s_0},\end{aligned}$$

and we have $1/m^{s-s_0} - 1/r^{s-s_0} = (s-s_0)\int_m^r dx/x^{s-s_0+1}$. Since all partial sums $q_r(s_0) = \sum_{n=1}^{n=r} a_n/n^{s_0}$ are bounded by some positive constant, when $|s-s_0|$ is bounded and $\Re e(s) \geqslant s_0 + \varepsilon$ with positive ε, the sum $\sum_{n=m}^{n=r} a_n/n^s$ tends uniformly to 0 when $m, r \to +\infty$.

Assume that $|\sum_{n=1}^{n=r} a_n| \leqslant r$. Then for the Dirichlet series $\sum_{n\geqslant 1} a_n/n^s$ we have

$$|q_r(s) - q_m(s)| \leqslant \sum_{n=m+1}^{n=r-1} ns \int_n^{n+1} dx/x^{s+1} + 1/r^{s-1},$$

and $\sum_{n=m+1}^{n=r-1} n \int_n^{n+1} dx/x^{s+1} \leqslant \int_{m+1}^{r} dx/x^s$. Thus, under the assumption this Dirichlet series is a holomorphic function on $\Re e(s) > 1$.

In particular, the Dirichlet series for $\zeta_{\mathbb{Q}}(s)$ diverges at $s = 1$ and converges absolutely and uniformly on compact subsets for $\Re e(s) > 1$ and there $\zeta_{\mathbb{Q}}(s) = \sum_{n\geqslant 1} \frac{1}{n^s} = \prod(1-p^{-s})^{-1}$. We deduce

$$\log \zeta_{\mathbb{Q}}(s) = \sum_{m\geqslant 1} \sum_{p} (mp^{ms})^{-1} \quad \text{for } \Re e(s) > 1.$$

We also deduce that for real $s > 1$

$$1/(s-1) = \int_1^{\infty} 1/x^s \leqslant \zeta_{\mathbb{Q}}(s) \leqslant 1 + 1/(s-1).$$

We use the notation $f \sim g$ for two functions defined for real $s > 1$, with singularity at $s = 1$, and whose difference does not have singularity at $s = 1$. Then we get

$$\zeta_{\mathbb{Q}}(s) \sim 1/(s-1).$$

Since $\sum_{m\geqslant 2} \sum_p (mp^{ms})^{-1}$ converges uniformly and absolutely for $\Re e(s) > 1/2+\varepsilon$, we deduce

$$\sum_p p^{-s} \sim \log \zeta_{\mathbb{Q}}(s) \sim -\log(s-1).$$

For a number field F the index of a maximal ideal P of $\mathcal{O}_F$ is its norm $N(P) = p^{f(P|p\mathbb{Z})}$ where $p\mathbb{Z}$ is the ideal of $\mathbb{Z}$ lying under P. Since there are at most $n = |F : \mathbb{Q}|$ maximal ideals of $\mathcal{O}_F$ over $p\mathbb{Z}$, for $\Re e(s) > 1$ we have

$$\log \prod_P (1 - N(P)^{-s})^{-1} = \sum_{m\geqslant 1} \sum_P m^{-1} N(P)^{-ms} \leqslant n \sum_{m\geqslant 1} \sum_p m^{-1} p^{-ms} = n \log \zeta_{\mathbb{Q}}(s).$$

Therefore, on $\Re e(s) > 1$

$$\zeta_F(s) = \prod_P (1 - N(P)^{-s})^{-1} = \sum_I N(I)^{-s},$$

where I runs through non-zero ideals of $\mathcal{O}_F$, converges absolutely and uniformly on compact subsets. Similarly to $\zeta_{\mathbb{Q}}(s)$, we obtain

$$\sum N(P)^{-s} \sim \sum_{N(P) \text{ is prime}} N(P)^{-s} \sim \log \zeta_F(s)$$

where P runs through maximal ideals of $\mathcal{O}_F$ whose residue field has prime cardinality.

Maximal ideals of $\mathbb{F}_q[t]$ are principal ideals generated by monic irreducible polynomials f over $\mathbb{F}_q$, so for $\Re e(s) > 1$ we have

$$\prod_P (1-N(P)^{-s})^{-1} = \prod_f (1-q^{-s\deg(f)})^{-1} = \sum_g q^{-s\deg(g)}$$

where g runs through all monic polynomials in $\mathbb{F}_q[t]$, their number of degree m is q^m, so the latter sum $= \sum_{m\geqslant 0} q^m q^{-sm} = (1-q^{-s+1})^{-1}$. Taking into account the valuation $-\deg$, we obtain

$$\zeta_{\mathbb{P}^1(\mathbb{F}_q)}(s) = (1-q^{-s})^{-1}(1-q^{-s+1})^{-1}$$

converges absolutely and uniformly on compact subsets on $\Re e(s) > 1$ and the only poles are at s such that $q^s = 1$ or $q^{s-1} = 1$. We also have

$$\log \zeta_{\mathbb{P}^1(\mathbb{F}_q)}(s) \sim -\log(1-q^{-s+1}) \sim \log(s-1).$$

Let F be a global field F of characteristic p, a separable extension of $\mathbb{F}_q(t)$ of degree n. Since there are at most n maximal ideals P of $\mathscr{O}_F$ over a maximal ideal P_v of $\mathbb{F}_q[t]$ corresponding to a discrete valuation v of $\mathbb{F}_q(t)$, on $\Re e(s) > 1$ we have

$$\begin{aligned}\log \prod_P (1-N(P)^{-s})^{-1} &= \sum_{m\geqslant 1}\sum_P m^{-1} N(P)^{-ms} \\ &\leqslant n \sum_{m\geqslant 1}\sum_v m^{-1} |k(v)|^{-ms} = n \log \zeta_{\mathbb{F}_q(t)}(s).\end{aligned}$$

Therefore, the zeta function of F converges absolutely and uniformly on compact subsets of $\Re e(s) > 1$, and on $\Re e(s) > 1$ we have

$$\begin{aligned}\zeta_F(s) &= \prod_{w|-\deg} (1-|k(w)|^{-s})^{-1} \prod_P (1-N(P)^{-s})^{-1} \\ &= \prod_{w|-\deg} (1-|k(w)|^{-s})^{-1} \sum_I N(I)^{-s},\end{aligned}$$

where I runs through non-zero ideals of $\mathscr{O}_F$.

5.2. Denote by

$$j_v \colon F_v^\times \longrightarrow J_F$$

the homomorphism sending $\alpha \in F_v^\times$ to the idele all of whose components are 1 except the v-component which is equal α.

Now we define *twists of zeta functions by characters*, they are traditionally called *L-functions*.

Let χ be a non-trivial character of J_F.

For example, for number fields such characters may come from characters of the ideal class group I_F/P_F, using the homomorphism $J_F/F^\times \longrightarrow I_F/P_F$ and for function fields from characters of the Picard group using $J_F/F^\times \longrightarrow \operatorname{Pic}^0(F)$, as in 4.7.

According to the first paragraph of 4.3, the kernel of the restriction of χ on the subgroup of ideles with component 1 at all infinite places is its open subgroup, hence the kernel of χ contains $j_v(U_v)$ for almost all v.

DEFINITION. Define

$$\chi(v) = \begin{cases} 0 & \text{if } \chi(j_v(U_v)) \neq 1 \\ \chi(j_v(\pi_v)) & \text{if } \chi(j_v(U_v)) = 1. \end{cases}$$

In the second case which happens for almost all finite v, the value $\chi(j_v(\pi_v))$, where π_v is any prime element of F_v, does not depend on the choice of π_v.

Let C be a finite subset of finite places v of F. Define

$$L_C(s,\chi) = \prod_{v \notin C} (1 - \chi(v)|k(v)|^{-s})^{-1},$$

the product is taken over finite places v not in C.

The product of finitely many factors $(1-\chi(v)|k(v)|^{-s})^{-1}$ does not affect the behaviour of $L_C(s,\chi)$ near $s=1$.

Then

$$L_C(s,1) = \prod_{v \notin C} (1 - |k(v)|^{-s})^{-1}$$

which, when multiplied with the finitely many Euler factors for $v \in C$, is $\zeta_F(s)$.

Except possibly finitely many factors corresponding to places in positive characteristic over $-\deg$, the product $\prod_{v \notin C}(1-\chi(v)|k(v)|^{-s})^{-1}$ equals the product $\prod_{v \notin C}(1-\chi(P_v)N(P_v)^{-s})^{-1}$ where P_v runs through maximal ideals of $\mathcal{O}_F$ and $\chi(P_v) = \chi(v)$. For a non-zero ideal $I = \prod P_i^{n_i}$ of $\mathcal{O}_F$ put $\chi(I) = \prod \chi(P_i)^{n_i}$. By the same reasons as for $\zeta_F(s)$, the product converges absolutely and uniformly on compact subsets of $\Re e(s) > 1$. On this half-plane we have, similarly to 5.1 in characteristic zero, we get

$$\sum_{v \notin C} \chi(v)|k(v)|^{-s} \sim \sum_{v \notin C, |k(v)| \text{ is prime}} \chi(v)|k(v)|^{-s} \sim \log L_C(s,\chi).$$

5.3. We have seen that the additive and multiplicative group of local fields with finite residue field and of adeles are abelian locally compact groups. Hence, in accordance with basic theory of locally compact abelian groups, for each of these groups one has a non-trivial translation invariant measure. For these groups one can construct such measures directly, see e.g. [**14**]. Each of such measures is defined up to multiplication by a positive real constant.

Translation invariant measures μ_v on the additive group of a local field F_v with finite residue field with the ring of integers $\mathcal{O}_v$ and maximal ideal $\mathcal{M}_v$ have an especially easy description. Counting indices and using the virtual index, similarly to 4.4, we immediately deduce the formula for the measure/volume of closed balls

$$\mu_v(\alpha+\mathcal{M}_v^n)=\mu_v(\mathcal{M}_v^n)=|\mathcal{O}_v:\mathcal{M}_v^n|^{-1}\mu_v(\mathcal{O}_v),$$

thus one only needs to fix the value $\mu_v(\mathcal{O}_v)\in\mathbb{R}_{>0}$.

We will use the local and adelic characters ψ_v^0 of F_v and ψ^0 of A_F defined in 4.4.

DEFINITION. By the first paragraph of 4.3, the kernel of ψ_v^0 is an open subgroup of F_v. For a finite v denote by d_v the maximal integer such that ψ_v^0 sends the group $\mathcal{M}_v^{-d_v}$ to 1. In other words,

$$\mathrm{Tr}_{F_v/k_{v_0}}(\mathcal{M}_v^{-d_v})\subset\mathcal{O}_{k_{v_0}},\quad \mathrm{Tr}_{F_v/k_{v_0}}(\mathcal{M}_v^{-d_v-1})\not\subset\mathcal{O}_{k_{v_0}}$$

where v_0 is the place of k (equal to $\mathbb{Q}$ or $\mathbb{F}_p(t)$) under v. For unramified v in the extension F_v/k_{v_0} we have $d_v=0$. The numbers d_v are zero for almost all v, since only finitely many places ramify in F/k by Proposition 4.6.

DEFINITION. Choose normalised measures μ_v as the usual Lebesque measure on $\mathbb{R}$, twice the usual Lebesque measure on $\mathbb{C}$, and for finite v the normalisation is

$$\mu_v(\mathcal{O}_v)=|k(v)|^{-d_v/2}.$$

Choose the translation invariant measure $\mu_{\mathrm{A}_F}=\mu=\prod_v\mu_v$ on A_F, it is well defined since $\mu_v(O_v)=1$ for almost all v.

One immediately checks that the normalised absolute values $|\ |_v$ defined in 4.5 are the module functions associated to μ_v, i.e. for every $\alpha\in F_v^\times$ we have $|\alpha_v|_v=\mu_v(\alpha_vA)/\mu_v(A)$ for any measurable subset A of F_v of non-zero volume. For finite places this comparison follows right away from the displayed formula above.

Hence $|\ |$ is the module function associated to μ, i.e. $|\alpha|=\mu(\alpha A)/\mu(A)$ for any measurable subset A of A_F of non-zero volume.

DEFINITION. On the multiplicative group $F_v^\times$ define the translation invariant measure $\mu_v^\times$ by the formula

$$\mu_v^\times=(1-|k(v)|^{-1})^{-1}\mu_v/|\ |_v$$

in the non-archimedean case and $\mu_v^\times=\mu_v/|\ |_v$ in the archimedean case.

Choose the translation invariant measure $\mu_{J_F}=\mu^\times=\prod_v\mu_v^\times$ on $\mathrm{A}_F^\times$, it is well defined since $\mu_v^\times(O_v^\times)=1$ for almost all v

5.4. We now define certain spaces of functions convenient to compute Fourier transforms.

DEFINITION. Define a space $S(F_v)$ of locally constant functions on F_v with compact support in the non-archimedean case; and of smooth functions on F_v such that the product of it and every of its derivatives with any polynomial function tends to 0 when the absolute value of the argument tends to infinity in the archimedean case. Functions in $S(F_v)$ are integrable with respect to μ_v.

Define $S(\mathrm{A}_F)$ as the space spanned by functions $\otimes_v g_v$ with $g_v \in S(F_v)$ such that $g_v|_{\mathscr{O}_v} = 1$ for almost all v. Functions in $S(\mathrm{A}_F)$ are integrable with respect to μ, and $\int_{\mathrm{A}_F} \otimes g_v\, \mu = \prod \int_{F_v} g_v\, \mu_v$ for $\otimes g_v \in S(\mathrm{A}_F)$.

Define the *local Fourier transforms* of $g_v \in S(F_v)$ as

$$\mathscr{F}_v(g_v)(\alpha_v) = \int_{F_v} g_v(\beta_v)\, \psi_v^0(\alpha_v\beta_v)\, \mu_v(\beta_v).$$

It gives a linear map $\mathscr{F}_v \colon S(F_v) \longrightarrow S(F_v)$.

Define the *adelic Fourier transforms* of $g \in S(\mathrm{A}_F)$ as

$$\mathscr{F}(g)(\alpha) = \int_{\mathrm{A}_F} g(\beta)\, \psi^0(\alpha\beta)\, \mu(\beta).$$

The adelic transform is well defined and it gives a linear map

$$\mathscr{F} \colon S(\mathrm{A}_F) \longrightarrow S(\mathrm{A}_F).$$

The definitions imply $\mathscr{F}(\otimes_v g_v) = \otimes \mathscr{F}_v(g_v) \in S(\mathrm{A}_F)$ for $\otimes_v g_v \in S(\mathrm{A}_F)$.

LEMMA. $\mathscr{F} \circ \mathscr{F}(g)(\alpha) = g(-\alpha)$ *for any* $g \in S(\mathrm{A}_F)$.

Proof. By general harmonic analysis there is a constant c such that $\mathscr{F} \circ \mathscr{F}(g)(\alpha) = cg(-\alpha)$ for all $g \in S(\mathrm{A}_F)$. To show that $c = 1$, it is sufficient to check the formula for at least one non-zero function.

DEFINITION. Choose the following functions in $S(F_v)$

$f_v(\alpha) = \exp(-\pi|\alpha|_v^2)$ when v is real,
$f_v(\alpha) = \exp(-2\pi|\alpha|_v)$ when v is complex,
$f_v = \mathrm{char}_{\mathscr{O}_v}$ when v is finite.

We easily deduce that $\mathscr{F}_v(f_v) = f_v$ for infinite v and

$$\mathscr{F}_v(f_v)(\alpha_v) = |\delta_v|_v^{1/2} f_v(\delta_v\alpha_v)$$

where $\delta_v \in F_v^\times$ is such that $|\delta_v|_v = |k(v)|^{-d_v}$.

Thus, these functions f_v are eigenfunctions of $\mathscr{F}_v$ with eigenvalue 1 for all v except finitely many finite v. Despite their very different look, the archimedean f_v

are similar to the non-archimedean f_v in the sense of being eigenfunctions of the corresponding Fourier transform.

Then for $f = \otimes f_v \in S(\mathbb{A}_F)$ we have $\mathscr{F}(f)(\alpha) = |\delta|^{1/2} f(\delta\alpha)$ where $\delta \in J_F$ has components δ_v at finite places and 1 at infinite places (in the number field case), so

$$|\delta| = \prod_v |k(v)|^{-d_v}.$$

If $g \in S(\mathbb{A}_F)$ then for every $\beta \in J_F$ the function $g_\beta : \alpha \mapsto g(\alpha\beta)$ belongs to $S(\mathbb{A}_F)$. We have

$$\begin{aligned}\mathscr{F}(g_\beta)(\alpha) &= \int_{\mathbb{A}_F} g(\beta\gamma)\psi^0(\alpha\gamma)\,\mu_{\mathbb{A}_F}(\gamma)\\ &= |\beta|^{-1}\int_{\mathbb{A}_F} f(\gamma')\psi^0(\gamma'\beta^{-1}\alpha)\,\mu_{\mathbb{A}_F}(\gamma') = |\beta|^{-1}\mathscr{F}(g)(\beta^{-1}\alpha),\end{aligned}$$

where $\gamma' = \gamma\beta$. Thus, $\mathscr{F}(g_\beta) = |\beta|^{-1}\mathscr{F}(g)_{\beta^{-1}}$.

For $\beta \in J_F$ with infinite components 1 we now deduce

$$\mathscr{F} f_\beta = |\delta|^{1/2}|\beta|^{-1} f_{\delta\beta^{-1}}.$$

Hence, $\mathscr{F}\circ\mathscr{F}(f)(\alpha) = |\delta|^{1/2}|\delta|^{1/2}|\delta|^{-1} f(\alpha) = f(-\alpha)$. □

Thus, the measures μ_v and μ are the self-dual measures with respect to the characters ψ_v^0 and ψ^0, i.e. such that the double Fourier transform of a function $g(x)$ is $g(-x)$.

REMARK. In characteristic zero it is easy to show that $|\delta| = |d_F|^{-1}$ where d_F is the discriminant of F. In positive characteristic, when F is the function field of a proper smooth geometrically irreducible curve of $\mathbb{F}_q$, 4.4 implies that the image of $\delta \in J_F$ with respect to $\rho\colon J_F \longrightarrow \mathrm{Div}(F)$ of 4.7 is a canonical divisor $\kappa = \sum d_v[v]$ of F and $|\delta| = q^{-\deg\kappa} = q^{2-2g}$ where q is the cardinality of the constant subfield of F and g is the genus of the curve $\mathscr{C}$. See Exercise 1.7.

5.5. The additive group F is a discrete locally compact group, its translation invariant measure is an atomic measure where each point have volume $c > 0$. Choose the measure μ_F which is the counting measure, i.e. $c = 1$. Since the measure on F is atomic counting, we have

$$\int_F g(a)\,\mu_F(a) = \sum_{a\in F} g(a).$$

Using harmonic analysis or even directly, see [**14**], define the measure $\mu_{\mathbb{A}_F/F}$ on $\mathbb{A}_F/F$ such that

$$\mu_{\mathbb{A}_F} = \mu_{\mathbb{A}_F/F}\otimes\mu_F,$$

i.e. for all $f \in S(\mathrm{A}_F)$ the equality

$$\int_{\mathrm{A}_F} f\,\mu_{\mathrm{A}_F} = \int_{\mathrm{A}_F/F} \left(\int_F f(\beta + a)\,\mu_F(a) \right) \mu_{\mathrm{A}_F/F}(\overline{\beta})$$

holds where $\overline{\beta} = \beta + F$.

Recall that the orthogonal complement of F with respect to ψ^0 is F and the group of characters of A_F/F is isomorphic to F, see Corollary 2 of 4.3. When applying inverse Fourier transform, one uses the dual measure on the group of characters. The following Proposition shows in particular that the measure $\mu_{\mathrm{A}_F/F}$ is dual to the counting measure μ_F.

PROPOSITION. *The volume of* A_F/F *with respect to* $\mu_{\mathrm{A}_F/F}$ *is 1, and* $\mu_{\mathrm{A}_F/F}$ *is dual to* μ_F.

Let $g \in S(\mathrm{A}_F)$ *and* $\beta \in J_F$. *Then*

$$\int_F g(a)\,\mu_F(a) = \int_F \mathscr{F}(g)(a)\,\mu_F(a),$$

which can be called generalised Gauß–Cauchy–Poisson summation formula.

For every $\beta \in \mathrm{A}_F^\times$ *we have*

$$\int_F g(\beta a)\,\mu_F(a) = |\beta|^{-1} \int_F \mathscr{F}(g)(\beta^{-1}a)\,\mu_F(a).$$

Proof. For $g \in S(\mathrm{A}_F)$ denote by $\hat{g}$ the function $\alpha \mapsto \int_F g(\alpha + a)\,\mu_F(a)$ on A_F/F.

Denote by $\mathscr{F}_{\mathrm{A}_F/F}$ the Fourier transform of functions on compact A_F/F using the character induced by ψ^0 and the property $\psi^0(F) = 1$. Then for $b \in F$

$$\begin{aligned} \mathscr{F}_{\mathrm{A}_F/F}(\hat{g})(b) &= \int_{\mathrm{A}_F/F} \hat{g}(\overline{\beta})\psi^0(b\beta)\,\mu_{\mathrm{A}_F/F}(\overline{\beta}) \\ &= \int_{\mathrm{A}_F/F}\int_F g(\beta + a)\,\mu_F(a)\psi^0(b\beta)\,\mu_{\mathrm{A}_F/F}(\overline{\beta}) \\ &= \int_{\mathrm{A}_F/F}\int_F g(\beta + a)\,\mu_F(a)\psi^0(b(\beta + a))\,\mu_{\mathrm{A}_F/F}(\overline{\beta}) \\ &= \int_{\mathrm{A}_F} g(\gamma)\psi^0(\gamma b)\,\mu_{\mathrm{A}_F}(\gamma) = \mathscr{F}(g)(b), \end{aligned}$$

where $a \in F$, $\gamma = \beta + a$. So the functions $\mathscr{F}_{\mathrm{A}_F/F}(\hat{g})$ and $\mathscr{F}(g)$ coincide on F.

Denote by m the volume of A_F/F with respect to $\mu_{\mathrm{A}_F/F}$. Applying the inverse Fourier transform to the function $\mathscr{F}_{\mathrm{A}_F/F}(\hat{g})$ on F, we obtain

$$\hat{g}(\overline{\beta}) = m^{-1} \int_F \mathscr{F}(g)(a)\,\overline{\psi^0(a\beta)}\,\mu_F(a).$$

Thus,

$$\int_F g(a)\,\mu_F(a) = \hat{g}(0) = m^{-1} \int_F \mathscr{F}(g)(a)\,\mu_F(a).$$

Since $g \in S(\mathbb{A}_F)$, all the computations are justified. Using Lemma 5.4, applying this formula to $\mathscr{F}(g)$, we deduce $m = 1$ and the generalised Gauß–Cauchy–Poisson formula.

The second formula follows from it and 5.4. □

REMARKS.

1. For $F = \mathbb{Q}$ the second formula of the Proposition for $g = f$ is the functional equation of the classical theta-function of real variable $\theta(x) = \sum_{n\in Z} \exp(-\pi n^2 x)$.

2. In positive characteristic the second formula of the Proposition implies another proof of the Riemann–Roch formula. Namely, for a divisor d of F, let $\beta \in J_F$ be such that the map ρ defined in 4.7 sends it to d. Then for the specific function f defined in 5.4, the last formula of the previous Proposition and the observation

$$|F \cap \mathbb{A}_F(d)| = \int_F f(\beta a)\, \mu_F(a)$$

together with Remark 5.4 imply the summation formula Riemann–Roch formula *Riemann–Roch formula*

$$\dim_{\mathbb{F}_q} H^0(\mathscr{A}_F(d)) = \deg_{\mathbb{F}_q} d + \chi_{\mathscr{A}_F}(0) + \dim_{\mathbb{F}_q} H^0(\mathscr{A}_F(\kappa - d)),$$

proved differently in 4.4.

5.6. For a character $\chi = \otimes \chi_v$ of J_F denote by χ_v the character $\chi \circ j_v$ of F_v. Then $\chi_v(U_v) = 1$ for almost all v by 5.2, and so $\chi = \otimes \chi_v$.

For $g = \otimes g_v \in S(\mathbb{A}_F)$ and a character $\chi = \otimes \chi_v$ of J_F we have

$$\int_{J_F} g\chi |\;|^s \mu_{J_F} = \prod \int_{F_v^\times} g_v \chi_v |\;|_v^s \mu_v^\times$$

when one of the sides converges.

We will use the counting measure $\mu_{F^\times}$ on the discrete group $F^\times$, so

$$\int_F g\, \mu_F = g(0) + \int_{F^\times} g\, \mu_{F^\times}.$$

DEFINITION. Define the translation invariant measure $\mu_{J_F/F^\times}$ such that

$$\mu_{J_F} = \mu_{J_F/F^\times} \otimes \mu_{F^\times}.$$

Hence for all $h = g\chi$ with $g \in S(\mathbb{A}_F)$, χ is a character of J_F that sends $F^\times$ to 1, the equality

$$\int_{J_F} h\, \mu_{J_F} = \int_{J_F/F^\times} \left(\int_{F^\times} h(\beta a)\, \mu_{F^\times}(a) \right) \mu_{J_F/F^\times}(\overline{\beta})$$

holds when the integral on one of the sides of the equation converges, $\overline{\beta} = \beta F^\times$.

Recall that $|J_F| = \mathbb{R}_{>0}^\times$ in the number field case and $|J_F| = q^{\mathbb{Z}}$ in the global function case when the constant field of F is $\mathbb{F}_q$ (see Corollary 3 of 4.9).

DEFINITION. Choose a subgroup M of J_F such that

$$J_F = M \times J_F^1.$$

Hence $M \simeq |J_F|$. Endow M with the standard multiplicative measure $\mu_{\mathbb{R}^\times} = \mu_{\mathbb{R}}/|\ |_{\mathbb{R}}$ of positive reals or with the counting discrete measure in positive characteristic case.

Define the translation invariant measure $\mu_{J_F^1}$ such that

$$\mu_{J_F} = \mu_{J_F^1} \otimes \mu_M.$$

This can be viewed as a generalisation of the use of spherical coordinates $\mathbb{R}^\times_{>0} \times S^1$.

Define the translation invariant measure $\mu_{J_F^1/F^\times}$ such that

$$\mu_{J_F^1} = \mu_{J_F^1/F^\times} \otimes \mu_{F^\times}.$$

Recall (see 4.8) that in the number field case, using the log map, $J_F^1/F^\times$ fits into an exact sequence

$$1 \longrightarrow \mu_F \longrightarrow J_F^1/F^\times \longrightarrow \mathbb{R}^{s-1}/T \longrightarrow 0,$$

where $s = r_1 + r_2$ and T is a lattice of $\mathbb{R}^{s-1}$. Using it, one can explicitly compute the volume $\mu_{J_F^1/F^\times}(J_F^1/F^\times)$, see Exercise 3.10.

DEFINITION. For $g \in S(\mathbb{A}_F)$, $s \in \mathbb{C}$ and a character χ of J_F that vanishes on $F^\times$, the *zeta integral* is

$$\zeta(g,s,\chi) = \int_{J_F} g(\alpha)|\alpha|^s\chi(\alpha)\,\mu_{J_F}(\alpha).$$

By the first paragraph of 4.3 and Proposition 4.7, the restriction of χ on J_F^1 gives a character of finite order of $J_F^1/F^\times$.

The zeta integral involves the factor g which has information about the integral structures of almost all completions of F and the multiplicative factor $\chi|\ |^s$.

There are two ways to compute it, and this will prove the equality of the results of the two computation.

The first computation.

We will use $J_F = \prod' F_v^\times$, compute local zeta integrals and take their product to get $\zeta(f,s,1)$, and then do the same for $\zeta(\mathcal{F}(f,)s,1)$. In order to handle the case of characters χ different from the trivial character, we modify the function f to a function f^χ, and again do local zeta integrals to get $\zeta(f^\chi,s,\chi)$ and $\zeta(\mathcal{F}(f^\chi),s,\chi)$.

Let's start with the case of $\chi = 1$ and let g be the function f defined in 5.4. Then

$$\zeta(f,s,1) = \prod_v \zeta_v(f_v,s,1), \quad \zeta_v(f_v,s,1) = \int_{F_v^\times} f_v(\alpha)|\alpha|_v^s \mu_{F_v^\times}(\alpha).$$

For example,

$$\begin{aligned}\zeta_v(\mathrm{char}_{\mathcal{O}_v},s,1) &= (1-|k(v)|^{-1})^{-1}\int_{\mathcal{O}_v} |\alpha|_v^{s-1}\,\mu_{F_v}(\alpha)\\ &= \sum_{n\geqslant 0} |k(v)|^{-ns} = (1-|k(v)|^{-s})^{-1},\end{aligned}$$

since $\mathcal{O}_v = (\mathcal{O}_v \setminus \mathcal{M}_v) \cup (\mathcal{M}_v \setminus \mathcal{M}_v^2) \cup \cdots$. Calculations, involving Gaussian integrals, immediately show that

$$\zeta_v(f_v,s,1) = \begin{cases} |k(v)|^{-d_v/2}(1-|k(v)|^{-s})^{-1} & \text{if } v \text{ is finite,}\\ \Gamma_{\mathbb{R}}(s) = \pi^{-s/2}\Gamma(s/2) & \text{if } v \text{ is real,}\\ \Gamma_{\mathbb{C}}(s) = (2\pi)^{1-s}\Gamma(s) & \text{if } v \text{ is complex,}\end{cases}$$

d_v was defined in 5.3. Recall that $\Gamma(s)$ is defined as $\int_0^\infty y^s \exp(-y)dy/y$ for $\Re e(s) > 0$, it has a meromorphic continuation to the complex plane, has no zeros there and has simple poles at non-positive integers. We have $\Gamma_{\mathbb{R}}(1) = \Gamma_{\mathbb{C}}(1) = 1$.

Thus,

$$\zeta(f,s,1) = \zeta_F(s)\zeta_{F,\infty}(s)\prod_v |k(v)|^{-d_v/2},$$

where $\zeta_{F,\infty}(s) = \Gamma_{\mathbb{R}}(s)^{r_1}\Gamma_{\mathbb{C}}(s)^{r_2}$ in the number field case and $\zeta_{F,\infty}(s) = 1$ in positive characteristic. Note that the function $\zeta_{F,\infty}(s)\prod_v |k(v)|^{-d_v/2}$ is a meromorphic function on the complex plane and it does not have zeros there. Using this equality on $\Re e(s) > 1$ where $\zeta_F(s)$ absolutely and uniformly converges, we deduce that the zeta integral $\zeta(f,s,1)$ is a holomorphic function on $\Re e(s) > 1$, and it absolutely and uniformly converges for $\Re e(s) > 1$.

In the case of $F = \mathbb{Q}$, we have

$$\zeta_{\mathbb{Q}}(f,s,1) = \zeta_{\mathbb{Q}}(s)\pi^{-s/2}\Gamma(s/2).$$

By 5.4 the local components of $\mathcal{F}(f)$ are equal to $|\delta_v|_v^{1/2} f_{v\delta_v}$, so this is f_v at all finite places where $d_v = 0$. Using 5.4, we obtain

$$\zeta_v(\mathcal{F}(f_v),s,1) = |k(v)|^{-d_v s}(1-|k(v)|^{-s})^{-1}$$

at finite places. Hence

$$\zeta(\mathcal{F}(f),s,1) = \zeta_F(s)\zeta_{F,\infty}(s)\prod_v |k(v)|^{-d_v s}.$$

Now let χ be a non-trivial character. Let V_χ be the finite set of all finite places v where $\chi(j_v(U_v)) \neq 1$, j_v is defined in 5.2. Denote $U_{0,F_v} = U_{F_v}$.

DEFINITION. For a finite v the kernel of $\chi \circ j_v$ is an open subgroup of U_v by the first paragraph of 4.3. Define the v-conductor $c_v = c_v(\chi)$ of χ as the smallest non-negative integer such that $\chi(j_v(U_{c_v,F_v})) = 1$. Thus, $v \in V_\chi$ if and only if $c_v(\chi) \neq 0$.

The definition in 5.2 shows that $\chi(v) = 0$ when $v \in V_\chi$. We also have

$$L_C(s,\chi) = L_{C\cup V_\chi}(s,\chi).$$

Note that $\zeta_v(f_v,s,\chi) = 0$ when $c_v > 0$, since the sum of the values of a non-trivial character of a finite group $U_v/U_{c_v,F_v}$ on all of its elements is 0. We will modify f_v at $v \in V_\chi$ to get non-zero local zeta integrals.

As a side remark which we will not use, for $g_1, g_2 \in S(F_v)$ and $0 < \Re e(s) < 1$ one can easily show that

$$\zeta_v(\mathscr{F}(g_1), 1-s, \chi^{-1})\zeta_v(g_2,s,\chi) = \zeta_v(\mathscr{F}(g_2), 1-s, \chi^{-1})\zeta_v(g_1,s,\chi),$$

hence the quotient $\zeta_v(\mathscr{F}(g), 1-s, \chi^{-1})/\zeta_v(g,s,\chi)$ when the denominator is non-zero does not depend on the choice of $g \in S(F_v)$.

If v is a real place, the composite character $\chi \circ j_v$ is a character of $\mathbb{R}^\times$ and so there is a uniquely determined number a which is 0 or 1, such that this character sends $\alpha \in \mathbb{R}^\times$ to $(\alpha/|\alpha|)^a$; define $\Gamma_\mathbb{R}(s,\chi) = \Gamma_\mathbb{R}(s+a)$. If v is complex, for the composite character $\chi \circ j_v$ of $\mathbb{C}^\times$ there is a uniquely determined number $n \in \mathbb{Z}$ such that this character sends $\alpha \in \mathbb{C}^\times$ to $(\alpha/|\alpha|)^n$, then define $\Gamma_\mathbb{C}(s,\chi) = \Gamma_\mathbb{C}(s+|n|/2)$.

Now define, following Tate's choice,

$$f^\chi = \otimes_v f_v^\chi, \quad f_v^\chi(\alpha) = \begin{cases} \alpha^a f_v(\alpha) & \text{if } v \text{ is real}, \\ \overline{\alpha}^n f_v(\alpha) & \text{if } v \text{ is complex and } n \geqslant 0, \\ \alpha^{-n} f_v(\alpha) & \text{if } v \text{ is complex and } n < 0, \\ f_v(\alpha) & \text{if finite } v \notin V_\chi, \\ \psi_v^0(\alpha)\,\mathrm{char}_{\mathscr{M}_v^{d_v-c_v}}(\alpha) & \text{if finite } v \in V_\chi. \end{cases}$$

Then $f^1 = f$. At finite places $f_v^\chi = f_v^{\chi^{-1}}$. One calculates

$$\zeta_v(f_v^\chi, s, \chi) = \begin{cases} \Gamma_\mathbb{R}(s,\chi) & \text{if } v \text{ is real}, \\ \Gamma_\mathbb{C}(s,\chi) & \text{if } v \text{ is complex}, \\ |k(v)|^{-d_v/2}(1-\chi(v)|k(v)|^{-s})^{-1} & \text{if finite } v \notin V_\chi, \\ |k(v)|^{(c_v+d_v)s} \times \text{non-zero constant} & \text{if finite } v \in V_\chi. \end{cases}$$

Note that $\zeta_v(f_v^\chi, s, \chi)$ has no complex zeros.
We have

$$\mathscr{F}(f_v^\chi)(\alpha) = \begin{cases} i^a f_v^\chi(\alpha) & \text{if } v \text{ is real,} \\ i^{|n|} f_v^{\chi^{-1}}(\alpha) & \text{if } v \text{ is complex,} \\ |\delta_v|_v^{1/2} f_v(\delta_v \alpha), & \text{if finite } v \notin V_\chi, \\ |k(v)|^{d_v/2+c_v} \mathrm{char}_{U_{c_v,F_v}} & \text{if finite } v \in V_\chi. \end{cases}$$

Then

$$\zeta_v(\mathscr{F}(f_v^\chi), s, \chi) = \begin{cases} i^a \Gamma_{\mathbb{R}}(s,\chi) & \text{if } v \text{ is real,} \\ i^{|n|} \Gamma_{\mathbb{C}}(s,\chi) & \text{if } v \text{ is complex,} \\ \chi(v)^{d_v} |k(v)|^{-d_v s}(1-\chi(v)|k(v)|^{-s})^{-1} & \text{if finite } v \notin V_\chi, \\ \text{non-zero constant} & \text{if } v \in V_\chi. \end{cases}$$

For a finite set of places C the function $L_C(s,\chi)$ is defined in 5.2. We obtain that for $\Re e(s) > 1$

$$\zeta(f^\chi, s, \chi) = L_C(s,\chi)\zeta_{F,\infty}(s,\chi) \prod_{v \in C \cup V_\chi} \zeta_v(f_v^\chi, s, \chi) \prod_{v \notin C \cup V_\chi} |k(v)|^{-d_v/2},$$

and

$$\begin{aligned} &\zeta(\mathscr{F}(f^\chi), s, \chi) \\ &= L_C(s,\chi) i^b \zeta_{F,\infty}(s,\chi) \prod_{v \in C \cup V_\chi} \zeta_v(\mathscr{F}(f_v^\chi), s, \chi) \prod_{v \notin C \cup V_\chi} \chi(v)^{d_v} |k(v)|^{-d_v s}, \end{aligned}$$

where in the number field case $\zeta_{F,\infty}(s,\chi) = \Gamma_{\mathbb{R}}(s,\chi)^{r_1} \Gamma_{\mathbb{C}}(s,\chi)^{r_2}$, integer b depends on the numbers a, n for real and complex places, and $\zeta_{F,\infty}(s,\chi) = 1$ in positive characteristic. The function $\zeta_{F,\infty}(s,\chi) \prod_{v \in C \cup V_\chi} \zeta_v(f_v^\chi, s, \chi)$ is a holomorphic function on $\Re e(s) > 0$, therefore the zeta integral $\zeta(f^\chi, s, \chi)$ is a holomorphic function on $\Re e(s) > 1$.

The second computation.

The second way to compute the zeta integral is to use the filtration $J_F > J_F^1 > F^\times$, like a generalisation of spherical coordinates. We use the equality of sets $F = F^\times \cup \{0\}$, to pass from the multiplicative situation to the additive situation and apply the summation formula of Proposition 5.5. This is a global computation.

For $m \in M$ denote

$$\zeta_m(g, s, \chi) = |m|^s \int_{J_F^1} g(m\gamma)\chi(m\gamma)\, \mu_{J_F^1}(\gamma).$$

Using Proposition 5.5 to pass from the third to the fourth line, we get

$$\begin{aligned}
&\zeta_m(g,s,\chi)+|m|^s g(0)\int_{C_F^1}\chi(m\gamma)\,\mu_{C_F^1}(\gamma)\\
&=|m|^s\int_{C_F^1}\chi(m\gamma)\int_{F^\times}g(m\gamma a)\,\mu_{F^\times}(a)\mu_{C_F^1}(\gamma)+|m|^s g(0)\int_{C_F^1}\chi(m\gamma)\,\mu_{C_F^1}(\gamma)\\
&=|m|^s\int_{C_F^1}\chi(m\gamma)\int_F g(m\gamma a)\,\mu_F(a)\,\mu_{C_F^1}(\gamma)\\
&=|m|^{s-1}\int_{C_F^1}\chi(m\gamma)\int_F\mathscr{F}(g)(m^{-1}\gamma^{-1}a)\,\mu_F(a)\,\mu_{C_F^1}(\gamma)\\
&=|m|^{s-1}\int_{C_F^1}\chi(m^{-1}\gamma)^{-1}\int_F\mathscr{F}(g)(m^{-1}\gamma a)\,\mu_F(a)\,\mu_{C_F^1}(\gamma)\\
&=\zeta_{m^{-1}}(\mathscr{F}(g),1-s,\chi^{-1})+|m|^{s-1}\mathscr{F}(g)(0)\int_{C_F^1}\chi^{-1}(m^{-1}\gamma)\,\mu_{C_F^1}(\gamma).
\end{aligned}$$

Thus,

$$\begin{aligned}
&\zeta_m(g,s,\chi)+|m|^s g(0)\int_{C_F^1}\chi(m\gamma)\,\mu_{C_F^1}(\gamma)\\
&\quad=\zeta_{m^{-1}}(\mathscr{F}(g),1-s,\chi^{-1})+|m|^{s-1}\mathscr{F}(g)(0)\int_{C_F^1}\chi^{-1}(m^{-1}\gamma)\,\mu_{C_F^1}(\gamma).
\end{aligned}$$

Now represent the measure space M as $M_-\cup M_+$ where M_-, M_+ correspond to $(0,1]$ and $[1,+\infty)$ with their measures in the number field case and M_-, M_+ correspond to $\{q^n : n<0\}\cup\{1\}$ and $\{q^n : n>0\}\cup\{1\}$ where q^n is given volume 1 when $n\neq 0$ and $\{1\}$ in both sets is given volume $1/2$. We have

$$\begin{aligned}
\zeta(g,s,\chi)&=\int_M\zeta_m(g,s,\chi)\,\mu_M(m)\\
&=\int_{M_-}\zeta_m(g,s,\chi)\,\mu_{M_-}(m)+\int_{M_+}\zeta_m(g,s,\chi)\,\mu_{M_+}(m).
\end{aligned}$$

Suppose that $g=f^\chi$. Then both integrals on the right hand side of the last equality converge for $\Re e(s)>1$. The second integral converges even better when $\Re e(s)$ gets smaller since $m\in M_+$, hence the second integral extends to an entire function $\xi(g,s,\chi)$ on the complex plane. For the first integral, using the previous computation for $\zeta_m(g,s,\chi)$, we get

$$\begin{aligned}
\int_{M_-}\zeta_m(g,s,\chi)\,\mu_{M_-}(m)&=\int_{M_-}\zeta_{m^{-1}}(\mathscr{F}(g),1-s,\chi^{-1})\,\mu_{M_-}(m)+\Delta(g,s,\chi)\\
&=\int_{M_+}\zeta_m(\mathscr{F}(g),1-s,\chi^{-1})\,\mu_{M_+}(m)+\Delta(g,s,\chi)\\
&=\xi(\mathscr{F}(g),1-s,\chi^{-1})+\Delta(g,s,\chi)
\end{aligned}$$

where

$\Delta(g,s,\chi) =$

$$\int_{M_-} (\mathscr{F}(g)(0)|m|^{s-1}\int_{C_F^1}\chi(m^{-1}\gamma)^{-1}\mu_{C_F^1}(\gamma) - g(0)|m|^s\int_{C_F^1}\chi(m\gamma)\mu_{C_F^1}(\gamma))\mu_{M_-}(m).$$

We have

$$\int_{M_-} |m|^s\,\mu_{M_-}(m) = \begin{cases} \dfrac{1}{s} & \text{when } \operatorname{char}(F)=0, \\ \dfrac{1}{1-q^{-s}} - \dfrac{1}{2} & \text{when } \operatorname{char}(F)>0. \end{cases}$$

When $\chi = 1$ we get $\int_{C_F^1}\chi(\gamma)\mu_{C_F^1}(\gamma) = \mu_{C_F^1}(C_F^1)$ and in characteristic zero

$$\zeta(g,s,1) = \xi(g,s,1) + \xi(\mathscr{F}(g),1-s,1) - \mu_{C_F^1}(C_F^1)\big(g(0)/s + \mathscr{F}(g)(0)/(1-s)\big),$$

and in positive characteristic

$$\begin{aligned}\zeta(g,s,1) &= \xi(g,s,1) + \xi(\mathscr{F}(g),1-s,1)\\ &\quad - \mu_{C_F^1}(C_F^1)\left(g(0)\frac{1+q^s}{2(1-q^s)} + \mathscr{F}(g)(0)\frac{1+q^{1-s}}{2(1-q^{1-s})}\right).\end{aligned}$$

If $\chi(C_F^1) \neq 1$ then since by 4.3 χ induces a non-trivial character of finite order of compact C_F^1, $\int_{C_F^1}\chi(\gamma)^{-1}\mu_{C_F^1}(\gamma)$ is zero. Therefore, in this case

$$\zeta(g,s,\chi) = \xi(g,s,\chi) + \xi(\mathscr{F}(g),1-s,\chi^{-1})$$

extends to an entire function on the complex plane. Since $\mathscr{F}\circ\mathscr{F}(g)(\alpha) = g(\alpha)$ for $g = f^{\chi}$, we get the functional equation for the zeta integral

$$\zeta(g,s,\chi) = \zeta(\mathscr{F}(g),1-s,\chi^{-1}).$$

When $\chi = 1$, $\zeta(g,s,1)$ extends to a meromorphic function on the complex plane. We have $\Delta(g,s,1) = \Delta(\mathscr{F}(g),1-s,1)$ and hence the functional equation for the zeta integral

$$\zeta(g,s,1) = \zeta(\mathscr{F}(g),1-s,1).$$

Taking $g = f$, so $f(0)$ and $\mathscr{F}(f)(0)$ are non-zero, we obtain that $\mu_{C_F^1}(C_F^1) < \infty$. We have $\mu_{C_F^1}(C_F^1) > 0$ since otherwise $\mu_{J_F^1} = 0$, $\mu_{J_F} = 0$ and $\zeta(f,s,1) = 0$, contradicting the first computation of the zeta integral. Therefore, the poles of $\zeta(f,s,1)$ are at $s = 0$ and $s = 1$ in characteristic zero and at s satisfying $q^s = 1$ and $q^{1-s} = 1$ in positive characteristic.

THEOREM. *The zeta integral $\zeta(f,s,1)$ extends to a meromorphic function on the complex plane and its only poles are at $s = 0$ and $s = 1$ in characteristic zero* *and at s such that $q^s = 1$ and $q^{1-s} = 1$ in positive characteristic. It satisfies the functional equation*

$$\zeta(f,s,1) = \zeta(\mathscr{F}(f),1-s,1).$$

For a character χ of J_F such that $\chi(J_F) \neq 1 = \chi(F^\times)$, the zeta integral $\zeta(f^\chi,s,\chi)$ extends to an entire function on the complex plane and satisfies the functional equation

$$\zeta(f^\chi,s,\chi) = \zeta(\mathscr{F}(f^\chi),1-s,\chi^{-1}).$$

Therefore, the zeta function $\zeta_F(s)$ extends to a meromorphic function on the complex plane, with the only poles at $s=0$ and $s=1$ in characteristic zero and at s such that $q^s = 1$ and $q^{1-s} = 1$ in positive characteristic.

Define $\widehat{\zeta}_F(s)$ as $(\pi^{-s/2}\Gamma(s/2))^{r_1}((2\pi)^{1-s}\Gamma(s))^{r_2}\zeta_F(s)$ in characteristic zero and as $\zeta_F(s)$ in positive characteristic. Then $\widehat{\zeta}_F(s) = \zeta(f,s,1)|\delta|^{-1/2}$.

It satisfies the functional equation

$$\widehat{\zeta}_F(s) = |\delta|^{-1/2+s}\widehat{\zeta}_F(1-s),$$

i.e. using Remark 5.4

$$\widehat{\zeta}_F(s) = |d_F|^{1/2-s}\widehat{\zeta}_F(1-s) \text{ in characteristic zero,}$$
$$\zeta_F(s) = (q^{2g-2})^{1/2-s}\zeta_F(1-s) \text{ in positive characteristic.}$$

If $\chi \neq 1$, for a finite set C of finite places the function $L_C(s,\chi)$ extends to an entire function on the complex plane and it satisfies the functional equation relating $L_C(s,\chi)$ and $L_C(1-s,\chi^{-1})$.

Proof. Use the computations above.

From the comparison of the entire function $\zeta(f^\chi,s,\chi)$ and $L_C(s,\chi)$ and the fact that the function $\zeta_{F,\infty}(s,\chi)\prod_{v\in C\cup V_\chi}\zeta_v(f_v^\chi,s,\chi)$ has no complex zeroes, we obtain that $L_C(s,\chi)$ extends to an entire function on $\mathbb{C}$. The functional equation for $L_C(s,\chi)$ follows from the two displayed lines in the last paragraph of the first computation. □

For the computation of the residue of the zeta function at $s=1$ see Exercise 3.10.

COROLLARY. *For a finite abelian extension L/F the group $J_F/(F^\times N_{L/F}J_L)$ is finite by Corollary 2 of (4.7). Let χ be a character of J_F, trivial on $F^\times$, that induces a non-trivial character of the finite group $J_F/(F^\times N_{L/F}J_L)$. Then for a finite set C of finite places the function $L_C(s,\chi)$ extends to an entire function on the complex plane and in particular the order of its zero at $s=1$ is non-negative.*

REMARKS.

1. The proof the Theorem uses local compactness of the additive and multiplicative groups of completions of a global fields and 4.1–4.3. It does not use any other non-trivial results from Chapter 2 or from Sections 1–4 of this chapter.

2. The computation of the zeta integral in the proof of the Theorem proves that $\mu_{C_F^1}(C_F^1) < \infty$. Since every locally compact abelian group of finite measure is compact, we deduce that C_F^1 is compact. This proof is different from the proof in 4.7 and in Chapter 1.

3. There are classical analytic ways without involving zeta integrals to prove the statement of the Theorem about the zeta and L-functions. Hecke's classical proof of the functional equation of the L-functions of number fields was historically the first. The proof included in this section, due to Iwasawa and Tate, [**19**], [**34**], derives the functional equation of the zeta function and its twist by a character using the adelic zeta integral, self-duality of adeles and how global elements sit there, the Fourier transform of functions on adeles and the right mixture of the additive and multiplicative structures and topological properties.

4. Generalisations of the zeta integral play key roles in two generalisations of class field theory: the Langlands program and in higher zeta integrals theory, see [**9**]; while the second computation of the archimedean local zeta integral, ie. Gaussian integral, has certain similarities to aspects of the IUT theory.

5. For further properties of $\zeta_{\mathbb{Q}}$ and the Prime Number Theorem see Exercise 3.15.

5.7. Now let's look at the use of L-function to extend Corollary 1 of 4.8 and complete a proof of the class field theory axioms A1, A2 for global fields,

THEOREM. *The index of $N_{L/F}C_L$ in C_F for a global field F and a cyclic extension L/F of prime degree does not exceed the degree of the extension. Hence, in view of Corollary* 1 *of* (4.8),

$$|C_F : N_{L/F}C_L| = |L : F|, \quad \ker N_{L/F} = C_L^{1-\sigma}.$$

Proof. Denote by C the set of all finite places v for which $e(w|v) > 1$ in L/F. Due to to the last displayed formula in 4.1 for all such v we have $e(w|v) = |L : F|$ since the latter is a prime number. The set C is finite due to Proposition 4.6. So finite $v \notin C$ are unramified in L/F.

Denote $r = |J_F : F^\times N_{L/F}J_L|$, it is finite and divisible by $n = |L : F|$, due to Corollary 1 of 4.8. Moreover, every elements of $J_F/(F^\times N_{L/F}J_L)$ is of order dividing n, since $J_F^n \subset N_{L/F}J_F \subset N_{L/F}J_L$. Hence, every character of $J_F/(F^\times N_{L/F}J_L)$ different from 1 has order n.

A non-trivial character χ of the finite abelian group $J_F/(F^\times N_{L/F}J_L)$ gives a character of J_F, using the quotient homomorphism $J_F \longrightarrow J_F/(F^\times N_{L/F}J_L)$, denote it also by χ. Denote by $n(\chi)$ the order of zero of $L_C(s,\chi)$ at $s = 1$. By Corollary of 5.6, $n(1) = -1$ and $n(\chi) \geqslant 0$ for characters χ different from the trivial character 1.

For number fields, by 5.2 and 5.1, we have for real $s > 1$

$$\sum_v \chi(v)|k(v)|^{-s} \sim \sum_{v\notin C} \chi(v)|k(v)|^{-s} \sim \log L_C(s,\chi),$$

but in order to handle the general case of global fields, we will not use this property.

By 5.2 and 5.1, we have for real $s > 1$

$$\sum_{m\geqslant 1}\sum_{v\notin C} m^{-1}\chi(v)^m|k(v)|^{-ms} = \log L_C(s,\chi) \sim n(\chi)\log(s-1).$$

Hence, summing over all characters of the finite abelian group $J_F/(F^\times N_{L/F}J_L)$, we get

$$\sum_\chi\sum_{m\geqslant 1}\sum_{v\notin C} m^{-1}\chi(v)^m|k(v)|^{-ms} = \sum_\chi \log L_C(s,\chi) \sim -(1-\sum_{\chi\neq 1} n(\chi))\log(s-1).$$

We have

$$\sum_{m\geqslant 1}\sum_{v\notin C} m^{-1}\chi(v)^m|k(v)|^{-ms} \sim \sum_{n\nmid m}\sum_{v\notin C} m^{-1}\chi(v)^m|k(v)|^{-ms}, \qquad (*)$$

since

$$\sum_{n|m}\sum_{v\notin C} m^{-1}\chi(v)^m|k(v)|^{-ms} = n^{-1}\sum_{m\geqslant 1}\sum_{v\notin C} m^{-1}|k(v)|^{-mns} = n^{-1}\log L_C(ns,1) \sim 0,$$

as $\log L_C(ns,1)$ has no singularity at $s = 1$ by 5.6.

Letting α run through all elements of the finite abelian group $J_F/(F^\times N_{L/F}J_L)$ we get

$$\sum_v \chi(v)|k(v)|^{-s} = \sum_\alpha \chi(\alpha) \sum_{v:\overline{j_v(\pi_v)}=\alpha} |k(v)|^{-s}$$

where $\overline{j_v(\pi_v)}$ is the coset with representative $j_v(\pi_v)$.

When $\chi \neq 1$ and $n \nmid m$, the character χ^m is of order n. Summing over all characters χ we obtain

$$\sum_{\chi}\sum_{n\nmid m}\sum_{v\notin C} m^{-1}\chi(v)^m|k(v)|^{-ms} = \sum_{\chi}\sum_{n\nmid m}\sum_{\alpha}\chi^m(\alpha)\sum_{v\notin C,\overline{j_v(\pi_v)}=\alpha} m^{-1}|k(v)|^{-ms}$$
$$= r\sum_{n\nmid m}\sum_{v\notin C,\overline{j_v(\pi_v)}=1} m^{-1}|k(v)|^{-ms},$$

since for a character χ^m of order n, the sum $\sum_{\chi}\chi^m(\alpha)$ equals zero if α is different from the identity of the quotient group and equals its order r otherwise.

Denote by $S(L/F)$ the set of finite places of F which completely split in L/F. These are exactly the places outside C with $f(w|v) = 1$. For each of them there are n places w of L over v. Since $j_v(F_v^{\times}) \subset N_{L/F}J_L$, i.e. $\overline{j_v(\pi_v)} = 1$, for real $s > 1$ we get

$$\sum_{n\nmid m}\sum_{v\notin C,\overline{j_v(\pi_v)}=1} m^{-1}|k(v)|^{-ms} \geqslant \sum_{n\nmid m}\sum_{v\in S(L/F)} m^{-1}|k(v)|^{-ms}$$
$$= n^{-1}\sum_{n\nmid m}\sum_{w:f(w|v)=1,v\notin C} m^{-1}|k(w)|^{-ms}.$$

Using (*) we obtain

$$\sum_{n\nmid m}\sum_{w:f(w|v)>1,v\notin C} m^{-1}|k(w)|^{-ms} \leqslant \sum_{n\nmid m}\sum_{v\notin C} m^{-1}|k(v)|^{-2ms}$$
$$\sim \sum_{m\geqslant 1}\sum_{v\notin C} m^{-1}|k(v)|^{-2ms} = \log L_C(2s,1) \sim 0.$$

Using (*) again, we deduce

$$\sum_{n\nmid m}\sum_{w:f(w|v)=1,v\notin C} m^{-1}|k(w)|^{-ms} \sim \sum_{m\geqslant 1}\sum_{w\notin C'} m^{-1}|k(w)|^{-ms}$$
$$= \log L_{C'}(s,1) \sim -\log(s-1)$$

where C' consists of places w of L over places $v \in C$.

We conclude

$$1 - \sum_{\chi\neq 1} n(\chi) \geqslant \frac{r}{n}.$$

Thus, $r = n$ and $n(\chi) = 0$ for every non-trivial character χ. □

REMARKS.

1. The method of working with L-series and using the properties of the summation over all characters of a finite abelian group and over all elements of the group is a long tradition starting from Dirichlet's proof of his theorem about primes in arithmetic progressions. See Exercises 3.13 and 3.14.

2. A purely algebraic proof (by Chevalley) of the first statement of the Theorem can be obtained using Kummer theory in characteristic zero (see [**1**] or [**30**]) and Artin–Schreier–Witt theory in positive characteristic p for Galois extensions of degree p^n, so without using L-functions. The proof above, essentially due to Weber, but in adelic language, is an example of an L-functions technique for establishing various theorems in algebraic number theory.

6. Global Class Field Theory

DEFINITION. For the field of real numbers define the reciprocity map

$$\Psi_{\mathbb{R}}\colon \mathbb{R}^\times \longrightarrow \mathrm{Gal}(\mathbb{C}/\mathbb{R})$$

as $r \mapsto \tau^{(1-\mathrm{sign}(r))/2}$ where $\tau \in \mathrm{Gal}(\mathbb{C}/\mathbb{R})$ is the complex conjugation. It induces an isomorphism $\Psi_{\mathbb{C}/\mathbb{R}}\colon \mathbb{R}^\times/N_{\mathbb{C}/\mathbb{R}}\mathbb{C}^\times \longrightarrow \mathrm{Gal}(\mathbb{C}/\mathbb{R})$. We can denote the inverse to it by $\Upsilon_{\mathbb{C}/\mathbb{R}}$. Of course, we can identify $\mathrm{Gal}(\mathbb{C}/\mathbb{R}) \xrightarrow{\sim} \mathbb{Z}/2\mathbb{Z}$ with the group $\{\pm 1\}$.

For the field of complex numbers define the reciprocity map

$$\Psi_{\mathbb{C}}\colon \mathbb{C}^\times \longrightarrow \mathrm{Gal}(\mathbb{C}/\mathbb{C}) = \{1\}$$

as the map which sends everything to 1.

For a trivial extension of archimedean completions define Υ is the trivial map sending 1 to 1.

Even though we do not have profinite extensions of archimedean completions with Galois groups isomorphic to $\widehat{\mathbb{Z}}$ and hence we do not have Frobenius elements in the sense of abstract class field theory in 2.1 and no hence analog of the map Υ of Section 2 defined using Frobenius elements, one checks immediately that for infinite places and the reciprocity maps defined above we have analogs of the first commutative diagram in 2.5 and commutative diagrams of Theorem 2.9, using the fact that the Galois groups involved are either trivial or $\mathrm{Gal}(\mathbb{C}/\mathbb{R})$. In particular, if $E/L/F$, $E/M/F$ are finite extension of archimedean completions, then $\Psi_{E/M}(\beta)|_L = \Psi_{L/F}(N_{M/F}(\beta))$ for $\beta \in M^\times$.

6.1. For abelian extensions the decomposition group $\mathrm{Gal}(L/F)_w$ of a place w of L over a place v of F depends on v only, due to the equality $\mathrm{Gal}(L/F)_w = \sigma^{-1}\,\mathrm{Gal}(L/F)_w\sigma = \mathrm{Gal}(L/F)_{\sigma w}$. Keeping in mind 4.1, for abelian L/F we will denote $\mathrm{Gal}(L/F)_v := \mathrm{Gal}(L/F)_w$, $L_v := L_w$, $i_v := i_w$

$$i_v\colon \mathrm{Gal}(L_v/F_v) \longrightarrow \mathrm{Gal}(L/F), \quad i_v(\mathrm{Gal}(L_v/F_v)) = \mathrm{Gal}(L/F)_v.$$

DEFINITION. Let F be a global field. Using the local reciprocity maps for all completions of F_v, define for a finite abelian extension L/F the homomorphism

$$\Phi_{L/F}\colon J_F \longrightarrow \mathrm{Gal}(L/F), \quad \Phi_{L/F}(\alpha) = \prod_v i_v \circ \Psi_{L_v/F_v}(\alpha_v)$$

where v runs through all places of F, $\Psi_{L_v/F_v}\colon F_v^\times \longrightarrow \mathrm{Gal}(L_v/F_v)$ is the local reciprocity map. The product is well defined, since for almost all v the element $\alpha_v \in U_{F_v}$ and the extension L_v/F_v is unramified by Proposition 4.6, so $\Psi_{L_v/F_v}(\alpha_v) = 1$. Note that in this definition only one local field L_v for a place v of F is used.

PROPOSITION. *Let $M/F, E/L$ be finite separable extensions of global fields and L/F and E/M be finite abelian extensions. Then the diagram*

$$\begin{array}{ccc} J_M & \xrightarrow{\Phi_{E/M}} & \mathrm{Gal}(E/M) \\ {\scriptstyle N_{M/F}}\downarrow & & \downarrow \\ J_F & \xrightarrow{\Phi_{L/F}} & \mathrm{Gal}(L/F) \end{array}$$

is commutative, where the right vertical map is the restriction of Galois automorphisms and the left vertical map is the norm map $N_{M/F}$.

Proof. For an idele (β_w) of J_M and $w|v$ for a place v of F we know from Theorem 2.9, Section 3 and the Definition preceding 6.1 that

$$\Psi_{E_w/M_w}(\beta_w)|_{L_v} = \Psi_{L_v/F_v}(N_{M_w/F_v}(\beta_w)), \quad w|v.$$

Since $N_{M/F}((\beta_w))_v = \prod_{w|v} N_{M_w/F_v}(\beta_w)$ by 4.2, we get

$$\begin{aligned} \Phi_{L/F}(N_{M/F}((\beta_w))) &= \prod_v i_v \circ \Psi_{L_v/F_v}(N_{M/F}(\beta_w)_v) \\ &= \prod_v \prod_{w|v} i_v \circ \Psi_{L_v/F_v}(N_{M_w/F_v}(\beta_w)) = \prod_v \prod_{w|v} i_v \circ \Psi_{E_w/M_w}(\beta_w)|_{L_v} \\ &= \Phi_{E/M}((\beta_w))|_L. \end{aligned}$$

□

DEFINITION. For an infinite abelian extension L/F define

$$\Phi_{L/F}\colon J_F \longrightarrow \mathrm{Gal}(L/F)$$

as the inverse limit of $\Phi_{R/F}(\alpha)$ for finite subextensions R/F of L/F, using the previous Proposition for $M = F$.

DEFINITION. For an infinite abelian extension L/F and a place v of F define L_v as the direct limit with respect to inclusions of completions R_w for all finite subextensions R/F of L/F, where w is an extension of v chosen in compatible way in all finite subextensions of L/F.

COROLLARY. *The equality* $\Phi_{L/F}(\alpha) = \prod_v i_v \circ \Psi_{L_v/F_v}(\alpha_v)$ *remains valid for infinite abelian extensions* L/F. *The previous Proposition remains true for infinite* L/F *and* E/M.

Proof. The product $\prod_v i_v \circ \Psi_{L_v/F_v}(\alpha_v)$ converges to $\Phi_{L/F}(\alpha)$ in $\mathrm{Gal}(L/F)$. Indeed, for a finite subextension R/F of L/F let $\sigma_R = \prod_v i_v \circ \Psi_{R_v/F_v}(\alpha_v)$. By the previous Proposition for any finite subextension S/F of R/F, $\sigma_R|_S = \prod_v i_v \circ \Psi_{S_v/F_v}(\alpha_v) = \Phi_{S/F}(\alpha) = \sigma_S$, so $\{\sigma_R\}_R$ converge to $\Phi_{L/F}(\alpha)$ in the profinite topology of $\mathrm{Gal}(L/F)$. The second assertion of the Corollary follows immediately. □

6.2. In characteristic zero, the maximal cyclotomic extension $\mathbb{Q}^{\mathrm{cycl}}$ is the composite of all finite cyclotomic extensions $\mathbb{Q}(\zeta_m)$ of $\mathbb{Q}$, and

$$\mathrm{Gal}(\mathbb{Q}^{\mathrm{cycl}}/\mathbb{Q}) \simeq \varprojlim (\mathbb{Z}/n\mathbb{Z})^\times \simeq \widehat{\mathbb{Z}}^\times.$$

We have $\widehat{\mathbb{Z}}^\times = \prod_p \mathbb{Z}_p^\times$ and from the description of the units of local number fields we know that $\mathbb{Z}_p^\times \simeq \mathbb{Z}/(p-1)\mathbb{Z} \times \mathbb{Z}_p$ for odd prime p and $\mathbb{Z}_2^\times \simeq \mathbb{Z}/2\mathbb{Z} \times \mathbb{Z}_2$. Hence

$$\widehat{\mathbb{Z}}^\times \simeq T \times \widehat{\mathbb{Z}}, \quad T = \mathbb{Z}/2\mathbb{Z} \times \prod_{p>2} \mathbb{Z}/(p-1)\mathbb{Z}.$$

Since $\widehat{\mathbb{Z}}$ has no non-trivial torsion, the torsion subgroup of $\mathrm{Gal}(\mathbb{Q}^{\mathrm{cycl}}/\mathbb{Q})$ coincides with the torsion subgroup of T. The latter contains $\mathbb{Z}/2\mathbb{Z} \oplus \oplus_{p>2}\mathbb{Z}/(p-1)\mathbb{Z}$ whose closure in $\widehat{\mathbb{Z}}^\times$ coincides with T.

DEFINITION. For $k = \mathbb{Q}$ denote by $\tilde{k}$ the fixed field of T, it is a $\widehat{\mathbb{Z}}$-extension of k.

In positive characteristic, the field $k = \mathbb{F}_p(t)$ has the $\widehat{\mathbb{Z}}$-extension $\tilde{k} = \mathbb{F}_p^{\mathrm{sep}}(t)$.

We get the surjective homomorphism

$$\deg \colon G_k \longrightarrow \mathrm{Gal}(\tilde{k}/k) \longrightarrow \widehat{\mathbb{Z}}.$$

For every finite separable extension F of k we get, similar to Section 2, the surjective homomorphism

$$\deg_F = f_F^{-1} \deg \colon G_F \longrightarrow \mathrm{Gal}(\widetilde{F}/F) \longrightarrow \widehat{\mathbb{Z}},$$

where $f_F = |F \cap \tilde{k} : k|$, $\widetilde{F} = F\tilde{k}$. It is continuous, since the restriction of Galois automorphisms is continuous.

LEMMA. *Let* l *be a prime number. Denote by* A_l *the subextension of* $\widetilde{\mathbb{Q}}/\mathbb{Q}$ *with the Galois group* $\mathbb{Z}_l$. *Then* $\widetilde{\mathbb{Q}} = \prod_l A_l$. *For a finite extension* K *of* $\mathbb{Q}$ *let* $\breve{K} = KA_l$ *be the* $\mathbb{Z}_l$*-subextension of* $\widetilde{K}/K$. *Then for every finite extension* E *of* $\mathbb{Q}_p$ *containing* K, *the image of* $\mathrm{Gal}(E\breve{K}/E) \simeq \mathrm{Gal}(\breve{K}/E \cap \breve{K})$ *in* $\mathrm{Gal}(\breve{K}/K)$ *is a non-trivial open subgroup of the latter and the intersection* $E \cap \breve{K}$ *is of finite degree over* K.

Proof. Put $l' = l$ if l is odd and $l' = 4$ if $l = 2$. The field A_l is linearly disjoint with $\mathbb{Q}(\zeta_{l'})$ and their composite is the maximal l-cyclotomic extension $\mathbb{Q}(\zeta_{l^\infty})$ of $\mathbb{Q}$. Since the finite extension $E(\zeta_{l'})$ of E does not include $E(\zeta_{l^\infty})$, the extension $E\check{K}/E$ is non-trivial. Hence the image of $\mathrm{Gal}(E\check{K}/E)$ in $\mathrm{Gal}(\check{K}/K)$ is a subgroup of finite index. □

We denote the element of $\mathrm{Gal}(\widetilde{F}/F)$ that is sent by $\deg_F$ to $1 \in \widehat{\mathbb{Z}}$ as φ_F. This is the Frobenius element in abstract class field theory in the sense of 2.1, but we will not use this name in the case of global fields in order not to confuse it with the Frobenius automorphisms of completions of global fields.

THEOREM. *For a global field F let*

$$w_F = \deg_F \circ \Phi_{\widetilde{F}/F} \colon J_F \longrightarrow \widehat{\mathbb{Z}}.$$

Then $w_F(F^\times) = 1$. The homomorphism w_F induces the continuous homomorphism

$$v_F \colon C_F \longrightarrow \widehat{\mathbb{Z}}.$$

Proof. Since $\Phi_{\widetilde{F}/F}(\alpha) = \Phi_{\widetilde{k}/k}(N_{F/k}(\alpha))$ by Corollary of Proposition 6.1, it is sufficient to prove the statement for $k = \mathbb{Q}$ and $k = \mathbb{F}_p(t)$.

In characteristic zero, it suffices to show that $\Phi_{\mathbb{Q}(\zeta)/\mathbb{Q}}(a) = 1$ for every root ζ and $a \in \mathbb{Q}^\times$. If ζ_1, ζ_2 are roots of orders m_1, m_2 and $(m_1, m_2) = 1$, then $\zeta = \zeta_1\zeta_2$ is of order m_1m_2. Using Proposition 6.1 we deduce $\Phi_{\mathbb{Q}(\zeta)/\mathbb{Q}}(a)|_{\mathbb{Q}(\zeta_2)} = \Phi_{\mathbb{Q}(\zeta_2)/\mathbb{Q}}(a)$ and $(\zeta^{\Phi_{\mathbb{Q}(\zeta)/\mathbb{Q}}(a)-1})^{m_1} = (\zeta_2^{\Phi_{\mathbb{Q}(\zeta_2)/\mathbb{Q}}(a)-1})^{m_1}$.

Hence it is sufficient to show $\zeta^{\Phi_{\mathbb{Q}(\zeta)/\mathbb{Q}}(a)-1} = 1$ for every root ζ of order $l^n > 2$, l a prime number. When l is different from a prime p, the extension $\mathbb{Q}_p(\zeta)/\mathbb{Q}_p$ is unramified. Therefore we obtain $\Psi_{\mathbb{Q}_p(\zeta)/\mathbb{Q}_p}(a)(\zeta) = \zeta^{p^{v_p(a)}}$ by Remark 18.2 of Chapter 2. When $p = l$ then by Proposition 3.3 we know $\Psi_{\mathbb{Q}_p(\zeta)/\mathbb{Q}_p}(a)(\zeta) = \zeta^{u^{-1}}$ where $a = l^{v_l(a)}u$ with $u \in \mathbb{Z}_l^\times$. When v is infinite then $\mathbb{R}(\zeta) = \mathbb{C}$ and $\Psi_{\mathbb{R}(\zeta)/\mathbb{R}}(a)(\zeta) = \zeta^{\mathrm{sign}(a)}$. Since $u = \mathrm{sign}(a)\prod_{p \neq l} p^{v_p(a)}$, we deduce $\Phi_{\mathbb{Q}(\zeta)/\mathbb{Q}}(a) = 1$.

In positive characteristic p, for a root ζ of order prime to p and $a \in k^\times$, $k_v(\zeta)/k_v$ is unramified for all places v of k and $\Psi_{k_v(\zeta)/k_v}(a)(\zeta) = \zeta^{|k(v)|^{v(a)}}$. Since $1 = |a^{-1}| = \prod_v |k(v)|^{v(a)}$, we obtain $\Phi_{k(\zeta)/k}(a) = 1$. Hence $\Phi_{\widetilde{k}/k}(a) = 1$

Thus, $\Phi_{\widetilde{F}/F}$ induces the homomorphism $C_F \longrightarrow \mathrm{Gal}(\widetilde{F}/F)$ and we have the homomorphism $v_F \colon C_F \longrightarrow \widehat{\mathbb{Z}}$.

The map $\Phi_{\widetilde{F}/F}$ is continuous, since the preimage of $\mathrm{Gal}(\widetilde{F}/L)$ for a finite subextension L/F of $\widetilde{F}/F$ contains $N_{L/F}C_L$, since it contains $F^\times$ and the image of the norms of L_v/F_v for places v of F by 4.2, so it also contains $N_{L/F}J_L$. The group $N_{L/F}C_L$ is an open subgroup in C_F by Corollary 4.6. □

REMARK. In positive characteristic v_F has a simple description. Denote by k_F the finite field of constants of F. Note that the restriction of the local Frobenius automorphism of F_v^{ur}/F_v on $\widetilde{F} = F \otimes_{k_F} k_F^{\mathrm{sep}}$ is $\varphi_F^{|k(v):k_F|}$ and by local class field theory $\Psi_{F_v^{\mathrm{ur}}/F_v}(\alpha_v) = \varphi_{F_v}^{v(\alpha_v)}$. Hence $\Phi_{\widetilde{F}/F}(\alpha) = \varphi_F^{\sum_v v(\alpha_v)|k(v):k_F|} = \varphi_F^{-\log_{|k_F|}|\alpha|}$ and we obtain

$$v_F(\alpha) = -\log_{|k_F|}|\alpha|.$$

In particular, $\Phi_{\widetilde{F}/F}(\alpha) = 1$ if and only if $\alpha \in J_F^1$.

PROPOSITION. *In characteristic zero $v_F(C_F) = \widehat{\mathbb{Z}}$. In positive characteristic $v_F(C_F)$ is isomorphic to the group $\mathbb{Z}$. For every finite separable extension L/F we have*

$$v_F(N_{L/F}C_L) = |L \cap \widetilde{F} : F|^{-1} v_L(C_L).$$

Proof. To prove the first assertion, note that for every finite subextension L/F of $\widetilde{F}/F$ the image $\Phi_{L/F}(J_F)$ contains all the decomposition groups $\mathrm{Gal}(L/F)_v = i_v(\mathrm{Gal}(L_v/F_v))$ for all all places v of F, since $\Phi_{L_v/F_v}(F_v^\times) = \mathrm{Gal}(L_v/F_v)$. Denote by M the fixed field of $\Phi_{L/F}(J_F)$, then $M_v = F_v$ for all places v of F. By Corollary 2 of 4.9 we deduce $M = F$. Thus, by finite Galois theory $\Phi_{\widetilde{F}/F}(J_F)|_L = \mathrm{Gal}(L/F)$ for every finite subextension L/F of $\widetilde{F}/F$. Therefore, the image $\Phi_{\widetilde{F}/F}(C_F)$ is dense in $\mathrm{Gal}(\widetilde{F}/F)$.

Note that infinitely divisible elements of a group go to the identity element of a profinite group with respect to any homomorphism from the former to the latter.

In characteristic zero $C_F/C_F^1 \simeq \mathbb{R}_{>0}^\times$ which is an infinitely divisible group, hence $\Phi_{\widetilde{F}/F}(C_F) = \Phi_{\widetilde{F}/F}(C_F^1)$. Since C_F^1 is compact and $\Phi_{\widetilde{F}/F}$ is continuous, $\Phi_{\widetilde{F}/F}(C_F^1)$ is closed. Thus, $\Phi_{\widetilde{F}/F}(C_F) = \mathrm{Gal}(\widetilde{F}/F)$.

In positive characteristic, for every F_v the image $\Psi_{F_v}(F_v^\times)$ restricted on $\widetilde{F} = F\mathbb{F}_q^{\mathrm{sep}}$ is an infinite cyclic subgroup of the infinite cyclic subgroup generated by φ_F, hence $v_F(C_F) \simeq \mathbb{Z}$.

Finally, using Theorem, Corollary of Proposition 6.1 and the property of deg, we deduce

$$\begin{aligned} v_F(N_{L/F}C_L) &= \deg_F \circ \Phi_{\widetilde{F}/F}(N_{L/F}J_L) = \deg_F \circ \Phi_{\widetilde{L}/L}(J_L)|_{\widetilde{F}} \\ &= |L \cap \widetilde{F} : F|^{-1} \deg_L \circ \Phi_{\widetilde{L}/L}(J_L) = |L \cap \widetilde{F} : F|^{-1} v_L(C_L). \end{aligned}$$

□

6.3. The map $\deg_k\colon G_k \longrightarrow \widehat{\mathbb{Z}}$ for condition C1 of 2.1 of abstract class field theory is the surjective homomorphism $\deg_k\colon G_k \longrightarrow \mathrm{Gal}(\tilde{k}/k) \simeq \widehat{\mathbb{Z}}$.

DEFINITION. Put $A = \varinjlim C_E$ where E runs through all finite separable extensions of k, with the inductive limit topology of the topologies of C_E. This is a G_k-module. Then $A_F = C_F$ by Lemma 4.6. Since every element of A belongs to some finite separable extension F/k, $A = \varinjlim A_{G_F}$.

The map $v = v_k\colon A_k \longrightarrow \mathbb{Z}$ is defined in the Theorem and Proposition 6.2. The A_k and v satisfy condition C2 of 2.4 of abstract class field theory, as established in Proposition 6.2.

Properties A1 and A2 of 2.7, i.e. for cyclic extensions L/F of prime degree the kernel of the norm map $N_{L/F}\colon C_L \longrightarrow C_F$ equals $C_L^{1-\sigma}$, σ is a generator of $\mathrm{Gal}(L/F)$, and the index of the norm group $N_{L/F}C_L$ equals to the degree, hold true by Theorem 5.7.

Thus, using Section 2 we obtain

THEOREM. *For a finite Galois extension L/F of global fields we have the homomorphism*

$$\Upsilon_{L/F}\colon \mathrm{Gal}(L/F) \longrightarrow C_F/N_{L/F}C_L,$$

its kernel is $[\mathrm{Gal}(L/F), \mathrm{Gal}(L/F)]$ *and it is surjective. All the properties stated in Section* 2 *hold.*

Using the inverse homomorphism provides the surjective homomorphism

$$\Psi_{L/F}\colon C_F \longrightarrow \mathrm{Gal}(L/F)^{\mathrm{ab}}$$

with kernel is $N_{L/F}C_L$.

Therefore, we have the global reciprocity map

$$\Psi_F\colon C_F \longrightarrow G_F^{\mathrm{ab}}$$

with all the properties in 2.8 *and* 2.9 *taking place. The map Ψ_F is continuous.*

Proof. Continuity of Ψ_F follows from $\Psi_F^{-1}(\mathrm{Gal}(L/F)) = N_{L/F}C_L$ for a finite abelian extension L/F and the openness of the norm group in Corollary 4.6. □

COROLLARY. *For a finite cyclic extension L/F an element $a \in F^\times$ is in the norm group $N_{L/F}L^\times$ if and only if for every place v of F its image in the completion $F_v^\times$ belongs to the local norm group N_{L_v/F_v}.*

Proof. If a is in the image of the local maps N_{L_v/F_v} for all v, then $a = N_{L/F}\beta$ for an idele $\beta \in J_L$. Hence $N_{L/F}(\beta L^\times) = 1$ in C_F. Therefore by Corollary 2.7 we obtain $\beta = \gamma^{1-\sigma}b$ for some $\gamma \in J_L$, $\sigma \in \mathrm{Gal}(L/F)$ and $b \in L^\times$. Thus, $a = N_{L/F}b$. □

6.4. In the approach to class field theory in this book, compatibility of the local reciprocity maps and the global reciprocity map needs to be established.

THEOREM. *For every finite abelian extension L/F and every place v of F we have the commutative diagram*

$$\begin{array}{ccc} F_v^{\times} & \xrightarrow{\Psi_{L_v/F_v}} & \mathrm{Gal}(L_v/F_v) \\ {\scriptstyle j_v}\downarrow & & \downarrow{\scriptstyle i_v} \\ C_F & \xrightarrow{\Psi_{L/F}} & \mathrm{Gal}(L/F) \end{array}$$

where j_v sends an element $\alpha \in F_v^{\times}$ to the class of the idele with components 1 everywhere except at v where its component is α.

Proof. Let F be a number field.

First consider infinite places. If $F_v^{\times}$ is infinitely divisible the diagram commutes. If $L_v = F_v$ then $j_v(F_v^{\times}) \in N_{L/F}C_L$ and the diagram commutes. If v is a real place and $\alpha \in F_v^{\times}$ is not infinitely divisible, then it is -1 modulo the subgroup $\mathbb{R}_{>0}^{\times}$ of infinitely divisible elements; if L_v/F_v is non-trivial then $L_v \simeq \mathbb{C}$, hence $|L:F|$ is even. Then $\Psi_{L/F}(j_v(-1))^2 = 1$ (identity automorphism). Since $i_v \circ \Psi_{L_v/F_v}(-1)$ is of order 2, we only need to check that $\Psi_{L/F}(j_v(-1)) \neq 1$. Consider the special case $L = F(\zeta_4)$ where $\zeta_4^2 = -1$. If $\Psi_{L/F}(j_v(-1)) = 1$ then $j_v(-1) \in N_{L/F}C_L$, i.e. $j_v(-1) = N_{L/F}(\beta)b$ for some $\beta \in J_L$ and $b \in F^{\times}$. Then (i) $b \in N_{L_{v'}/F_{v'}}L_{v'}^{\times}$ for $v' \neq v$, and (ii) $-b \in N_{L_v/F_v}L_v^{\times}$. On the other hand, $w_F(b) = 1$ by Theorem 6.2, so from (i) we deduce $b \in N_{L_v/F_v}L_v^{\times}$. But then from (ii) $-1 \in N_{L_v/F_v}L_v^{\times}$, a contradiction. Thus, for the special case $L = F(\zeta_4)$ we have $\Psi_{L/F}(j_v\alpha) = i_v \circ \Psi_{L_v/F_v}(\alpha)$. In the general case of real v, define $L' = L(\zeta_4)$ and choose F' as the fixed field of the restriction of the complex conjugation to L'. Then L' is an extension of F' of degree 2, L'/F' is the special case as above, $L' \supset L$, $F' \supset F$, $F'_v \simeq \mathbb{R}$ and $L'_v \simeq \mathbb{C}$. Therefore, $F_v = F'_v$, $L_v = L'_v$. For L'/F' we already know that $\Psi_{L'/F'}(j'_v\alpha) = i_v \circ \Psi_{L_v/F_v}(\alpha)$, where $j'_v \colon F_v = F'_v \longrightarrow C_{F'}$. Due to formula for the norm map on ideles in 4.2, $j_v(\alpha) = N_{F'/F}(j'_v(\alpha))$. Using the first Proposition of 2.5 we conclude $\Psi_{L/F}(j_v(\alpha)) = i_v \circ \Psi_{L_v/F_v}(\alpha)$.

Now we deal with finite places v in characteristic zero. By Theorem 2.9 $\deg_F \circ \Psi_{\widetilde{F}/F} = v_F$. Since $w_F = \deg_F \circ \Phi_{\widetilde{F}/F}$, in the special case of a finite subextension L/F of $\widetilde{F}/F$ we get $\Psi_{L/F}(\alpha) = \prod_v i_v \circ \Psi_{L_v/F_v}(\alpha_v)$ and, in particular, the diagram is commutative. We will reduce the general case to this special case, similar how in the study of Υ one reduces the general case of finite Galois extensions to the case of finite Galois extensions inside $\widetilde{F}/F$.

We have the diagram

$$\begin{array}{ccc} \mathrm{Gal}(L_v/F_v) & \xrightarrow{\Upsilon_{L_v/F_v}} & F_v^\times/N_{L_v/F_v}L_v^\times \\ {\scriptstyle i_v}\big\downarrow & & \big\downarrow{\scriptstyle j_v^*} \\ \mathrm{Gal}(L/F) & \xrightarrow{\Upsilon_{L/F}} & C_F/N_{L/F}C_L, \end{array}$$

where j_v^* is induced by j_v, and the horizontal maps are isomorphisms. Its commutativity is equivalent to the commutativity of the diagram in the statement of the Theorem. So we need to show that for every element $\sigma \in \mathrm{Gal}(L_v/F_v)$ its image in the lower right corner does not depend on the path.

Since finite abelian groups are generated by their elements of prime power order, we can assume that the order of σ is l^m for a prime l and a positive integer m. We can also assume that $i_v(\sigma)$ generates $\mathrm{Gal}(L/F)$ by passing to its fixed field and using the functorial properties as above. Then $i_v\colon \mathrm{Gal}(L_v/F_v) \to \mathrm{Gal}(L/F)$ is an isomorphism.

We use the notation $\breve{\mathbb{Q}}$ for the $\mathbb{Z}_l$-extension of $\mathbb{Q}$, similar to Lemma 6.2. We use the notation $\breve{F} = F\breve{\mathbb{Q}}$, $\breve{F}_v = F_v\breve{\mathbb{Q}}$ and similarly for finite extensions of F and F_v. The restriction map gives the homomorphism $G_{F_v} \longrightarrow \mathrm{Gal}(\breve{F}_v/F_v) \longrightarrow \mathrm{Gal}(\breve{F}/F)$. By Lemma 6.2, $n_l = |F_v \cap \breve{F} : F|$ is a positive integer. So there is an isomorphism $\mathrm{Gal}(\breve{F}_v/F_v) \simeq \mathrm{Gal}(\breve{F}/\breve{F} \cap F_v) \simeq \mathbb{Z}_l$ and we have the surjective homomorphism

$$\mathrm{deg}_{F_v}^{\smile}\colon G_{F_v} \longrightarrow \mathrm{Gal}(\breve{F}_v/F_v) \simeq \mathbb{Z}_l$$

which is different from the $\deg_{F_v}$ in local class field theory.

For the local fields extension L_v/F_v and $\sigma \in \mathrm{Gal}(L_v/F_v)$ we can use $\mathrm{deg}_{F_v}^{\smile}$ as in Remark 2.2. Hence, there is an element ϕ of $\mathrm{Frob}^{\smile}(L_v/F_v) = \{\tau \in \mathrm{Gal}(\breve{L}_v/F_v) : \mathrm{deg}_{F_v}^{\smile}(\tau) \in \mathbb{Z}_{>0}\}$ such that $\phi|_{L_v} = \sigma$. Denote by ϕ' the image of ϕ in $\mathrm{Gal}(\breve{L}/F)$ with respect to the restriction $\mathrm{Gal}(\breve{L}_v/F_v) \longrightarrow \mathrm{Gal}(\breve{L}/F)$. We have $\deg_F(\phi') = n_l \, \mathrm{deg}_{F_v}^{\smile}(\phi) \in \mathbb{Z}_{>0}$. Denote by R the fixed field of ϕ, it is of finite degree over F_v. Denote $K = R \cap \breve{L}$, it is the fixed field of ϕ'. By 2.2 K is of finite degree over F. Denote $M = KL$, by 2.2 it is of finite degree over K and L and is inside $\breve{K} = \breve{L}$. Denote by M_w the completion of M with respect to a place w of M over v of L, then $M_w \supset L_v$. Denote by the same notation w the place of K under the place w of M. The field R contains K and F_v, therefore it contains K_w. So $\phi|_{M_w}$ is in $\mathrm{Gal}(M_w/K_w)$. We deduce that the restriction map $\mathrm{Gal}(M_w/K_w) \longrightarrow \mathrm{Gal}(L_v/F_v)$ sends $\phi|_{M_w}$ to σ. The extension M/K is of the special type, so the preceding diagram is commutative for M/K, M_w/K_w.

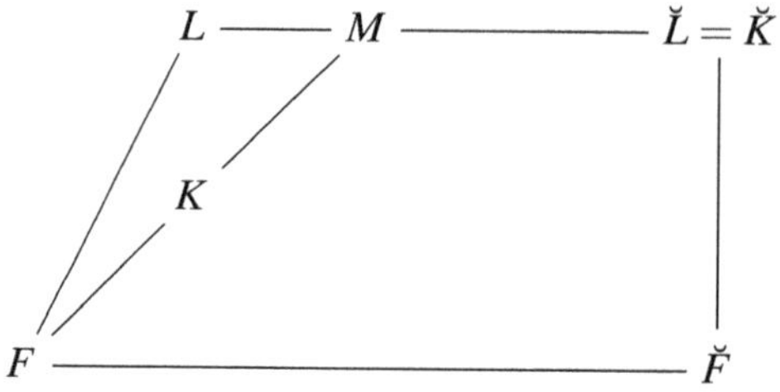

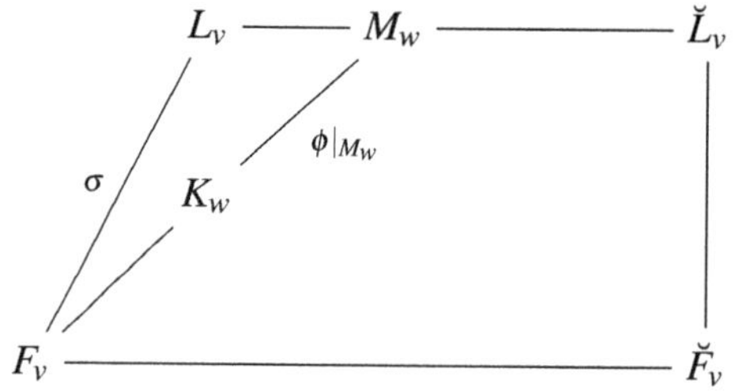

It remains to use the following cube diagram, all side diagrams except the bottom square are commutative. Hence the bottom square is commutative for σ.

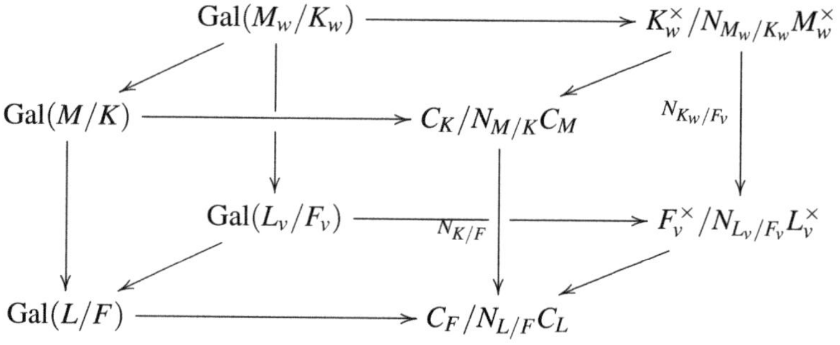

Finally, in positive characteristic $\widetilde{F} = F \otimes_{\mathbb{F}_q} \mathbb{F}_q^{\mathrm{sep}}$ where $\mathbb{F}_q$ is the constant subfield of F, and each completion F_v is the compositum of F_v and $\widetilde{F}$. We argue similarly to the characteristic zero case argument, using the usual $\deg_{F_v}$ in local class field theory of local fields of positive characteristic with finite residue field. □

REMARK. In the last part of the proof for number fields it would be more satisfying to work with local extensions $F_v\widetilde{F}/F_v$. However unlike in Lemma 6.2 for $\mathbb{Z}_l$-extensions, the intersection $\widetilde{\mathbb{Q}} \cap \mathbb{Q}_p$ is not of finite degree over $\mathbb{Q}$. Indeed, for odd primes l different from p and a primitive lth root ζ_l it is easy to check that the degree of the unramified extension $\mathbb{Q}_p(\zeta_l)/\mathbb{Q}_p$ is r_l where r_l is the minimal positive integer such that $p^{r_l} \equiv 1 \mod l$. Hence the fixed field R_l of the decomposition group $\mathrm{Gal}(\mathbb{Q}_p(\zeta_l)/\mathbb{Q}_p)$ of p in $\mathrm{Gal}(\mathbb{Q}(\zeta_l)/\mathbb{Q})$ is of degree $(l-1)/r_l$ over $\mathbb{Q}$. By the last sentence in the proof of Theorem 5.7, there are infinitely many primes which split completely in $\mathbb{Q}(\sqrt{p})/\mathbb{Q}$, hence, by Theorem 3.5.9 in Chapter 1, there are infinitely many primes l such that p is a quadratic residue modulo l, and for them $(l-1)/r_l \geqslant 2$. So $\widetilde{\mathbb{Q}} \cap \mathbb{Q}_p$ contains disjoint non-trivial extensions R_l of $\mathbb{Q}$ for infinitely many l.

6.5. RELATION OF THE GLOBAL AND LOCAL RECIPROCITY MAPS. *For every abelian extension L/F of global fields and $\alpha = (\alpha_v) \in J_L$*

$$\Psi_{L/F}(\alpha) = \prod_v i_v \circ \Psi_{L_v/F_v}(\alpha_v).$$

For every principal idele $a \in F^\times$ the reciprocity law holds

$$\prod_v i_v \circ \Psi_{L_v/F_v}(a) = 1.$$

Proof. The first formula for idele $j_v(b)$ and every place v is the content of Theorem 6.4. Hence it holds for the subgroup of ideles which have almost all of their components equal to 1. This subgroup is a dense subgroup of ideles. The reciprocity map $\Psi_{L/F}$ is continuous by Theorem 6.3, $\prod_v i_v \circ \Psi_{L_v/F_v}(\alpha_v)$ is continuous, since the local reciprocity maps are continuous by Theorem 3.2. Hence we obtain the first statement. The second statement follows. □

REMARK. The relation of the global and local reciprocity maps was first used by Chevalley to define the global reciprocity map using the local reciprocity maps. This is different from Artin's approach which did not use non-existing one hundred years ago local class field theory, and it is different from the approach of this book.

COROLLARY. *For every finite abelian extension L/F and every place v of F*

$$j_v(F_v^\times) \cap F^\times N_{L/F} J_L = j_v(N_{L_v/F_v} L_v^\times).$$

The places of L over the place v of F are in one-to-one correspondence with elements of the finite group $J_F/(F_v^\times N_{L/F} J_L)$.

Proof. For the first assertion, the $\supset$ inclusion follows from the description of the norm map in 4.2. Let $j_v(\alpha) \in F^\times N_{L/F} J_L$ for $\alpha \in F_v^\times$, i.e. $j_v(\alpha) = a N_{L/F}(\beta)$ for some $a \in F^\times$ and $\beta \in J_L$. This implies $\Psi_{L_{v'}/F_{v'}}(a) = 1$ for all places $v' \neq v$, hence by the Theorem, $\Psi_{L_v/F_v}(a) = 1$, and therefore a and α are in $N_{L_v/F_v} L_v^\times$.

The places of L over v correspond the cosets of $\mathrm{Gal}(L/F)_v = i_v(\mathrm{Gal}(L_v/F_v))$ in $\mathrm{Gal}(L/F)$, and since $\Psi_{L/F}$ and Ψ_{L_v/F_v} are isomorphisms, we deduce the last statement. □

6.6. To state the next Theorem we need to make several definitions and observations.

The Hilbert symbol $(\ ,\ )_{n,F_v}\colon F_v^\times \times F_v^\times \longrightarrow \mu_n$ for local fields F_v with finite residue field containing a primitive nth root of unity was defined and studied in 3.5. Similarly we can define it for archimedean completions F_v using the same formula $(\alpha,\beta)_{n,F_v} = \gamma^{-1}\Psi_{F_v}(\alpha)(\gamma)$ where $\gamma^n = \beta$ and F_v contains a primitive nth root of 1. Then $(\alpha,\beta)_{n,\mathbb{C}} = 1$ for all non-zero complex α,β since $\mathbb{C}^\times$ is infinitely divisible; $(\alpha,\beta)_{2,\mathbb{R}} = 1$ if $\alpha > 0$ or $\beta > 0$ and $= -1$ otherwise.

DEFINITION. For a finite v let integer $n > 1$ be such that such that a primitive nth root of unity is in F_v and $n \in \mathcal{O}_v^\times$. Then the group μ_n of roots of order dividing n is isomorphic to its residue image in $\mathcal{O}_v/\mathcal{M}_v$ and we identify them.

The nth power v-residue symbol is defined as

$$\left(\frac{-}{v}\right)_{n,F_v} : \mathcal{O}_v^\times \longrightarrow \mu_n, \quad \left(\frac{\alpha}{v}\right)_{n,F_v} := \alpha^{(|k(v)|-1)/n} \mod \mathcal{M}_v.$$

So $\left(\frac{\alpha}{v}\right)_{n,F_v} = 1$ if and only if $\overline{\alpha} \in k(v)^{\times n}$, which explains the name.

Below in this subsection let an integer $n > 1$ be such that a primitive nth root of unity is in F.

Let I be a non-zero fractional ideal of F with factorisation $I = \prod P_{v_i}^{n_i}$ in the product of maximal ideals P_i of $\mathcal{O}_F$ with non-zero integer n_i. Let, in addition, $n \in \cap \mathcal{O}_{v_i}^\times$. Let $a \in F^\times$ be such that $a \in \cap \mathcal{O}_{v_i}^\times$. Define *the nth power I-residue symbol*

$$\left(\frac{a}{I}\right)_n := \prod \left(\frac{a}{v_i}\right)_{n,F_{v_i}}^{n_i}.$$

For $a,b \in F^\times$, such that $I = b\mathcal{O}_F$ and a satisfy the restrictions above, we see that for every finite place v corresponding to a maximal ideal of $\mathcal{O}_F$, at most one of $v(a), v(b)$ is not zero. We call such a,b relatively prime. For relatively prime $a,b \in F^\times$ such that also ab, n are relatively prime define the *nth power symbol*

$$\left(\frac{a}{b}\right)_n := \left(\frac{a}{b\mathcal{O}_F}\right)_n.$$

This is equivalent to

$$\left(\frac{a}{b}\right)_n = \prod_{v(b)\neq 0} \left(\frac{a}{v}\right)_{n,F_v}^{v(b)},$$

where v runs through finite places of F and not over the place $-\deg$ in positive characteristic.

It is easily checked that when $F = \mathbb{Q}$ and $n = 2$, for coprime positive odd integers a, b the symbol $\left(\frac{a}{b}\right)_2$ is the Legendre quadratic symbol.

THEOREM. (*Reciprocity Law for nth power residue symbols*). *Let integer $n > 1$ be such that such that a primitive nth root of unity is in F. Denote by S' the set of archimedean places of F in characteristic zero and the set of places over $-\deg$ in positive characteristic. Let $a, b \in F^\times$. Assume that for every finite place v of F not over $-\deg$ in positive characteristic, if one of $v(a), v(b)$ is not 0 then the other and $v(n)$ are 0. Then*

$$\left(\frac{a}{b}\right)_n \left(\frac{b}{a}\right)_n^{-1} = \prod_{v(n)>0 \text{ or } v\in S'} (a,b)_{n,F_v}.$$

Proof. The previous Theorem implies that for $a, b \in F^\times$, $\mu_n \subset F$, and $\gamma^n = b$

$$\prod_v (a,b)_{n,F_v} = \gamma^{-1}\Big(\prod_v \Psi_{F_v}(a)\Big)(\gamma) = \gamma^{-1}\Psi_F(a)(\gamma) = 1,$$

here v runs through all places of F.

For finite v such that $v(n) = v(a) = 0$ we know from the proof of the Proposition about the tame symbol in 3.5 that $(b,a)_{n,F_v} = \left(\frac{a}{v}\right)_{n,F_v}^{v(b)}$. Similarly we obtain, $(b,a)_{n,F_v} = \left(\frac{b}{v}\right)_{n,F_v}^{-v(a)}$ if $v(b) = v(n) = 0$.

So

$$\begin{aligned}\left(\frac{a}{b}\right)_n \left(\frac{b}{a}\right)_n^{-1} &= \prod_{v(b)\neq 0} \left(\frac{a}{v}\right)_{n,F_v}^{v(b)} \prod_{v(a)\neq 0} \left(\frac{b}{v}\right)_{n,F_v}^{-v(a)} = \prod_{v(ab)\neq 0} \left(\frac{a}{v}\right)_{n,F_v}^{v(b)} \left(\frac{b}{v}\right)_{n,F_v}^{-v(a)} \\ &= \prod_{v(ab)\neq 0} (b,a)_{n,F_v} = \prod_{v(n)=0} (b,a)_{n,F_v},\end{aligned}$$

where $v \notin S'$. Applying the first sentence of the proof, the proof is completed. □

REMARKS.

1. For number fields, using not easy to deduce explicit formulas for the nth Hilbert symbol, n is divisible by the residue characteristic, see 3.5, one answers Hilbert's Problem 9 about an explicit description of

$$\left(\frac{a}{b}\right)_n \left(\frac{b}{a}\right)_n^{-1}.$$

2. One immediately checks that

$$(a,b)_{2,\mathbb{Q}_2} = (-1)^{(a-1)(b-1)/4}, \quad \text{if } a,b \in \mathbb{Z}_2^\times.$$

Hence the partial case of the Theorem for $F = \mathbb{Q}$, $n = 2$ gives a proof of *Gauß quadratic reciprocity law* for coprime positive odd integers a,b. It is the only proof, in the list of currently known 330 proofs included several by Gauß, which really explains why this law holds. The auxiliary formula for $\left(\frac{2}{b}\right)_2$ also follows immediately.

3. In positive characteristic p, if a primitive nth root of unity is in F then n is not divisible by p and $v(n) = 0$ for all places v of F; so in this case the factors on the right hand side run through places v over $-\deg$.

6.7. EXISTENCE THEOREM. *The kernel of the reciprocity map Ψ_F coincides with the intersection of all open subgroups of finite index in C_F. It is surjective in characteristic zero. In positive characteristic its image is everywhere dense, and it sends C_F^1 isomorphically onto* $\mathrm{Gal}(F^{\mathrm{ab}}/\widetilde{F})$.

The correspondence between open subgroups N of finite index in C_F and the norm subgroups of finite abelian extensions L/F:

$$N \longleftrightarrow L/F,$$

$N = N_{L/F}C_L = \Psi_F^{-1}(\mathrm{Gal}(F^{\mathrm{ab}}/L))$, *is an order reversing bijection between the lattice of open subgroups of finite index in C_F (with respect to the intersection $N_1 \cap N_2$ and the product N_1N_2) and the lattice of finite abelian extensions of F (with respect to the compositum L_1L_2 and intersection $L_1 \cap L_2$).*

Proof. By Theorem 2.9 the image of Ψ_F is dense in $\mathrm{Gal}(F^{\mathrm{ab}}/F)$. In characteristic zero $C_F = M \times C_F^1$ where $M \simeq \mathbb{R}_{>0}^\times$ is an infinite divisible group. Hence $\Psi_F(C_F) = \Psi_F(C_F^1)$. Since C_F^1 is compact and Ψ_F is continuous, $\Psi_F(C_F)$ is closed, so Ψ_F is surjective. In positive characteristic, due to Remark 6.2 the image $\Psi_F(C_F^1)$ is in $\mathrm{Gal}(F^{\mathrm{ab}}/\widetilde{F})$, it is dense and closed, hence $\Psi_F(C_F^1) = \mathrm{Gal}(F^{\mathrm{ab}}/\widetilde{F})$, and the cokernel of the reciprocity map is isomorphic to $\widehat{\mathbb{Z}}/\mathbb{Z}$.

To verify that an open subgroup N of finite index in C_F coincides with the norm subgroup $N_{L/F}C_L$ of some finite abelian extension L/F, it suffices to verify that N contains the norm subgroup $N_{M/F}C_M$ of some finite separable extension M/F. Indeed, in this case N contains $N_{E/F}C_E$, where E/F is a finite Galois extension, $E \supset M$. Then by Proposition 2.8 we deduce that $N = N_{R/F}C_R$, where R is the fixed field of $\Psi_{E/F}(N)$ and R/F is abelian. The argument in this paragraph is identical to the argument in the proof of Existence theorem in local class field theory in 3.2.

Denote by n the index of N in C_F (in fact, it suffices to consider the case when n is a power of prime number, but the argument there is the same as below). Assume first that n is not divisible by characteristic of F. The preimage of N in J_F is an open subgroup of J_F of index n, so it contains the product of $F^\times$ and the subgroup

$$N_{S_0} = \prod_{v \notin S_0} U_v \times \prod_{v \in S_0} F_v^{\times n}$$

for some finite subset S_0 of places of F, containing all infinite places in characteristic zero.

Denote $E = F(\zeta_n)$ for a primitive nth root ζ_n. We will use Kummer theory for E. Enlarge S_0 so that it contains all ramified places in E/F (their number is finite by Proposition 4.6) and all places dividing n. Denote by S the set of all places of E over places in S. Further enlarge finite S_0 so that the set S of all places of E over places in S_0 has the property $J_E = E^\times J_E(S)$ (see Corollary 1 of 4.7). Consider the Kummer extension M of E obtained by extracting all nth roots from all elements of $E^\times(S)$. Then $\mathrm{Gal}(M/E)^n = \{1\}$. By Proposition 4.8 the group $E^\times(S)$ is isomorphic to the product of a free abelian group of rank $s-1$, $s = |S|$, and the finite group of roots in E. Since $\mu_n \subset E$, we obtain $|E^\times(S) : E^\times(S)^n| = n^s$ and by Kummer theory the extension M/E has degree n^s. Each place $w \notin S$ is unramified in M/E, so the group U_w of units of the ring of integers of E_w is in the norm group $N_{M_w/E_w} M_w^\times$. For $w \in S$ the nth powers $E_w^{\times n}$ are in $N_{M_w/E_w} M_w^\times$ since $\mathrm{Gal}(M_w/E_w)^n = 1$. Hence $N_{M/E} J_M$ contains

$$N_S = \prod_{w \notin S} U_w \times \prod_{w \in S} E_w^{\times n}.$$

Note that $N_S \cap E^\times = E^\times(S)^n$. To show the non-trivial inclusion, for an element $a \in N_S \cap E^\times$ consider the cyclic Kummer extension $K = E(\sqrt[n]{a})$. Then $K_w = E_w$ for all $w \in S$ and K_w/E_w is unramified for all $w \notin S$. Hence every idele in $J_E(S)$ is in $E^\times N_{K/E} J_K$. Since $E^\times J_E(S) = J_E$, we deduce $C_E = N_{K/E} C_K$ and therefore $K = E$ and $a \in E^{\times n}$. Therefore, $N_S \cap E^\times \subset E^{\times n} \cap J_E(S) \subset E^\times(S)^n$.

For three finite abelian subgroups $G \supset H$ and K of a larger group, the order of G/H is the product of the order of GK/HK and the order of $(G \cap K)/(H \cap K)$. We have $J_E/(E^\times N_S) \simeq E^\times J_E(S)/(E^\times N_S)$. Its order is the quotient of the order r of the group $J_E(S)/N_S$ by $n^s =$ the order of $(J_E(S) \cap E^\times)/(N_S \cap E^\times) = E^\times(S)/E^\times(S)^n$. We also have $J_E(S)/N_S \simeq \prod_{w \in S} E_w^\times / E_w^{\times n}$ and due to the description in 18.3 of Chapter 2 in the non-archimedean case and the obvious description in the archimedean case, we obtain $|E_w^\times : E_w^{\times n}| = n^2 |n|_w^{-1}$ for all places w of E. Since $|n|_w = 1$ for $w \notin S$, we obtain $r = n^{2s} \prod_w |n|_w^{-1} = n^{2s}$. Thus, the order of $J_E/(E^\times N_S)$ is $n^s = |M : E|$ and hence using Theorem 6.3 we derive

$E^\times N_S = E^\times N_{M/E} J_M$. Thus, the image of N_S in C_E equals to $N_{M/E} C_M$. Therefore, $N \supset N_{M/F} C_M$, as desired.

To handle the case when n is divisible by $\operatorname{char}(F) = p$, it is sufficient to show by induction on $m \geqslant 1$ that any open subgroup N of index p^m in C_F contains a norm group, and then, similarly to the proof of local Existence Theorem 3.2 one only needs to treat the case $m = 1$ where one can use Remark 2 below, working with the adelic version of the Artin–Schreier pairing of 3.6.

Everything else follows from Proposition 2.8. □

REMARKS.

1. For a description of the kernel of the reciprocity map see Exercise 3.11. For ray class fields see Exercise 3.12.

2. Let F be a finite separable extension of $\mathbb{F}_p(t)$. Using the local Artin–Schreier pairings from 3.6, define a pairing

$$(\, , \,] \colon J_F \times F \longrightarrow \mathbb{F}_p, \quad (\alpha, b] = \sum_v (\alpha_v, b]_v, \quad (\alpha_v, b]_v = \operatorname{Tr}_{k(v)/\mathbb{F}_p} \operatorname{res}_v (b\, d_t \alpha/\alpha)$$

where res_v is $\operatorname{res}_{\pi_v}$ for any prime element π_v of F_v as in 3.6, $d_t\alpha = dt\, d_{\pi_v}\alpha / d_{\pi_v} t$. Since only finitely many places ramify in finite separable extension $F/\mathbb{F}_p(t)$ by Proposition 4.6, the element t is a local parameter of F_v for almost all places v of F, and hence $(\, , \,]_v$ is the local Artin–Schreier pairing for almost all v.

If $(J_F, b] = 0$ then $b \in \wp(F_v)$ for almost all v by 3.6, hence the extension $F(\wp^{-1}(b))/F$ splits completely for almost all v, hence $F(\wp^{-1}(b)) = F$ by Corollary 2 of 4.9 and thus $b \in \wp(F)$. If $(\alpha, F] = 0$ then $d_\omega(F\alpha^{-1} d_t \alpha) = 0$ where $\omega = dt$ and d_ω is defined in 4.4, hence by 4.4 $\alpha^{-1} d_t \alpha = c\, dt$ for some $c \in F$. Let Der_t be the operator sending γ to $d_t\gamma/dt = d_{\pi_v}\gamma/d_{\pi_v} t$ for any prime element π_v of F_v, and let M_β be the operator of multiplication by β. The equality for α and c can be rewritten as $\operatorname{Der}_t + M_c = M_{\alpha^{-1}} \circ \operatorname{Der}_t \circ M_\alpha$. Hence $(\operatorname{Der}_t + M_c)^m = M_{\alpha^{-1}} \circ \operatorname{Der}_t^m \circ M_\alpha$. Since $\operatorname{Der}_t^p = 0$, there is a maximal $m < p$ for which $l = (\operatorname{Der}_t + M_c)^m (1) \neq 0$. Then $(\operatorname{Der}_t + M_c) l = 0$, $c = l \operatorname{Der}_t(l^{-1})$ and $\operatorname{Der}_t(\alpha l) = 0$. So each v-component of αl is in F_v^p and so $\alpha l \in J_F^p$, $\alpha \in J_F^p F^\times$.

Thus, we obtain the perfect continuous pairing $C_F/C_F^p \times F/\wp(F) \longrightarrow \mathbb{F}_p$ which induces, by Artin–Schreier theory, the continuous isomorphism $C_F/C_F^p \xrightarrow{\sim} \operatorname{Gal}(F_p/F)$ where F_p is the maximal abelian extension of F of exponent p. This implies that every open subgroup N of index p in C_F is the norm group of the Artin–Schreier extension $L = F(\wp^{-1}(b))$ of F where $b\mathbb{F}_p + \wp(F)$ is the complement of N with respect to the perfect pairing.

3. Similarly to Remark 2 and alternatively to the proof of the Existence theorem, when $\mu_n \subset F$, one can use the local Hilbert symbols to define the pairing

$$C_F/C_F^n \times F^\times/F^{\times n} \longrightarrow \mu_n$$

check its non-degenerate property and an adelic analog of Theorem 3.5, to prove that every open subgroup N of index n in C_F is the norm group of the Kummer extension $L = F(\sqrt[n]{b})$ of F and N is the complement of b with respect to the pairing.

The following Corollary which is crucial for special class field theory briefly introduced at the end of Chapter 1, is not used in this chapter on general class field theory.

COROLLARY. (*Kronecker–Weber Theorem*) *The maximal abelian extension* $\mathbb{Q}^{\mathrm{ab}}$ *of* $\mathbb{Q}$ *coincides with the maximal cyclotomic extension* $\mathbb{Q}^{\mathrm{cycl}}$.

Proof. By the previous Theorem it is sufficient to show that every open subgroup N of $C_\mathbb{Q}$ contains the norm group of a cyclotomic extension of $\mathbb{Q}$. Since N is open, for some positive integer m the group N contains $J_\mathbb{Q}(m)\mathbb{Q}^\times/\mathbb{Q}^\times$, where $m = \prod p^{n_p}$ and

$$J_\mathbb{Q}(m) = \mathbb{R}^\times_{>0} \times \prod_{p|m} U_{n_p,\mathbb{Q}_p} \times \prod_{p\nmid m} U_{\mathbb{Q}_p}.$$

Without loss of generality we can assume that $n_2 > 1$.

We will show that $J_\mathbb{Q}(m)\mathbb{Q}^\times/\mathbb{Q}^\times = N_{\mathbb{Q}(\zeta_m)/\mathbb{Q}} C_{\mathbb{Q}(\zeta_m)}$. We can use the computations of the norm groups of cyclotomic extensions of p-adic fields in the proof of Theorem 3.3 where it was shown that the norm group of $\mathbb{Q}_p(\zeta_{p^{n_p}})/\mathbb{Q}_p$ is $\langle p \rangle \times U_{n_p,\mathbb{Q}_p}$ if $p^{n_p} > 2$. The group $U_{n_p,\mathbb{Q}_p}$ is contained in the norm group of any unramified extension of $\mathbb{Q}_p$, so the norm group of $\mathbb{Q}_p(\zeta_m)/\mathbb{Q}_p$ contains $U_{n_p,\mathbb{Q}_p}$. Then $N_{\mathbb{Q}(\zeta_m)/\mathbb{Q}} C_{\mathbb{Q}(\zeta_m)} \supset J_\mathbb{Q}(m)\mathbb{Q}^\times/\mathbb{Q}^\times$. We have $J_\mathbb{Q}/\mathbb{Q}^\times \simeq \mathbb{R}^\times_{>0} \times \prod_p U_{\mathbb{Q}_p}$ and $J_\mathbb{Q}(m)\mathbb{Q}^\times/\mathbb{Q}^\times \simeq \mathbb{R}^\times_{>0} \times \prod_{p\nmid m} U_{\mathbb{Q}_p} \times \prod_{p|m} U_{n_p,\mathbb{Q}_p}$, so the quotient $J_\mathbb{Q}/(J_\mathbb{Q}(m)\mathbb{Q}^\times)$ is isomorphic to $\prod_{p|m} U_{\mathbb{Q}_p}/U_{n_p,\mathbb{Q}_p} \simeq (\mathbb{Z}/m\mathbb{Z})^\times$. Hence, the index of the subgroup $J_\mathbb{Q}(m)\mathbb{Q}^\times/\mathbb{Q}^\times$ in $C_\mathbb{Q}$ equals the degree of $\mathbb{Q}(\zeta_m)/\mathbb{Q}$. Theorem 6.3 now implies $N_{\mathbb{Q}(\zeta_m)/\mathbb{Q}} C_{\mathbb{Q}(\zeta_m)} = J_\mathbb{Q}(m)\mathbb{Q}^\times/\mathbb{Q}^\times$. □

6.8. REMARKS.

1. There is a certain analogy between Neukirch's approach to class field theory and the zeta integral theory of Iwasawa–Tate: in both cases one extends the original math setting (finite Galois extensions/zeta functions) to something much

larger and complete in appropriate sense where one has richer structures (infinite Galois groups/ideles and adeles) that can be used to produce simpler proofs than the classical ones.

2. Using 6.2, 6.3, 6.6, one immediately deduces all the statements in 5.4 of Chapter 1 for class field theory of rational numbers. One can show that the only algebraic number field F for which the equality $F^{\rm ab} = F^{\rm cycl}$ holds is the field of rational numbers.

3. Historically, without using abstract class field theory, one develops special class field theory for $\mathbb{Q}$, called cyclotomic class field theory (Kronecker, Weber and others), using explicit cyclotomic methods. Special in the sense of using more information about Galois action on torsion elements than the abstract general class field theory of Section 2 does and not functorial with respect to finite separable base change. Another special class field theory is complex multiplication theory for quadratic imaginary fields (it started from Kronecker–Weber and continued till Shimura), see Chapter 2 of [**33**]. More generally, there is special class field theory for totally imaginary extensions of totally real fields (Shimura). There is no special class field theory for arbitrary algebraic number fields. See also 5.4.3 of Chapter 1. General functorial class field theory such as in this text is very much different from those special theories, both conceptually and technically, see [**9**].

4. Other approaches to class field theory of global fields include general class field theories such as

– for number fields by Artin, almost 100 years old, building on the previous work of German mathematicians and Takagi, using L-functions and Chebotarev's Density Theorem (stated in Exercise 3.14),

– for number fields by Hasse, using central division algebras and the computation of the Brauer group of the field to define a canonical pairing of the group of characters of the field with, in the modern language, the idele class group, using a cup product, [**15**]

– algebraic approach by Chevalley, using ideles and not using L-functions, [**37**]

– Galois cohomology approach, [**1**]. This approach derives the local and global reciprocity maps using certain cup products of H^1 cohomology groups, hence, similarly to the Hasse's approach, not quite explicitly. However, for applications one often needs to know explicit descriptions of the reciprocity maps or explicit formulas for the local Hilbert symbol.

In positive characteristic only:

– an explicit approach using Artin–Schreier–Witt pairing, [**23**]

– using a theory of Hayes–Drinfeld modules of rank 1.

See [**15**] for a short history of class field theory before 1940.

5. Higher adelic theory studies adelic structures associated to two-dimensional arithmetic schemes. There are two main adelic structures there: one of more geometric (1-cocyles) nature (its use leads to an adelic proof of the Riemann–Roch Theorem for surfaces and a two-dimensional version of the homomorphism ρ of 4.7 and one of more arithmetic (0-cycles) nature crucial for a two-dimensional version of the Iwasawa–Tate theory and applications to the study of meromorphic continuation and functional equation of the zeta function of the scheme and its poles, [**9**].

6. Three main generalisations of class field theory are higher class field theory, Langlands program, anabelian geometry. See [**9**] for various discussions and open questions.

It is remarkable that Neukirch's mechanism of class field theory was motivated by his previous work in anabelian geometry of number fields and he developed his approach with the eye towards potential applications in the Langlands program. Maybe, one of the readers of this book will realise his dream.

Chapter 4

Exercises

1. Algebraic Numbers Exercises

1.1. Let A be an integral domain and K is its fraction field. Prove that A is a Dedekind ring if and only if every non-zero proper ideal of A can be written as a product of prime ideals if and only if every non-zero ideal I of A satisfies $A = \{a \in K : aI \subset A\}I$.

1.2.

(a) Let F be an algebraic number field of degree d. Let m be a positive integer. For $a_i \in F^\times$ and independent variables $X_1, \ldots, X_m$ put

$$\begin{aligned} f(X_1, \ldots, X_m) &= N_{F/\mathbb{Q}}(a_1X_1 + \cdots + a_mX_m) \\ &= \prod_{\sigma \in \mathrm{Hom}_{\mathbb{Q}}(F,\mathbb{C})} (\sigma(a_1)X_1 + \cdots + \sigma(a_m)X_m). \end{aligned}$$

Show that $f(X_1, \ldots, X_m)$ is a homogeneous polynomial of degree d (i.e. every monomial expression is a monomial of total degree d) with coefficients from $\mathbb{Q}$.

(b) Show that f defined in (a) is irreducible over $\mathbb{Q}$.

(c) Let $g(X_1, \ldots, X_r)$ be a homogeneous polynomial of degree d with rational coefficients. Assume that g is irreducible over $\mathbb{Q}$. Assume also that there exists an algebraic number field L such that g splits into the product of linear polynomials over L. Show that then there is an algebraic number field F, a positive integer m and elements $a_i \in F^\times$, $1 \leqslant i \leqslant m$, such that $g = N_{F/\mathbb{Q}}(f)$ as in (a).

1.3. Let $b > 1$ be an odd number and let $m > 1$ be an integer. Suppose that $d = b^m - 1$ is square-free.

(a) Show that $d \equiv 2 \mod 4$.

(b) Show that $(b)^m = (1+d)$ factorises into the product of ideals $(1+\sqrt{-d})$ and $(1-\sqrt{-d})$ of $\mathbb{Z}[\sqrt{-d}]$.

(c) Show that if a proper non-zero ideal I of $\mathbb{Z}[\sqrt{-d}]$ divides both $(1+\sqrt{-d})$ and $(1-\sqrt{-d})$, then 2 is contained in I and therefore $2^2 = 4$ is contained in the product $(1+\sqrt{-d})(1-\sqrt{-d})=(1+d)$. Deduce from (a) that this is impossible; thus, the ideals $(1+\sqrt{-d})$ and $(1-\sqrt{-d})$ don't have common factors.

(d) Prove that there are ideals I,J of $\mathbb{Z}[\sqrt{-d}]$ such that $(1+\sqrt{-d})=I^m$ and $(1-\sqrt{-d})=J^m$ and $IJ=(b)$.

(e) Let n be the minimal positive integer such that I^n is a principal ideal, say $(e+c\sqrt{-d})$ of $\mathbb{Z}[\sqrt{-d}]$ for some $e,c\in\mathbb{Z}$. Show that $c\neq 0$.

(f) Show that $b^n=e^2+dc^2\geqslant d=b^m-1$ and deduce that $n\geqslant m$. Conclude that the ideal class group of $\mathbb{Q}(\sqrt{-d})$ has an element (namely, I) of order m.

Example: $b=3$, $m=3$, $d=26$, the class number of $\mathbb{Q}(\sqrt{26})$ is $\geqslant 3$.

1.4. Let d be a positive square free integer, $d\neq 5$. Suppose that $4^n+1=da^2$ with integer a. Prove that $2^n+a\sqrt{d}$ is a fundamental unit of $\mathbb{Q}(\sqrt{d})$.

1.5. Let P be a maximal ideal of the ring of integers of an algebraic number field F, such that $P^n=a\mathcal{O}_F$ is a principal ideal. Prove that the ideal $P\mathcal{O}_L$, generated by P in $\mathcal{O}_L$, a a principal ideal of the ring $\mathcal{O}_L$ of integers of the field $L=K(\sqrt[n]{a})$.

1.6. Prove that each algebraic number field F has a finite extension L such that every ideal of the ring of integers of F generates a principal ideal of $\mathcal{O}_L$.

1.7. (to solve this exercise, knowledge of appropriate sections of Chapter 3 is needed). For a finite extension L/F of global fields the ideal $\{\alpha\in L: \mathrm{Tr}_{L/F}(\alpha\mathcal{O}_L)\subset\mathcal{O}_F\}$ is a fractional ideal of $\mathcal{O}_L$. The inverse to it is called the different $\mathcal{D}_{L/F}$ of L/F, it is an ideal of $\mathcal{O}_L$. Similarly, for a finite extension L_w/F_v of local fields the ideal $\{\alpha\in L_w:\mathrm{Tr}_{L_w/F_v}(\alpha\mathcal{O}_{L_w})\subset\mathcal{O}_{F_v}\}$ is a fractional ideal of $\mathcal{O}_{L_w}$. The inverse to it is called *the different* $\mathcal{D}_{L_w/F_v}$ of L_w/F_v, it is an ideal of $\mathcal{O}_{L_w}$.

For a discrete valuation v of a number field F over the p-adic valuation of $\mathbb{Q}$, $\mathcal{D}_{F_v/\mathbb{Q}_p}=\mathcal{M}_v^{d_v}$, where d_v is as in the first Definition of 5.3 of Chapter 3.

(a) Prove that for three global or local fields $L/M/F$ the different of L/F is the product of the differents of L/M and M/F.

(b) Using the formula in 4.1 of Chapter 3 for the trace $\mathrm{Tr}_{L/F}$ in relation to local traces and Approximation Theorem 2.8 of Chapter 2, prove that $\mathcal{O}_{L_w}\mathcal{D}_{L/F}=\mathcal{D}_{L_w/F_v}$.

(c) Prove that $\mathcal{D}_{L/F}=\prod_v\prod_{w|v}(\mathcal{D}_{L_w/F_v}\cap\mathcal{O}_L)$.

(d) Prove that the absolute value $|d_F|$ of the discriminant d_F of a number field F is $N(\mathcal{D}_{F/\mathbb{Q}})$.

(e) Deduce the formula $|d_F|=\prod|\delta_v|_v$ in Remark 5.4 of Chapter 3.

(f) Using 4.4 of Chapter 3 in positive characteristic deduce similarly the formula $q^{2-2g}=\prod|\delta_v|_v$ in Remark 5.4 of Chapter 3.

2. Local Fields Exercises

2.1. A subring $\mathcal{O}$ of a field F is said to be a valuation ring if $\alpha \in \mathcal{O}$ or $\alpha^{-1} \in \mathcal{O}$ for every nonzero element $\alpha \in F$. Show that the ring of integers of a valuation of F is a valuation ring. Conversely, for a valuation ring $\mathcal{O}$ in F one can order the group $F^\times/\mathcal{O}^\times$ as follows: $\alpha\mathcal{O}^\times \leqslant \beta\mathcal{O}^\times$ if and only if $\beta\alpha^{-1} \in \mathcal{O}$. Show that the map $F \to (F^\times/\mathcal{O}^\times) \cup +\infty$, which sends 0 to $+\infty$, is a valuation with the ring of integers $\mathcal{O}$.

2.2. Show that every isomorphism of $\mathbb{Q}_p$ onto a subfield of $\mathbb{Q}_p$ is continuous.

2.3. Let F be a field with a discrete valuation v and ring of integers $\mathcal{O}$ and maximal ideal $\mathcal{M}$. Show that the following conditions are equivalent:

(i) F is a Henselian discrete valuation field.

(ii) If $f(X) = X^n + \alpha_{n-1}X^{n-1} + \cdots + \alpha_0$ is an irreducible polynomial over F and $\alpha_0 \in \mathcal{O}$, then $\alpha_i \in \mathcal{O}$ for $0 \leqslant i \leqslant n-1$.

(iii) If $f(X) = X^n + \alpha_{n-1}X^{n-1} + \cdots + \alpha_0$ is an irreducible polynomial over F, $\alpha_{n-2}, \ldots, \alpha_0 \in \mathcal{O}$, then $\alpha_{n-1} \in \mathcal{O}$.

(iv) If $f(X) = X^n + \alpha_{n-1}X^{n-1} + \cdots + \alpha_0$ is an irreducible polynomial over F, $\alpha_{n-2}, \ldots, \alpha_0 \in \mathcal{M}, \alpha_{n-1} \in \mathcal{O}$, then $\alpha_{n-1} \in \mathcal{M}$.

2.4. Let F be a Henselian field with respect to each of non-trivial valuations $v, v' \colon F \to \mathbb{Q}$. Assume the topologies induced by v and v' are not equivalent.

(a) Show that if v is discrete, then v' is not.

(b) Deduce that F is separably closed.

2.5. Let π be a prime element of a discrete valuation field F, and let $\overline{F}^{\text{sep}}$ be of infinite degree over $\overline{F}$.

(a) Let F_i be finite unramified extensions of F, $F_i \subset F_j$, $F_i \neq F_j$ for $i < j$. Put

$$\alpha_n = \sum_{i=1}^{n} \theta_i \pi^i,$$

where $\theta_i \in \mathcal{O}_{F_{i+1}}, \notin \mathcal{O}_{F_i}$. Show that the sequence $\{\alpha_n\}_{n\geqslant 0}$ is a Cauchy sequence in F^{ur}, but $\lim \alpha_n \notin F^{\text{ur}}$.

(b) Show that F^{sep} is not complete, but the completion of F^{sep} is separably closed.

2.6. Prove that for every finite extension of complete discrete valuation fields L/F there is a finite extension K' of a maximal complete discrete valuation subfield K of F with perfect residue field such that $e(K'L|K'F)=1$ following the steps below:

(a) Let M_1/F, M_2/F be finite Galois subextensions of L/F. Show that the set of upper ramification jumps of M_1/F is a subset of upper ramification jumps of M_2/F. Denote by $B(L/F)$ the union of all upper ramification jumps of finite Galois subextensions of L/F.

(b) For a real x define $L(x)=\cup_M M(x)$ where M runs over all finite Galois extensions of F in L and $M(x)$ is the fixed field of $\mathrm{Gal}(M/F)(x)$ inside M. Show that if $x_1<x_2$, then $L(x_1)\neq L(x_2)$ if and only if $[x_1,x_2)\cap B(L/F)\neq\emptyset$.

(c) Show that if x is the limit of a monotone increasing sequence x_n, then $L(x)=\cup L(x_n)$.

(d) Show that if x is the limit of a monotone decreasing sequence x_n and $x\notin B(L/F)$, then $L(x)=\cap L(x_n)$.

(e) Let x be the limit of a strictly monotone decreasing sequence x_n. Define $L[x]=\cup_M(\cap_n M(x_n))$ where M runs over all finite Galois extensions of F in L. Show that $L[x]=\cap_n L(x_n)$. Show that $L[x]=L(x)$ is and only if $x\notin B(L/F)$.

(f) A subfield E of L, $F\subset E$ is called a ramification subfield if for every finite Galois subextension M/F of L/F there is y such that $E\cap M=M(y)$. Show that every ramifications subfield of L over F coincides either with some $L(x)$ or with some $L[x]$.

(g) Deduce that the set of all upper ramification jumps of L/F is the union of $B(L/F)$ and the limits of strictly monotone decreasing sequences of elements of $B(L/F)$.

2.7. Let L/F be a cyclic totally ramified extension of complete discrete valuation fields, $|L:F|=p^n$. Let $\mathrm{char}(F)=0$, $\mathrm{char}(\overline{F})=p$, and let $\overline{F}$ be perfect.

(a) Show that L/F has n ramification numbers $x_1<x_2<\cdots<x_n$.

(b) Show that if x_i are divisible by p, then $x_i=x_1+(i-1)e$ for $1\leqslant i\leqslant n$, where $e=e(F)$.

(c) For the rest of this Exercise assume that a primitive pth root of unity ζ belongs to F. Let $N_{L/F}(\alpha)=\zeta$ and $v_L(\alpha-1)=i$. Show that if $x_1<e/(p-1)$, then $x_1\leqslant i\leqslant h_{L/F}(e/(p-1))$ and if $x_1\geqslant e/(p-1)$, then $i=e/(p-1)$.

(d) Assume that M/F is cyclic of degree p^{n-1} and $L=M(\sqrt[p]{\alpha})$ with $\alpha\in M^*$. Let $\alpha^{-1}\sigma(\alpha)=\beta^p$ for a generator σ of $\mathrm{Gal}(L/F)$. Show that $N_{M/F}(\beta)$ is a primitive pth root of unity.

(e) Show that if $x_1\geqslant e/(p-1)$, then $x_i=x_1+(i-1)e$ for $1\leqslant i\leqslant n$.

(f) Let $n\geqslant 2$. Show that if $x_{n-1}\geqslant p^{n-2}e/(p-1)$, then $x_n=x_{n-1}+p^{n-1}e$, and if $x_{n-1}\leqslant p^{n-2}e/(p-1)$, then

$$(1+p(p-1))x_{n-1}\leqslant x_n\leqslant p^ne/(p-1)-(p-1)x_{n-1}.$$

2.8. Let L_n be a cyclic totally ramified extension of F of degree p^n, $p = \operatorname{char}(\overline{F})$ and $L_n \subset L_{n+1}$. Let $L = \cup L_n$. Show that $i(L_{n+1}|L_n) \geqslant i(L_n|L_{n-1}) + 1$. Deduce that L/F is arithmetically profinite.

2.9. Let F be a complete field with respect to some non-trivial valuation $v\colon F^\times \to \mathbb{Q}$. Let the perfect residue field $\overline{F}$ be of characteristic $p > 0$. Put $F^{(n)} = F$, and let $R^\times(F) = \varprojlim F^{(n)\times}$ with respect to the homomorphism of the raising to the pth power $F^{(n+1)} \xrightarrow{\uparrow p} F^{(n)}$. Put $R(F) = R^\times(F) \cup \{0\}$.

(a) Show that if $\mathrm{A} = (\alpha^{(n)}), \mathrm{B} = (\beta^{(n)}) \in R(F)$, then the sequence $(\alpha^{(n+m)} + \beta^{(n+m)})^{p^m}$ converges as $m \to +\infty$. Put $\gamma^{(n)} = \lim_{m\to+\infty}(\alpha^{(n+m)} + \beta^{(n+m)})^{p^m}$ and define $\mathrm{A} + \mathrm{B} = \Gamma = (\gamma^{(n)})$; put $\delta^{(n)} = \alpha^{(n)}\beta^{(n)}$ and define $\mathrm{A}\cdot\mathrm{B} = \Delta = (\delta^{(n)})$. Show that $R(F)$ is a perfect field of characteristic p.

(b) For $\mathrm{A} = (\alpha^{(n)})$ put $v(\mathrm{A}) = v(\alpha^{(0)})$. Show that v possesses the properties of a valuation. Let $\theta \in F$ be the multiplicative representative of $a \in \overline{F}$ and $\Theta = (\theta^{(n)})$ with $\theta^{(n)} = \theta^{1/p^n}$. Show that $R\colon a \to \Theta$ is an isomorphism of $\overline{F}$ onto a subfield in $R(F)$ which is isomorphic to the residue field of $R(F)$.

(c) Show that if $v\colon F^\times \to \mathbb{Z}$ is discrete, then $R(F)$ can be identified with $\overline{F}$.

(d) Show that if F is of characteristic p, then $\mathrm{A} = (\alpha^{(n)}) \mapsto \alpha^{(0)}$ is an isomorphism of $R(F)$ with the maximal perfect subfield in F.

2.10. Let L be an infinite arithmetically profinite extension of a local field F with residue field of characteristic p. Assume that the Hasse–Herbrand function $h_{L/F}$ grows relatively fast, namely that there exists a positive c such that $h_{L/F}(x_0)/h'_{L/F}(x_0) > c$ for all x_0 where the derivative is defined. Let C be the completion of the separable closure of F.

(a) For $(\alpha_E) \in N(L/F)$ show that there exists $\beta^{(n)} = \lim_E \alpha_E^{|E:L_1|/p^n} \in C$ where L_1/F is the maximal tamely ramified subextension of L/F and E runs over all finite extensions of L_1 in L. Show that $(\beta^{(n)})$ belongs to $R(C)$.

(b) Show that the homomorphism $N(L|F) \longrightarrow R(C)$ is a continuous (with respect to the discrete valuation topology on $N(L|F)$ and the topology associated with the valuation v defined in the previous exercise) field homomorphism.

(c) Let E be a separable extension of L. Let S be the completion of the (p-) radical closure of $N(E,L|F)$, i.e., the completion (with respect to the extension of the valuation) of the subfield of $N(E,L|F)^{\mathrm{alg}}$ generated by $\sqrt[p^n]{\alpha}$ for all n and $\alpha \in N(E,L|F)$. Show that there is a field isomorphism from S to $R(\widehat{E})$ where $\widehat{E}$ is the completion of E. Deduce that if F is of positive characteristic, then $\widehat{E}$ is a perfect field.

2.11. Let F be a discrete valuation field of characteristic 0 with residue field of characteristic p, and let C be the completion of the separable closure of F.

Define the map

$$g\colon W(\mathcal{O}_{R(C)}) \to \mathcal{O}_C$$

by the formula $g(\mathrm{A}_0, \mathrm{A}_1, \dots) = \sum_{n \geqslant 0} p^n \alpha_n^{(n)}$, where $\mathrm{A}_m = (\alpha_m^{(n)}) \in \mathcal{O}_{R(C)}$.

(a) Show that g is a surjective homomorphism. Show that its kernel is a principal ideal in $W(\mathcal{O}_{R(C)})$, generated by some element $(\mathrm{A}_0, \mathrm{A}_1, \dots)$ for which, in particular, $v(\alpha_0^{(0)}) = v(p)$.

(b) Let $W_F(R) = W(\mathcal{O}_{R(C)}) \otimes_{W(\overline{F})} F$. Show that g can be uniquely extended to a surjective homomorphism of K-algebras $g\colon W_F(R) \to C$.

(c) Show that the kernel I of this homomorphism is a principal ideal.

(d) Let B^+ be the completion of $W_F(R)$ with respect to I-adic topology and let B be its quotient field. Show that B does not depend on the choice of F and is a complete discrete valuation field with residue field C. The ring B plays a role in the theory of p-adic representations and p-adic periods.

2.12. For $n \geqslant 0$, find a local number field F such that $\mu_{p^n} \subset F, \mu_{p^{n+1}} \not\subset F$, and the extension $F(\mu_{p^{n+1}})/F$ is unramified.

2.13. Let L be a finite Galois extension of a local number field F with Galois group G. Show that L/F is tamely ramified if and only if the ring of integers $\mathcal{O}_L$ is a free $\mathcal{O}_F[G]$-module of rank 1.

2.14. Let F be a finite extension of $\mathbb{Q}_p$, $n = |F : \mathbb{Q}_p|$. Let L/F be a finite Galois extension, $G = \mathrm{Gal}(L/F)$. A field L is said to possess a normal basis over F, if the group $U_{1,L}$ of principal units decomposes, as a multiplicative $\mathbb{Z}_p[G]$-module, into the direct product of a finite group and a free $\mathbb{Z}_p[G]$-module of rank n.

(a) Show that if G is of order relatively prime to p, then L possesses a normal basis over F.

(b) Suppose the F has no roots of order p. Show that L possesses a normal basis over F if and only if L/F is tamely ramified.

Numerous further exercises on local fields can be found in [**12**].

3. Class Field Theory and Zeta Functions Exercises

3.1. Let L/F be a finite Galois totally ramified extension of local fields with finite residue field and E be the maximal abelian extension of F in L. Let $\alpha \in F^\times$

and $\alpha = N_{L^{\mathrm{ur}}/F^{\mathrm{ur}}}\beta$ for some $\beta \in L^{\mathrm{ur}}$. Let $\beta^{\varphi-1} = \prod_{i=1}^{m} \gamma_i^{\widetilde{\sigma}_i-1}$ with $\gamma_i \in L^{\mathrm{ur}*}$ and $\widetilde{\sigma}_i \in \mathrm{Gal}(L^{\mathrm{ur}}/F^{\mathrm{ur}})$. Show that

$$\Psi_{L/F}(\alpha)|_E = \widetilde{\sigma}^{-1}|_E$$

where $\widetilde{\sigma} = \widetilde{\sigma}_1^{v(\gamma_1)} \dots \widetilde{\sigma}_m^{v(\gamma_m)} \in \mathrm{Gal}(L^{\mathrm{ur}}/F^{\mathrm{ur}})$ and v is the discrete valuation of L^{ur}. Deduce that, in particular, if $\beta^{\varphi-1} = \pi^{\widetilde{\sigma}-1}$ for a prime element π of L^{ur}, then $\Psi_{L/F}(\alpha)|_E = \widetilde{\sigma}^{-1}|_E$.

3.2. Let p be an odd prime, and let ζ_p be a primitive pth root of unity.
(a) Show that $X^p - Y^p = \prod_{i=0}^{p-1} (\zeta_p^i X - \zeta_p^{-i} Y)$ and $\prod_{i=1}^{p-1} (\zeta_p^i - \zeta_p^{-i}) = p$.
(b) Put $c(\zeta_p) = \prod_{i=1}^{\frac{p-1}{2}} (\zeta_p^i - \zeta_p^{-i})$. Show that $c(\zeta_p)^2 = (-1)^{\frac{p-1}{2}} p$.
(c) For a positive integer b put

$$\left(\frac{b}{p}\right) = \begin{cases} 0 & \text{if } p|b, \\ 1 & \text{if } p\nmid b, b \equiv a^2 \mod p \text{ for} \\ -1, & \text{otherwise.} \end{cases}$$

Show that

$$\left(\frac{b}{p}\right) = \frac{c(\zeta_p^b)}{c(\zeta_p)}.$$

(d) Let q be an odd prime, $q \neq p$, and let ζ_q be a primitive qth root of unity. Show that

$$\left(\frac{q}{p}\right) = \prod_{i=1}^{\frac{p-1}{2}} \prod_{j=1}^{\frac{q-1}{2}} (\zeta_p^i \zeta_q^j - \zeta_p^{-i} \zeta_q^{-j}).$$

(e) Deduce one of the proofs of the quadratic reciprocity law: if p, q are odd primes, $p \neq q$, then

$$\left(\frac{p}{q}\right)\left(\frac{q}{p}\right) = (-1)^{\frac{p-1}{2}\frac{q-1}{2}}, \quad \left(\frac{2}{p}\right) = (-1)^{\frac{p^2-1}{8}}.$$

3.3. Let F be a local field with finite residue field, and let L be a totally ramified infinite arithmetically profinite extension of F. Let $N = N(L|F)$. Show that there is a homomorphism $\Psi\colon N^\times \to \mathrm{Gal}(L^{\mathrm{ab}}/L)$ induced by the reciprocity maps $\Psi_E\colon E^\times \mapsto \mathrm{Gal}(E^{\mathrm{ab}}/E)$ for finite subextensions E/F in L/F. Show that $\chi \circ \Psi = \Psi_N$, where the homomorphism $\chi\colon \mathrm{Gal}(L^{\mathrm{ab}}/L) \to \mathrm{Gal}(N^{\mathrm{ab}}/N)$ is defined similarly to the homomorphism $\tau \mapsto \mathrm{T}$ in 17.6 of Chapter 2.

3.4. Let ζ_p be a primitive pth root of unity, $p > 2$. Let $F = \mathbb{Q}_p(\zeta_p)$, $\pi = \zeta_p - 1$, $\mathrm{Tr} = \mathrm{Tr}_{F/\mathbb{Q}_p}$.

(a) Show that

$$\frac{1}{p}\mathrm{Tr}(\zeta_p \pi^i) \equiv \begin{cases} 1 \mod p & \text{if } \; i = p-1 \\ 0 \mod p & \text{if } \; i \neq p-1, i \geqslant 1, \end{cases}$$

(b) Let $\alpha \equiv 1 \mod \pi^2$, $\beta \equiv 1 \mod \pi$. If $\gamma = \sum a_i \pi^i$, $a_i \in \mathbb{Z}_p$, then let

$$d\log\gamma := \gamma^{-1}\left(\sum i a_i \pi^{i-1}\right),$$

this depends on the choice of expansion of β in a series in π. Let

$$\log\beta := (\beta - 1) - \frac{(\beta-1)^2}{2} + \frac{(\beta-1)^3}{3} - \cdots.$$

Prove an *Artin–Hasse formula*

$$(\alpha,\beta)_p = \zeta_p^{\mathrm{Tr}(\zeta_p \log\alpha \cdot d\log\beta)/p}$$

(c) Using a suitable expansion in a series in π, show that $d\log\zeta_p$ can be made equal to $-\zeta_p^{-1}$, $d\log\pi$ to π^{-1}. Prove *Artin–Hasse formulas*

$$(\zeta_p,\beta)_p = \zeta_p^{\mathrm{Tr}(\log\beta)/p} \qquad \text{for} \quad \beta \equiv 1 \mod \pi,$$

$$(\beta,\pi)_p = \zeta_p^{\mathrm{Tr}(\zeta_p \pi^{-1}\log\beta)/p} \qquad \text{for} \quad \beta \equiv 1 \mod \pi.$$

3.5. Let $F = \mathbb{Q}_p(\zeta_{p^n})$, where ζ_{p^n} is a p^nth primitive root of unity, $p > 2$. Denote $\mathrm{Tr} = \mathrm{Tr}_{F/\mathbb{Q}_p}$. Let $\pi_n = \zeta_{p^n} - 1$; then π_n is prime in F. Prove *Artin–Hasse formulas* for $\beta \equiv 1 \mod \pi_n$

$$(\zeta_{p^n},\beta)_{p^n} = \zeta_{p^n}^{\mathrm{Tr}(\log\beta)/p^n},$$

$$(\beta,\pi_n)_{p^n} = \zeta_{p^n}^{\mathrm{Tr}(\zeta_{p^n}\pi_n^{-1}\log\beta)/p^n}.$$

3.6. Let A be a commutative topological ring with unity containing a subfield F. Show that A is isomorphic to the ring of adeles $\mathbb{A}_F$ of a global field F if and only if A is locally compact but not compact and not discrete, F is discrete in A, A/F is compact, and the intersection of all closed maximal ideals of A is 0.

3.7. Let F be a global field. Using the proof of Proposition 4.8 of Chapter 3, prove

(a) For a collection of positive real numbers d_v for all places of F except one v', such that $d_v = 1$ for almost all places, there is $a \in F^\times$ such that $|a|_v \leqslant d_v$ for all $v \neq v'$.

(b) *Strong approximation theorem*: F is dense in $\mathbb{A}_F(\{v\})$ for every place v.

3.8. Let $g(x_1,\dots,x_n)$ be a quadratic form in several variables with coefficients in a number field F. Prove *Hasse's theorem*: the equation $g(x_1,\dots,x_n)=0$ has a solution $a_1,\dots,a_n \in F$ different from $0,\dots,0$ if and only if it has a solution different from $0,\dots,0$ in each completion of F.

3.9. For a number field F let L be the maximal abelian extension of F which is unramified at all finite places and in which real places stay real. Prove that the Galois group of L/F is isomorphic to the ideal class group of F. The field L is called the *Hilbert class field* for F.

3.10. (a) Let F be a number field. Following the notation in 5.6 of Chapter 3, prove that

$$\mu_{J_F^1/F^\times}(J_F^1/F^\times) = \frac{2^{r_1}(2\pi)^{r_2} c_F R_F}{r_F\sqrt{|d_F|}},$$

where c_F is the class number of F, r_F is the number of roots in F, d_F is the discriminant of F, R_F is the regulator of F, i.e. the volume of the fundamental domain of the lattice $T = \log_{S_\infty}(\mathcal{O}_F^\times)$ of $\mathbb{R}^{r_1+r_2-1}$ divided by $\sqrt{r_1+r_2}$.

(b) Deduce that the residue of the completed zeta function $\hat{\zeta}_F(s)$ and of the zeta function $\zeta_F(s)$ at $s=1$ is the right hand side of the displayed equation.

(c) Prove that in positive characteristic

$$\mu_{J_F^1/F^\times}(J_F^1/F^\times) = \frac{c_F}{q-1},$$

where c_F is the cardinality of $\mathrm{Pic}^0(F)$ and q is the cardinality of the constant subfield of F, and deduce that the residue of the zeta function $\zeta_F(s)$ at $s=1$ is $\dfrac{c_F}{(q-1)\log q}$.

3.11. Let D_F be the kernel of the reciprocity map for a global field F.

(a) Prove that D_F is an infinitely divisible group.

(b) Prove that $D_F = \{1\}$ in positive characteristic.

(c) In characteristic zero prove that

$$D_F \simeq (\mathbb{R}/\mathbb{Z})^{r_2} \times ((\prod \mathbb{Z}_p \times \mathbb{R})/\mathbb{Z})^r,$$

algebraically and topologically, where $r = r_1 + 2r_2$ are the standard numbers associated to the number field F.

3.12. Let F be an algebraic number field.

(a) For a cycle $z = \sum n_v[v]$, a linear combination with non-negative integer coefficients, almost all equal to 0, of classes of finite places v, define the z-ray idele class group $C_F^z := J_F^z F^\times/F^\times$ where $J_F^z := \prod U_{n_v,F_v} \times \prod U'_{F_v}$. Here the first product is over finite places, $U_{0,F_v} = U_{F_v}$, the second product is over infinite places and U'_{F_v} is the subgroup of all infinitely divisible elements of $F_v^\times$. Show that the set of open subgroups of finite index of C_F coincides with the set of closed subgroups of C_F which contain one of ray idele class groups. The finite abelian extension F^z/F corresponding to C_F^z by the existence theorem is called the ray class field for the cycle z.

(b) Denote by I_F^z the group of fractional ideals of F generated by maximal ideals whose places have coefficient 0 in $z = \sum n_v[v]$. Denote by P_F^z principal ideals generated by elements a such that $a - 1 \in \prod P_v^{n_v}$ and the image of a in each real completion F_v is in U'_{F_v}. Using Remark 5.1 of Chapter 3 show that $\rho\colon J_F \longrightarrow I_F$ of 5.3 of Chapter 3 induces an isomorphism

$$C_F/C_F^z \simeq I_F^z/P_F^z.$$

3.13. Let F be an algebraic number field.

(a) For a subset M of finite places of F its *Dirichlet's density* is

$$d(M) := \lim_{s\to 1+0} \frac{\sum_{v\in M} |k(v)|^{-s}}{\sum_v |k(v)|^{-s}}$$

if exists. Deduce from 5.6 of Chapter 3 that

$$d(M) := \lim_{s\to 1+0} \frac{\sum_{v\in M} |k(v)|^{-s}}{\log\frac{1}{s-1}}.$$

(b) For a cycle z let χ be a non-trivial character of I_F^z/P_F^z. By the previous exercise it corresponds to a non-trivial character of finite order of J_F/J_F^z. Let C be the support of z, i.e. those v for which $n_v \neq 0$. Show that

$$L_C(\chi, 1) \neq 1.$$

(c) Let R be a subgroup of I_F^z, $R \supset P_F^z$. Let M_{a+R} for $a \in I_F^z$ be the set of finite places whose maximal ideals belong to the coset $a+R$. Using the proof of Theorem 5.7 of Chapter 3 show that $d(M_{a+R}) = |I_F^z : R|^{-1}$.

(d) Deduce *Dirichlet's theorem on prime numbers in arithmetic progressions*: for a positive integer m and an integer a prime to m there are infinitely many prime numbers congruent to a modulo m.

3.14. Let F be an algebraic number field and L/F be a finite Galois extension.

(a) Let L/F be a cyclic extension. For a $\sigma \in \mathrm{Gal}(L/F)$ let M_σ be the set of all finite places v of F which are unramified in L/F and such that σ is the Frobenius automorphism of $\mathrm{Gal}(L_v/F_v) \subset \mathrm{Gal}(L/F)$. Using the proof of Theorem 5.7 of Chapter 3 show that $d(M_\sigma) = |L:F|^{-1}$.

(b) Let L/F be a finite Galois extension. For a $\sigma \in \mathrm{Gal}(L/F)$ let M_σ be the set of all finite places v of F which are unramified in L/F and such that the conjugate class Σ of σ in $\mathrm{Gal}(L/F)$ is the conjugate class of the Frobenius automorphism of $\mathrm{Gal}(L_w/F_v) \subset \mathrm{Gal}(L/F)$ for a place w of L over v. Deduce *Chebotarev's Density Theorem*:

$$|\Sigma|/|\mathrm{Gal}(L/F)| = d(M_\sigma).$$

3.15. *Prime Number Theorem* is not part of class field theory but it is an important property of the Euler–Riemann zeta function introduced in Chapter 3. Follow the steps below to obtain its short proof. For more details see [**25**].

For a positive real x denote

$$\psi(x) = \sum_{p \leqslant x} \log p,$$

where p runs through positive prime numbers.

(a) Prove that for $\Re e(s) > 1$

$$\frac{\zeta'_{\mathbb{Q}}(s)}{\zeta_{\mathbb{Q}}(s)} = \sum_p \frac{\log p}{1-p^s} = \sum_p \frac{\log p}{p^s(1-p^s)} - \xi(s), \quad \text{where} \quad \xi(s) = \sum_p \frac{\log p}{p^s}.$$

Deduce that $\xi(s)$ extends meromorphically to $\Re e(s) > 1/2$ and its poles there are at $s = 1$ and at zeros of $\zeta(s)$.

Prove that if $\zeta(s)$ has a zero of order m at $s = 1 + iy$ with a non-zero real y and a zero of order n at $1 + 2iy$, then

$$\lim_{\varepsilon \to 0+} \varepsilon\xi(1+\varepsilon \pm iy) = -m, \qquad \lim_{\varepsilon \to 0+} \varepsilon\xi(1+\varepsilon \pm 2iy) = -n.$$

Prove that

$$\sum_{\ell=-2}^{2} \binom{4}{2+\ell} \xi(1+\varepsilon+2i\ell y) = \sum_p \frac{\log p}{p^{1+\varepsilon}} (p^{iy} + p^{-iy})^4 \geqslant 0$$

and hence $m = 0$ and $\zeta(s)$ does not have zeros on the vertical line $\Re e(s) = 1$.

(b) Prove that

$$2n \log 2 > \log \binom{2n}{n} \geqslant \psi(2n) - \psi(n)$$

and deduce that there are positive real a, b such that $\psi(x) \leqslant ax + b$ for all positive real x.

(c) Show that on $\Re e(s) > 1$

$$\xi(s) = s \int_0^\infty e^{-st} \psi(e^t)\, dt.$$

Using (a) and (b), show that on $\Re e(s) > 0$

$$\frac{\xi(s+1)}{s+1} - \frac{1}{s} = \int_0^\infty (\psi(e^t)e^{-t} - 1)e^{-st}\,dt$$

extends holomorphically to $\Re e(s) \geqslant 0$.

(d) Deduce that the integral

$$\int_1^\infty \frac{\psi(x) - x}{x^2}$$

converges.

(e) Denote $\pi(x) = \sum_{p \leqslant x} 1$. Deduce that

$$\lim_{x \to \infty} \psi(x)/x = 1, \quad \text{and thus} \quad \lim_{x \to \infty} \pi(x)/x = 1.$$

Bibliography

[1] E. Artin, J.T. Tate, *Class Field Theory*, revised edition of 1967 edition, Amer. Math. Soc. 2009.

[2] I. Fesenko, The generalised Hilbert symbol in 2-adic case, Vestnik St. Petersburg Univ. 1985 issue 22, 112–114; English transl. in Vestnik St Petersburg Univ. Math. 18 (1985) 88–91.

[3] I. Fesenko, Class field theory of multidimensional local fields of characteristic zero, with residue field of positive characteristic, *Algebra i Analiz* (1991); English transl. in *St. Petersburg Math.* 3 (1992) 649–678; Multidimensional local class field theory II, *Algebra i Analiz* (1991); English transl. in *St. Petersburg Math. J.* 3 (1992), 1103–1126.

[4] I. Fesenko, Local class field theory: perfect residue field case, *Izvest. Russ. Acad. Nauk. Ser. Mat.* 1993; English transl. in *Russ. Acad. Scienc. Izvest. Math.* 43 (1994) 65–81.

[5] I. Fesenko, On general local reciprocity maps, *J. reine angew. Math.* 473 (1996) 207–222; available from `https://ivanfesenko.org/wp-content/uploads/2021/10/glr1.pdf`.

[6] I. Fesenko, Hasse–Arf property and abelian extensions, *Math. Nachr.* 174 (1995), 81–87; available from `https://ivanfesenko.org/wp-content/uploads/2021/10/hap.pdf`.

[7] I. Fesenko, On just infinite pro-p-groups and arithmetically profinite extensions, *J. Reine Angew. Math.* 517 (1999), 61–80.

[8] I. Fesenko, Noncommutative local reciprocity maps, In Class Field Theory - Its Centenary and Prospects, Adv. Studies in Pure Math., vol. 30, 63–78, Math. Soc. Japan 2001; available from `https://ivanfesenko.org/wp-content/uploads/2021/10/noncom.pdf`; On the image of non-commutative reciprocity map, Homology, *Homotopy and Applications*, 7 (2005) 53–62.

[9] I. Fesenko, Class field theory, its three main generalisations, and applications, *EMS Surveys* 8 (2021) 107–133.

[10] I. Fesenko, M. Kurihara (eds.), Invitation to Higher Local Fields, Geometry and Topology Monographs vol 3, 2000, available from `https://ivanfesenko.org/wp-content/uploads/2021/10/m3-partI.pdf`, `https://ivanfesenko.org/wp-content/uploads/2021/10/m3-partII.pdf`.

[11] I.B. Fesenko, S.V. Vostokov, S.H. Yoon, Generalised Kawada–Satake method for Mackey functors in class field theory, *Europ. J. Math.* 4 (2018), 953–987.

[12] I.B. Fesenko, S.V. Vostokov, Local Fields and Their Extensions, 2nd extended ed., Amer. Math. Soc., 2002, available from `https://ivanfesenko.org/wp-content/uploads/2021/10/vol.pdf`.

[13] J.-M. Fontaine, J.-P. Wintenberger, Le "corps des normes" de certaines extensions algébriques de corps locaux, *C. R. Acad. Sci. Paris Sér. A* 288 (1979) 367–370.

[14] D. Goldfeld, J. Hundley, *Automorphic Representations and L-Functions for the General Linear Group* Vol. 1 CPU 2011.

[15] H. Hasse, History of class field theory, In *Algebraic Number Theory: Proceed.* Instructional LMS Conference, Brighton, 1967.

[16] M. Hazewinkel, Local class field theory is easy, *Adv. Math.* 18 (1975) 148–181.

[17] L. Herr, $\Phi - \Gamma$-modules and Galois cohomology, pp. 263–272 in I. Fesenko, M. Kurihara (eds.), Invitation to higher local fields, Geometry and Topology Monographs vol 3, 2000, available from `https://ivanfesenko.org/wp-content/uploads/2021/10/m3-partII.pdf`.

[18] K.I. Ikeda, E. Serbest, Fesenko reciprocity map, *St. Petersburg Math. J.* 20 (2008) 407–445; Generalized Fesenko reciprocity map, *St. Petersburg Math. J.* 20 (2008) 593–624; Non-abelian local reciprocity law, *Manuscripta Math.* 132 (2010) 19–49; Ramification theory in non-abelian local class field theory, Acta Arithmetica, 144 (2010) 373–393.

[19] K. Iwasawa, A note on functions, In Proceed. ICM Cambridge, Mass., 1950, p. 322, AMS 1952; Letter to J. Dieudonné, 1952, reproduced in *Adv. Stud. Pure Math.*, 21, 1992, 445–450; Hecke's L-functions, Spring 1964, Springer 2019.

[20] K. Iwasawa, *Local class field theory*, Oxford Univ. Press and Clarendon Press 1986.

[21] M. Jarden, J. Ritter, On the characterization of local fields by their absolute Galois groups, *J. Number Theory* 11 (1979) 1–13.

[22] K. Kato, A generalization of local class field theory by using K-groups. I, *J. Fac. Sci. Univ. Tokyo Sect. IA Math.* 26 (1979) 303–376; II, *J. Fac. Sci. Univ. Tokyo Sect. IA Math.* 27 (1980) 603–683.

[23] Y. Kawada, I. Satake, Class formations. II , *J. Fac. Sci. Univ. Tokyo* 7 (1955) 453–490.

[24] H. Koch and E. de Shalit, Metabelian local class field theory, *J. reine angew. Math.* 478 (1996) 85–106.

[25] J. Korevaar, On Newman's quick way to the prime number theorem, *Math. Intelligencer* 4, issue 3 (1982) 108–115.

[26] M. Kurihara, Kato's higher local class field theory, pp. 53–60 in I. Fesenko, M. Kurihara (eds.), *Invitation to higher local fields, Geometry and Topology Monographs* vol 3, 2000, available from `https://ivanfesenko.org/wp-content/uploads/2021/10/m3-partI.pdf`.

[27] D.A. Marcus, Number Fields, 2nd edition, Springer 2018.

[28] S. Mochizuki, A version of the Grothendieck conjecture for p-adic fields, *Inter. J. Math.* 8 (1997) 499–506.

[29] J. Neukirch, *Class Field Theory*, Springer 1986.

[30] J. Neukirch, *Algebraic Number Theory*, Springer 1999.

[31] J. Ritter, $\mathfrak{p}$-adic fields having the same type of algebraic extensions, *Math. Ann.* 238 (1978) 281–288.

[32] I.R. Shafarevich, A general reciprocity law, *Sb. Math.* 68 (1950) 113–146; *Engl. transl. in Amer. Math. Soc. Transl. Ser.* 2, 4 (1956) 73–106.

[33] J.H. Silverman, *Advanced Topics in the arithmetic of Elliptic Curves*, Springer 1994.

[34] J. Tate, Fourier analysis in number fields and Hecke's zeta function, PhD thesis, Princeton Univ., 1950; reproduced In Algebraic Number Theory: Proceed. Instructional LMS Conference, Brighton, 1967

[35] S.V. Vostokov, Explicit form of the law of reciprocity, *Izv. Akad. Nauk SSSR Ser. Mat.* 42 (1978), 1288–1321; Engl. transl. in Math. USSR-Izv. 13 (1979).

[36] S.V. Vostokov, Explicit formulas for the Hilbert symbol, pp. 81–90 in I. Fesenko, M. Kurihara (eds.), *Invitation to higher local fields, Geometry and Topology Monographs* vol. 3, 2000, available from `https://ivanfesenko.org/wp-content/uploads/2021/10/m3-partI.pdf`.

[37] A. Weil, *Basic Number Theory*, 3rd edit., Springer 1974.

[38] J.-P. Wintenberger, Le corps des normes de certaines extensions infinies des corps locaux; applications, *Ann. Sci. École Norm. Sup.* 4, 16 (1983) 59–89.

[39] L. Xiao, I. Zhukov, Ramification of higher local fields, approaches and questions, *St. Petersburg Math. J.* 26 (2015) 695–740.

Index

www.ingramcontent.com/pod-product-compliance
Ingram Content Group UK Ltd.
Pitfield, Milton Keynes, MK11 3LW, UK
UKHW020316150926
8623IPUK00004B/4

9 789819 826810